Organic reactions and orbital symmetry

Cambridge Texts in Chemistry and Biochemistry

Organic reactions and orbital symmetry

SECOND EDITION

T. L. GILCHRIST and R. C. STORR

Department of Organic Chemistry
The Robert Robinson Laboratories
University of Liverpool

CAMBRIDGE UNIVERSITY PRESS
Cambridge
London · New York · Melbourne

CAMBRIDGE UNIVERSITY PRESS
Cambridge, New York, Melbourne, Madrid, Cape Town,
Singapore, São Paulo, Delhi, Tokyo, Mexico City

Cambridge University Press
The Edinburgh Building, Cambridge CB2 8RU, UK

Published in the United States of America by Cambridge University Press, New York

www.cambridge.org
Information on this title: www.cambridge.org/9780521293365

First published 1972
Second edition 1979
Re-issued 2011

A catalogue record for this publication is available from the British Library

Library of Congress Cataloguing in Publication data

Gilchrist, Thomas Lonsdale.
Organic reactions and orbital symmetry.
(Cambridge texts in chemistry and biochemistry)
Includes bibliographies and index.
1. Chemistry, Physical organic. 2. Chemical reaction,
Conditions and laws of. 3. Molecular orbitals. 4. Symmetry
(Physics) I. Storr, R. C., joint author. II. Title. III. Series.
QD476.G54 1979 547'.1'39 78-54578

ISBN 978-0-521-22014-9 Hardback
ISBN 978-0-521-29336-5 Paperback

Contents

Preface to the first edition

The application of the concept of orbital symmetry to organic chemistry by R. B. Woodward and R. Hoffmann has proved to be a major theoretical advance, in that it has succeeded in bringing together and rationalising diverse areas of the subject. In particular it has provided the basis for a unified mechanistic approach to cycloadditions and molecular rearrangements; partly as a result of the stimulus of the new theory, the importance of such reactions is now rightly recognised. For historical reasons these reactions have not usually been treated as fully as their importance warrants in student texts. Now that the concept of orbital symmetry control is well established it seems appropriate to present an account of these reactions within a modern mechanistic framework.

The major part of this book (chapters 3 to 7) is devoted to a descriptive account of rearrangements and cycloadditions. The aim has been to illustrate the scope and synthetic utility of the reactions as well as to discuss their mechanisms. Chapter 1 gives an introduction to the types of mechanisms by which such reactions can occur, and to the experimental methods available for establishing the mechanisms. In particular, the important distinction between stepwise and concerted processes is emphasised. The treatment in chapter 1 is elementary and descriptive, the aim being to provide a brief revision of certain terms and concepts which are used throughout the rest of the book. Chapter 2 is a comparative account of the various approaches to the theory of concerted reactions. In a brief final chapter (chapter 8) we have speculated on possible extensions of the theory to other types of concerted processes. Throughout the book, thermally induced reactions are given more detailed treatment than photochemical and catalysed reactions, for which the applications of the theory are, as yet, less firmly established. References are given at the end of each chapter to relevant reviews, and also to important original papers and to work which has appeared since the publication of the reviews.

We are indebted to our colleagues Dr D. Bethell and Dr M. J. P. Harger, and to the Series Editor, Dr K. Schofield, for their constructive criticisms. We are also grateful to Professor C. W. Rees, who not only made helpful criticisms of the text, but also was a source of advice and encouragement during its preparation.

August 1971

T. L. GILCHRIST
R. C. STORR

Preface to the second edition

This edition follows much the same pattern as the previous one, with the emphasis remaining on the descriptive aspects of pericyclic reactions. We have greatly expanded the treatment of frontier orbital theory, which has recently proved to be particularly useful in explaining the rates and selectivities of pericyclic processes. The chapters on cycloadditions and sigmatropic reactions have been expanded to cover this and other recent work. We have deleted the chapter on non-pericyclic reactions.

We are grateful to Professor K. Schofield for his helpful and constructive comments, and to our colleagues at Liverpool who have read parts of the manuscript.

April 1978

T. L. GILCHRIST
R. C. STORR

1 Classification and investigation of reaction mechanisms

The aim of this book is to describe the mechanisms and the synthetic applications of several important types of organic reactions, which include cycloadditions and many molecular rearrangements. The unifying feature of the reactions is that they can go by way of *cyclic* transition states.

Such processes involve the breaking and formation of more than one bond. These bonding changes can be *concerted*; that is, they occur by a mechanism in which the nature and type of new bond formation is coupled to and controlled by the movement of the electrons in the bonds which are being broken. Alternatively the bond making and breaking processes can be *stepwise*, in which case a discrete intermediate (a diradical or a zwitterion) is involved. The concerted or stepwise nature of a reaction is amenable to experimental investigation; the major part of this chapter is devoted to discussing the experimental criteria which have been used to determine the mechanisms, and the related terminology.

Experimental evidence supporting stepwise and concerted mechanisms for different types of cycloadditions and cyclic rearrangements has been available for several years. It was not until the mid 1960s, however, that any comprehensive rationalisation of these experimental observations was available. At that time R. B. Woodward and R. Hoffmann began to develop a general theory of concerted reactions which proceed through a cyclic transition state – processes which they termed *pericyclic.* They used the concept of orbital symmetry to predict which types of cyclic transition state are energetically feasible.[1] This and related theories are described in chapter 2, and form the mechanistic basis for the chemistry which follows in later chapters.

1.1. Free energy, enthalpy and entropy[2]

Two factors are important in determining the stability of a molecule: the stabilising energy resulting from the formation of bonds between the atoms and the destabilising energy due to the loss of freedom involved in constraining the atoms within the molecular structure. The thermodynamic function which embraces both of these factors is the *free energy* (G) now called Gibbs Function in SI terminology. Free energy is the fundamental quantity which controls the feasibility and the rates of all reactions.

The reactants in a chemical system usually have a higher free energy than the products and the forward reaction can then, in principle, proceed spontaneously. If the products have a higher free energy than the reactants, energy has to be supplied from some external source for the forward reaction to go. However, even when the reactants have a higher free energy than the products, the forward reaction is rarely spontaneous, but proceeds at a finite rate which may be extremely slow except at high temperatures. Theories which attempt to explain this use the concept of an *energy barrier* which has to be surmounted.

A reaction involves a gradual breaking of some bonds and/or formation of others. According to *transition state theory*, the reaction course can be considered as an infinite series of equilibria between one structure and another.

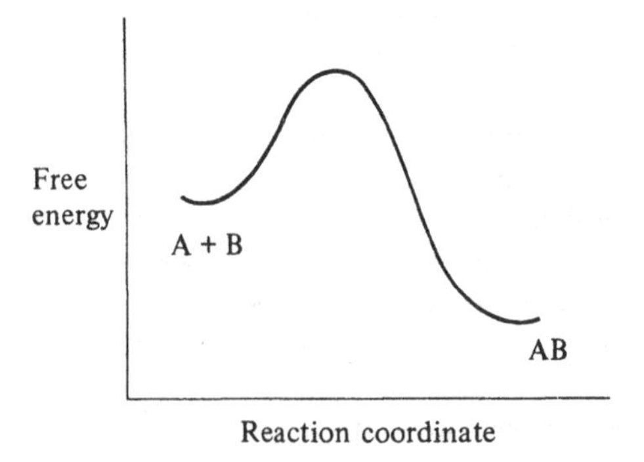

Fig. 1.1

The structure of highest free energy on the reaction path is called the *transition state* or *activated complex*. Although there is an infinite number of such routes from reactants to products, we are concerned only with the one of lowest energy – the so-called *reaction profile*. The free energy in such a process is illustrated in fig. 1.1 for the simple reaction.

$$A + B \rightleftharpoons AB$$

The free energy is shown as a function of *reaction coordinate*. This is a rather loose term which indicates progress of change from reactants to products: for the example shown, it could be the length of the A–B bond.

The equilibrium constant K for any reversible reaction is determined by the difference in free energy ΔG between reactants and products. For the system

$$A + B \rightleftharpoons AB$$

$$K = \frac{[AB]}{[A]\,[B]}$$

and $$\Delta G = -RT \ln K = -RT \ln \frac{[AB]}{[A]\,[B]}$$

The relative proportions of products and reactants at equilibrium are therefore determined by their difference in free energy.

Transition state theory also enables the rate of a reaction to be linked to free energy differences. The theory assumes an equilibrium between the reactants A + B and the activated complex A . . . B. The concentration of the activated complex is therefore determined by $\Delta G^{\ddagger}$, the difference in free energy between the reactants and the transition state. $\Delta G^{\ddagger}$ is called the *free energy of activation*, because it is the free energy barrier which must be surmounted for reaction to occur. In transition state theory the further assumption is made that all activated complexes break down at the same rate. This can be shown to be reasonable by a statistical argument which gives kT/h as the rate constant for the breakdown (k = Boltzmann's constant; h = Planck's constant). Thus, the rate of the forward reaction A + B → AB is

$$\frac{akT}{h}\,[A\ldots B]$$

where a is the fraction of activated complex passing on to product; it is assumed to be close to one. Putting

$$K^{\ddagger} = \frac{[A\ldots B]}{[A]\,[B]}$$

this expression becomes

$$\text{rate} = \frac{kT}{h} \cdot K^{\ddagger}[A]\,[B]$$

The rate constant for the reaction k_r is therefore

$$k_r = \frac{kT}{h} \cdot K^{\ddagger} = \frac{kT}{h} \cdot e^{-\Delta G^{\ddagger}/RT}$$

The rate constant at any particular temperature is thus determined by the free energy of activation $\Delta G^{\ddagger}$. The rate constant for the reverse reaction is similarly determined by the difference in free energy between products and transition state.

The discussion so far has concerned *reversible* reactions, but in fact not all reactions are found experimentally to be reversible. An *irreversible* reaction is one in which $\Delta G^{\ddagger}$ for the reverse reaction is large compared with $\Delta G^{\ddagger}$ for the forward reaction so that the rate of the reverse reaction is negligible in comparison with the rate of the forward reaction.

It is useful to split $\Delta G^{\ddagger}$ into its component enthalpy ($\Delta H^{\ddagger}$) and entropy ($\Delta S^{\ddagger}$) factors:

$$\Delta G^{\ddagger} = \Delta H^{\ddagger} - T\Delta S^{\ddagger}$$

Enthalpy is essentially a bond energy term and entropy reflects the ordering of the system, so that $\Delta H^{\ddagger}$ and $\Delta S^{\ddagger}$ give more insight into the nature of the transition state. This idea of the energy barrier for a reaction as a combination of two factors is derived naturally from an alternative, and pictorially simpler, approach to the theory of reaction rates - *collision theory*. This assumes that only that fraction of molecules which collide with sufficient energy to surmount the transition state barrier can react (the ΔH term), and that even sufficiently energetic collisions will only lead to reaction if the colliding molecules are correctly aligned (the ΔS term).

Experimental values of $\Delta H^{\ddagger}$, $\Delta S^{\ddagger}$ and $\Delta G^{\ddagger}$ are obtained by measuring the reaction rate constant at several different temperatures. The *Arrhenius equation*

$$k_r = Ae^{-E_a/RT} \quad (k_r = \text{rate constant})$$

is an empirical relationship between rate constant and temperature which fits well for most reactions. A and E_a are constants which are, to a first approximation, independent of temperature. E_a is called the *Arrhenius activation energy* and A the *pre-exponential factor*. It can be shown that for a reaction in solution

$$\Delta H^{\ddagger} = E_a - RT$$

and for a gas phase reaction

$$\Delta H^{\ddagger} = E_a - nRT \quad (n = \text{molecularity})$$

At room temperature RT is about 2.5 kJ mol^{-1} so there is little difference in the value of E_a and $\Delta H^{\ddagger}$.

Since $\Delta G^{\ddagger} = \Delta H^{\ddagger} - T\Delta S^{\ddagger}$, the expression for the rate constant for a reaction derived from transition state theory

$$k_r = \frac{kT}{h} \cdot e^{-\Delta G^{\ddagger}/RT}$$

can be rewritten in a form very similar to that of the empirical Arrhenius equation:

$$k_r = \left(\frac{kTe}{h} \cdot e^{\Delta S^{\ddagger}/R}\right) e^{-E_a/RT}$$

the expression within the brackets being equivalent to the pre-exponential factor A. Thus, $\Delta H^{\ddagger}$ and $\Delta S^{\ddagger}$ can be obtained from the experimentally determined values of E_a and A, respectively.

1.2. Concerted and stepwise reactions

Reactions in which more than one bond is broken or formed can be divided into two classes. The first is one in which all the bond forming and breaking processes occur simultaneously so that a one-step transformation of reactants to products occurs without the intervention of an intermediate. Such reactions are called *concerted* or *multicentre* since the bond changes occur in concert or at the same time at more than one centre. The energy curve for such processes is shown in fig. 1.1; it involves only one energy barrier and one transition state.

The second broad class of reactions is one in which the bond forming and breaking processes occur consecutively so that one or more intermediates is involved. These intermediates may be stable molecules capable of isolation, or they may be highly reactive species of only transient existence. Where the intermediate is a stable molecule it is often more convenient to consider the overall process as two or more consecutive concerted reactions. Where the intermediates are unstable the process is normally considered as one reaction which proceeds in a stepwise manner. This distinction is purely arbitrary and one of convenience.

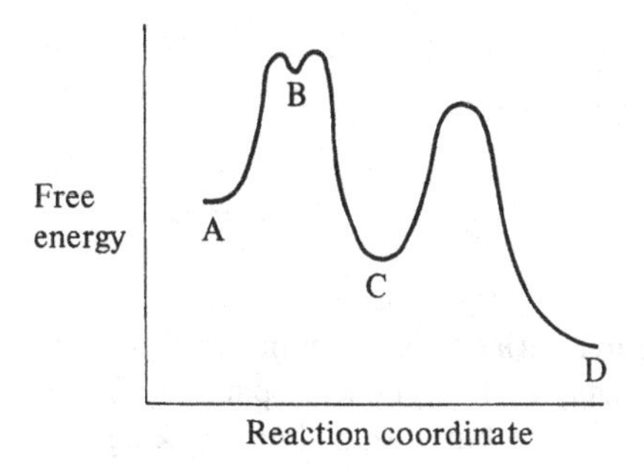

Fig. 1.2

The energy curve for a simple stepwise reaction could be as in fig. 1.2. In such a reaction C is formed from A via an unstable intermediate B which has a discrete, though short, lifetime. The instability of B is indicated by the fact that it lies in a shallow energy well. It has only a small energy barrier to surmount in order to pass over to products or to revert to reactants. The more stable such an intermediate is, the deeper the energy well. C is relatively stable and lies in a deep energy well.

Although C is the first isolable product to be formed, it may subsequently pass over to the more stable D in the reaction conditions. C is called the *kinetically controlled* product of reaction of A. D, which is the product isolated after the system reaches equilibrium, is called the *thermodynamically controlled* product. This is the type of situation where kinetic and thermodynamic control is most frequently encountered.

Kinetic and thermodynamic control can also operate in systems of the type $Q \rightleftharpoons P \rightleftharpoons R$. If the reactions $P \rightarrow Q$ and $P \rightarrow R$ have energy profiles as shown in fig. 1.3, Q is formed faster than R because the energy barrier for its formation is lower than that for R. However, if the reacting system reaches thermodynamic equilibrium, the proportions of products Q and R will be determined by their relative free energies, and R will predominate. In this type of situation, Q is the kinetically controlled product and R is the thermodynamically controlled product.

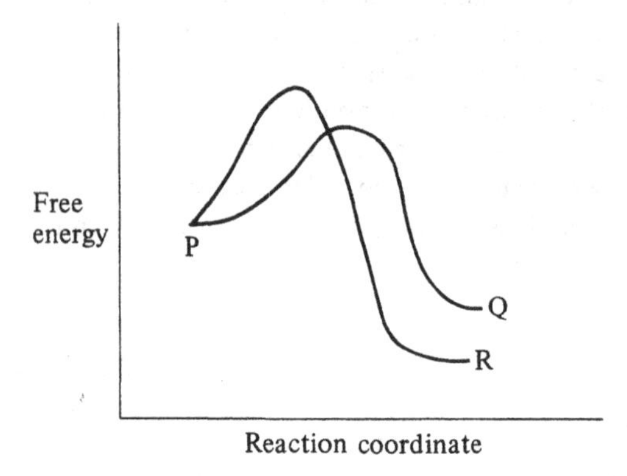

Fig. 1.3

If a reaction is reversible and a particular pathway is energetically the most favourable route from reactants to products, the lowest energy pathway for the reverse reaction will be along the same route but with all bond making and breaking processes reversed. Forward and reverse reactions will therefore pass through the same transition states and will involve the same mechanism. This is the *principle of microscopic reversibility.*

Throughout the book we are concerned with reactions in which cyclic structures are formed or cleaved or which, at least formally, involve a cyclic transition state. The extreme mechanisms by which such reactions can occur can conveniently be discussed with reference to a hypothetical fragmentation:

$$\begin{array}{c} a-b \\ |\quad| \\ c-d \end{array} \longrightarrow \begin{array}{c} a=b \\ \\ c=d \end{array}$$

Concerted mechanism

In the concerted fragmentation the breaking of the a—c bond is coupled with, and depends on, the breaking of the b—d bond, so that both processes are going on at the same time. The overall reaction does not involve an intermediate and the energy profile is of the type shown in fig. 1.1. There is only one energy barrier with a transition state in which both σ bonds are partially broken and the new π bonds in a=b and c=d are partially formed.

The two σ bonds can break simultaneously and at exactly the same rate (a *synchronous* process). It is reasonable that the bonds do not break at the same rate, however, especially if the reactant is unsymmetrical. In the latter case the two bonds will be broken to different extents in the transition state. However, no matter how lopsided the transition state may be, the process is still considered to be concerted if the breaking of the a—c bond is coupled with and controlled by the breaking of the b—d bond (fig. 1.4).

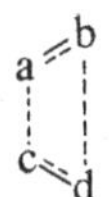

Fig. 1.4

Woodward and Hoffmann[1] have introduced the term *pericyclic* to cover all concerted reactions which involve a cyclic transition state, and define it as follows: 'a pericyclic reaction is a reaction in which all first order changes in bonding relationships take place in concert on a closed curve'.

Stepwise mechanism

The bonds a—c and b—d may break in two successive independent steps. The reaction then involves an intermediate in which only one of the σ bonds is broken. The energy profile will be as in fig. 1.5, with two transition states.

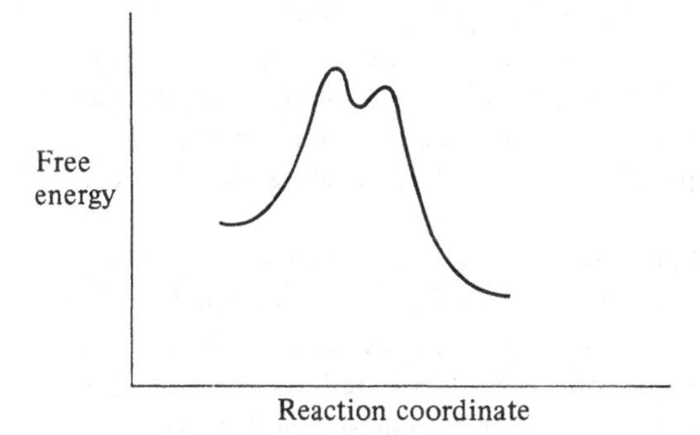

Fig. 1.5

Stepwise breaking and formation of bonds can be detected experimentally (§1.3). The intermediates fall into two categories; those which are highly polar, and those which are essentially non-polar. The first type is normally considered to be a *zwitterion*; that is, a species bearing both a positive and a negative charge; and the second type, a *diradical* (fig. 1.6).

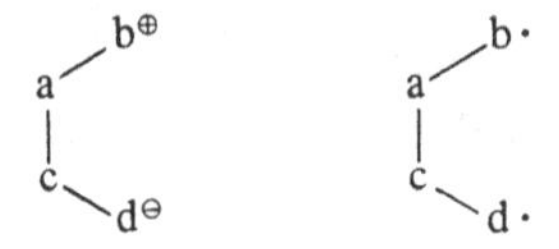

Fig. 1.6

Zwitterions. In the fragmentation shown a zwitterion results from heterolytic cleavage of one of the bonds. Such a process is favoured if the resulting charges are stabilised inductively or mesomerically. Collapse of the zwitterion to re-form the bond b—d gives back the starting material. Other courses are open to the zwitterion; the bond a—c may cleave, or the zwitterion may react with other species present. Bond rotation about the a—b, a—c, or c—d bonds in the intermediate may occur before re-closure or cleavage. It is the experimental observation of these competing processes which enables the reaction to be identified as a stepwise one.

Diradicals.[3] If the bond b—d cleaves homolytically, this leads to a diradical intermediate. Such a process is favoured if the structure is such that the separate monoradicals are stabilised by delocalisation. As the separation of the centres b and d increases, the correlation between the electrons decreases. At some point the separation is sufficiently large for the species to be considered as a diradical rather than as a vibrationally excited ring. Conversely, a diradical ·b—a—c—d· only becomes equivalent to a vibrationally excited ring when conformational changes bring b and d within bonding distance. The diradical is therefore a discrete intermediate; it can re-close, or fragment by cleavage of the bond a—c, both processes competing with bond rotation.

If the centres bearing the unpaired electrons in a diradical are well separated (for example, by a long carbon chain) there is little interaction between them and each might be expected to act as an independent monoradical function. Usually the centres are close enough for intramolecular interaction to preclude their reaction as separate chemical entities.

The concept of electron spin has proved very useful to organic chemists, and the question of relative spins of the unpaired electrons in a diradical is often alluded to. This is reasonable if the separation between the radical centres is small, but not otherwise. The species is referred to as a singlet if the electron spins are opposed, but as a triplet if the spins are parallel. As the bond b—d is cleaved, the electrons initially have antiparallel spins, so that the radical should be generated as a singlet. This singlet state may not be the ground state of the intermediate, however. The relationship between structure and reactivity of diradicals is a complex one, and further discussion is deferred until §6.3.

Photochemical reactions[4]

In a photochemical reaction, a reactant absorbs radiation (generally ultraviolet) which causes excitation of an electron to a higher vacant orbital. Most frequently, lone pair electrons or π electrons are promoted to low-lying π^* antibonding orbitals. σ Electrons are more tightly held and less easily excited. The type of transition can be controlled by the wavelength of light used, since the energy of the light quanta is given by $E = h\nu$.

In the excitation step the electron is promoted with retention of spin to give a singlet excited state. Excited states of higher energy rapidly decay to the lowest excited singlet. This may undergo chemical reaction or it can emit a quantum of light and collapse to the ground state (fluoresce). The singlet excited molecule may also undergo intersystem crossing, with spin inversion, to give a lower energy triplet state. This again may undergo chemical reactions, which are likely to be quite different from those of the singlet, or it may collapse to the ground state with emission of light (phosphoresce). Since this involves spin inversion it is a forbidden process and occurs more slowly than collapse of the excited singlet. The lifetime of an excited singlet is very short since both emission of light and intersystem crossing are fast. (The rate constants for these processes are of the order 10^7–10^8 s^{-1} and 10^8–10^{10} s^{-1} giving lifetimes of 10^{-7}–10^{-8} s and 10^{-8}–10^{-10} s, respectively.) The lowest triplet state has a relatively long lifetime ($\sim 10^{-3}$ s). Thus, intramolecular photochemical reactions may involve the singlet excited state, but intermolecular reactions are more likely to involve triplet states since intersystem crossing competes favourably with intermolecular collision.

Both singlet and triplet excited molecules may lose their energy by intermolecular transfer to other molecules. This type of process is utilised in photosensitised reactions where reactant molecules are excited not by direct absorption of light, but through energy transfer from photosensitisers. Light is absorbed by the sensitiser molecule which then transfers its energy to the reactant, if the latter has a lower-lying excited state available. If the sensitiser transfers its energy from a singlet state, a singlet excited reactant is produced. If the excited photosensitiser undergoes intersystem crossing to its triplet state and then transfers its energy, a triplet reactant molecule is produced.

Other processes are also possible; for example, triplet* + triplet → singlet* + singlet, as in the generation of singlet oxygen. Thus by careful choice of reaction conditions, the chemist can exercise considerable control over the nature of the excited species involved in a photochemical process.

1.3. Experimental investigation of mechanism[2]

The first step in any mechanistic investigation is to determine what all the reactants and products are. This may seem obvious, but incomplete knowledge of all products has often hampered investigations, and has led to wrong conclusions about a mechanism. Minor side products can sometimes provide clues to possible mechanisms.

It must be established that the observed products are actually formed in the reaction studied and not in some subsequent step. Thus, it must be known whether kinetic or thermodynamic control is operative. The fate of particular atoms in a reaction must be known. This is usually evident from the relative positions of substituents in the reactants and products but in more subtle cases atoms can be labelled by isotopes. Detailed stereochemistries of reactants and products must be known since the mechanism has to be able to account for the stereochemical course of the process. For example, we must know whether configuration is retained, inverted or lost at an atom to which a bond is broken and formed.

The second major type of information in a mechanistic investigation comes from rate measurements. Measurement of the rate at varying concentrations of reactants gives an empirical rate equation. In simple cases this gives directly the number of molecules of each type involved in the transition state. In other cases the rate equations are complex and can fit more than one mechanistic scheme. However, for any proposed reaction scheme it is possible to formulate a theoretical rate expression, and this must be compatible with the observed one.

The effects of changing substituents, replacement of atoms by their isotopes, and solvent polarity on the rate of a reaction all give additional useful information. Determination of the rate constant at several different temperatures leads to estimates of enthalpy and entropy of activation.

The information thus obtained can often enable a distinction to be made between concerted and stepwise mechanisms and if the reaction is stepwise it can indicate what sort of intermediate is involved. The following types of experiment have most commonly been used.

Isolation or detection of an intermediate[5]

Stepwise reactions involve intermediates which have widely differing stabilities. At one end of the stability scale these may be isolable; at the other end they may be extremely unstable and their presence only proved indirectly. If an intermediate can be detected by a method which does not affect the mechanism then the stepwise nature of the reaction is proved. The converse does not apply; if no intermediate is detectable the reaction is not necessarily concerted - it may be that the methods of detecting the intermediate are not sufficiently refined.

Sometimes intermediates can actually be isolated. If so, or if a suspected intermediate is available from any other source, it should be subjected to the reaction conditions and shown to undergo the postulated reactions. Intermediates which, although they cannot be isolated, have an appreciable lifetime, may be detected by observation of their infrared, ultraviolet, or nuclear magnetic resonance (n.m.r.) spectra. Transient radical intermediates can be detected by electron spin resonance spectroscopy provided that the intermediate is present in sufficient concentration ($\sim 10^{-6}$ M).

Certain products which are formed from radical pairs give unusual transient emission and absorption lines in their n.m.r. spectra when the reaction is carried out in the n.m.r. probe.[6] These appear in the same position as the normal absorption lines of the reaction product, and both absorption and emission intensities are often greatly enhanced. They are gradually replaced by the normal absorption peaks. This effect is known as chemically induced dynamic nuclear polarisation (CIDNP). A simplified explanation for this phenomenon is as follows. Radical pairs produced from a singlet precursor within a solvent cage are initially singlets, but there is interaction between the unpaired electrons and the adjacent nuclei in the radicals, the electron spin state being dependent upon the nuclear spin states in a particular radical pair. The interaction produces singlet–triplet mixing in the radical pairs. If the interaction produces radical pairs which are mainly triplet in character these tend to diffuse apart and out of the solvent cage, so that the remaining radical pairs have an excess of nuclear spin states which enhance the singlet character. The singlet pairs tend to collapse within the solvent cage to give covalent products. The excess of one type of nuclear spin states in these pairs leads to a polarised spectrum for the product, the magnitude and sign of the polarisation depending upon the hyperfine splitting constants and the g values of the component radicals. Products derived from the out-of-cage reactions will show a different type of polarisation. Rules have been formulated which allow the sign of the polarisation to be predicted.

The technique has been particularly useful in determining the mechanisms of some rearrangements: examples are given in chapter 7. Unfortunately it cannot normally be used to detect diradicals, so that it cannot distinguish between diradical and concerted mechanisms for cycloadditions. A problem with both electron spin resonance and CIDNP, which are very sensitive techniques, is that the radicals may not be involved in the major reaction pathway but may be part of an independent and minor competing reaction. Recent developments in the theory are aimed at obtaining a reliable estimate of the proportion of the total reaction which is detected by CIDNP.

Intermediates may also be intercepted by trapping with other reactants. These may be added deliberately to divert the intermediate from its normal mode of breakdown, or they may be solvent molecules or another molecule of starting material. The isolation of side products can often be explained by the involvement of an intermediate and so indicate its presence. Care has to be taken, however, since such side products may be formed in an independent competing reaction and not from branching of an intermediate, as shown. This problem can usually be resolved by a study of the kinetics of the reaction.[5]

reactants ⟶ intermediate ⟶ product A / product B or reactants ⟶ product A / product B

Crossover products resulting from intermolecular reaction should always be searched for carefully in any suspected intramolecular reaction. This is done by allowing closely related substrates to react in the presence of each other.

Determination of stereochemistry

The stereochemical course of a reaction can provide useful information about its mechanism. A reaction is said to be stereoselective if it proceeds predominantly by one of several possible stereochemical pathways.

In a concerted reaction, all bond forming and breaking processes are interconnected. Such reactions are therefore completely stereoselective. Stepwise reactions involve intermediates in which bond rotations may be able to compete with the second step of the process, and they are therefore often non-stereoselective. For example, in stepwise radical or ionic 2 + 2 additions of olefins (chapter 6) some loss of configuration results (1.1).

(1.1)

Stereoselectivity must be used cautiously as a criterion for concertedness, however. Apparent loss of stereochemistry in a reaction can be due to competition between two different but completely stereoselective pathways. Alternatively, a highly stereoselective reaction may take place by a stepwise mechanism in which the stereochemistry of the intermediates is maintained within a solvent cage, or in which bond rotations of the intermediates are prevented by steric or polar interactions. Consider as an example a stepwise cycloaddition to an isomeric pair of *cis*- and *trans*-olefins. If the olefinic substituents are bulky, steric interactions will cause rotamers in which the X groups are *trans* to predominate. Thus, addition to the *trans*-olefin may give only the adduct in which the X groups are *trans*. Addition to the *cis*-olefin will initially give rotamers with the X groups *cis*. There will then be a driving force for rotation; depending on the relative stabilities of *cis*- and *trans*-rotamers and on the relative rates of ring closure and rotation, some or all of the product will have the X groups *trans*. In such a case, addition to the *trans*-olefin is stereospecific and the lack of stereoselectivity is only apparent with the *cis*-isomer (1.2.). It is therefore important to test the stereoselectivity with both olefins, or at least with one for which rotation is more likely in the intermediate.

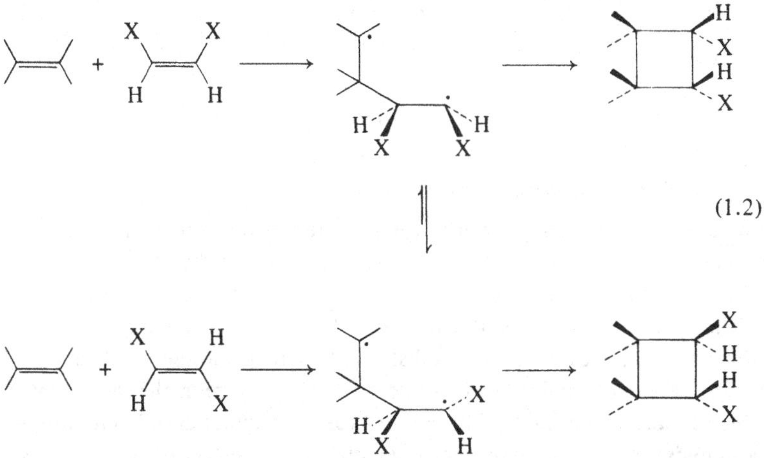

(1.2)

Optical activity provides a convenient way of following the stereochemical course of some reactions. The degree of retention or loss of configuration at an atom to which bonds are broken and formed can be determined: the classic work on the S_N1 and S_N2 mechanisms of aliphatic substitution is an example of the use of the technique. The Claisen rearrangement of **(1)** to **(2)** illustrates another application, involving the creation of a new asymmetric centre in the product. This is an example of *asymmetric induction*.

(1.3)

(1) **(2)**

If the reaction is concerted, the C-1 to C-6 bond forms as the C—O bond breaks, and complete retention of optical activity should be observed. A step-wise reaction involving cleavage of the C—O bond to give a pair of allylic radicals **(3)** should lead to loss of optical activity, but only if the radical pair has a sufficiently long lifetime for reorientation of the planar radicals to occur.

A concerted mechanism also requires *allylic inversion*; that is, formation of the product in which the migrating allyl group is attached at the carbon which was the terminal one in the starting material. The stepwise mechanism should lead to recombination through both C-1 and C-3 of the allyl radical, so that some of the isomer **(4)** should be produced. Asymmetric induction and allylic inversion are particularly useful in obtaining information about the mechanisms of molecular rearrangements of this type.

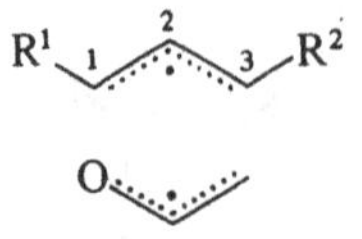

(3) 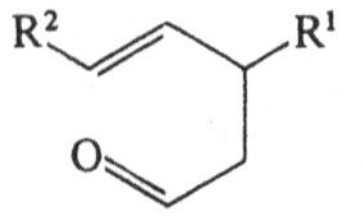 **(4)**

Determination of substituent effects

In a stepwise reaction involving a zwitterionic or diradical intermediate, substituents can stabilise or destabilise the intermediate and therefore profoundly affect the ease of its formation and hence the rate of the overall reaction. The pattern of substituent effects can distinguish between radical and zwitterionic processes since those substituents which stabilise a radical centre do not necessarily stabilise a cationic or anionic centre in the same way.

This simple picture does not apply in every case and under certain circumstances a stepwise reaction may be very insensitive to the effect of substituents. This is the case, for example, in a stepwise cycloaddition in which *closure* rather than formation of a zwitterionic intermediate $\overset{-}{b}$-a-c-$\overset{+}{d}$ is rate determining. The overall rate of such a reaction will be influenced little by the ability of different substituents to stabilise the negative charge at b and positive charge at d.

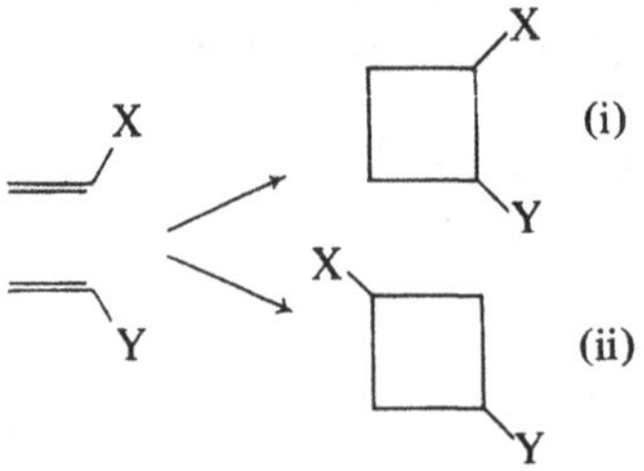

(i)

(ii)

(1.4)

In a stepwise cycloaddition, substituents can also control the *orientation* (i) or (ii) of the addition (1.4). Under kinetic control the product from the most stabilised intermediate predominates because the lower activation energy for its formation means that it is formed faster. Since the orientation of the most stabilised diradical is not always the same as that of the more stabilised zwitterion, the effects of substituents on orientation will often also serve to distinguish between radical and ionic processes.

Concerted reactions where no intermediate is involved might be expected to be much less sensitive to the nature of substituents and this is often the case. However, in many cases the effect of substituents on concerted reactions can be very significant both as regards the rate and orientation. It is only recently that a satisfactory picture of substituent effects on concerted reactions has begun to emerge (chapter 4). Before this the rather vague notion that concerted reactions involved some charge or radical character resulting from unequal bond formation in the transition state was commonly advanced.

The above discussion has considered only the electronic effect of substituents. It should be borne in mind that substituents can also influence the ease of a reaction by a steric effect. There is a very important proviso in the use of substituent effects to study a reaction mechanism, namely that the substituents do not have such a profound effect as to change the mechanism completely. As we shall see there are many examples where widely different substituents can change a reaction from a concerted to a stepwise one.

Substituent effects are thus a rather indefinite criterion for distinguishing stepwise and concerted processes. They must always be interpreted with caution; it is dangerous to argue simply that reactions which show large effects are stepwise and those which show small effects are concerted. Rather, the observed effects should be taken as just one piece of evidence which must be consistent with whatever model finally emerges for the mechanism.

Determination of solvent effects

A reaction going through a zwitterionic intermediate generally involves a transition state which is more polar than the reactants and so it should be accelerated by solvents which are more polar. Solvent polarity is a rather loose term involving the dielectric constant and hydrogen bonding ability. The rate of a concerted reaction, or of a stepwise reaction involving a diradical intermediate, where no significant charge build-up occurs, should be relatively independent of solvent. This is based on the Hughes-Ingold theory that polar species are better solvated by polar solvents than non-polar species, and are therefore more stabilised. For a concerted reaction in which some asymmetry in the transition state leads to a small charge build-up, the energy profiles for the reaction are as shown in fig. 1.7. A moderate solvent effect is to be expected. If the transition state has a higher dipole moment than the reactants the rate will be increased by changing to a more polar solvent. If the transition state has lower overall dipole moment than the reactants then the effect of a more polar solvent will be to decrease the rate.

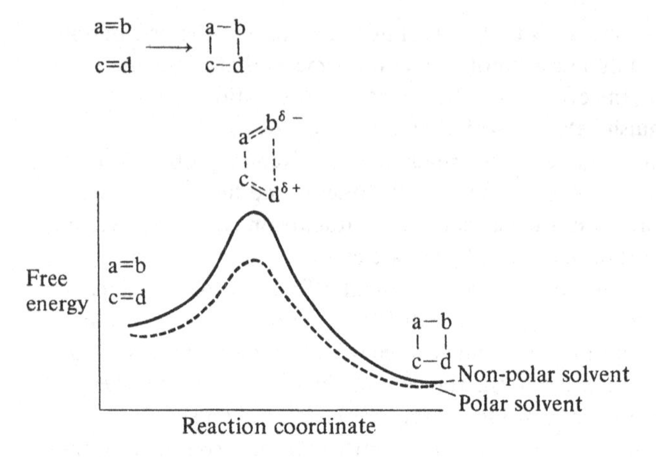

Fig. 1.7

A stepwise reaction via a dipolar intermediate involves two consecutive reactions and the effect of solvent on the activation energies for both reactions must be considered. For example, the effect of change of solvent on the energy profile for the stepwise ionic cycloaddition is illustrated in fig. 1.8.

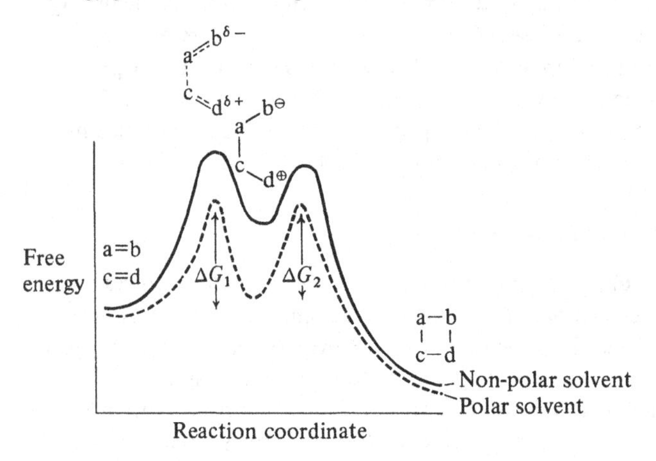

Fig. 1.8

If the first step is rate determining ($\Delta G_1 > \Delta G_2$) then changing from a non-polar to a polar solvent lowers ΔG_1 and increases the reaction rate. If the second step is rate determining ($\Delta G_2 > \Delta G_1$) the reverse applies. If $\Delta G_2 \triangleq \Delta G_1$, then small increases or decreases in rate occur depending on whether $\Delta G_1 > \Delta G_2$ or $\Delta G_2 > \Delta G_1$. Thus a concerted reaction involving an unsymmetrical transition state and a two-step ionic reaction can only be distinguished when the solvent effect is very large.

Detection of kinetic isotope effects[7]

Isotope effects give information concerning the bonds broken and formed in the rate-determining step of a reaction. They can be classed as primary and secondary isotope effects.

A primary isotope effect is the change observed in the rate of a reaction when one of the atoms of the bond being broken or formed in the rate-determining step is replaced by an isotope. The isotope effects most commonly studied in organic chemistry are those resulting from different rates of transfer of hydrogen and deuterium. Consider, as an example, the general reactions:

$$X{-}H + Y \rightarrow X...H...Y \rightarrow X + H{-}Y$$

$$X{-}D + Y \rightarrow X...D...Y \rightarrow X + D{-}Y$$

The zero point vibrational energy of X—D will be appreciably lower than that of X—H because of the difference in mass between hydrogen and deuterium. The zero point energies of the transition states for H and D transfer will depend on X and Y and on the extent of breaking of X—H (X—D) and formation of H—Y (D—Y). In general, X. . . D. . . Y will be lower in energy than X. . .H. . . Y and the difference in energy between the two states will usually be somewhat less than the energy difference between the reactants (X—H and X—D). The energy profiles for the two reactions will therefore be as in fig. 1.9 and X—H will react faster than X—D.

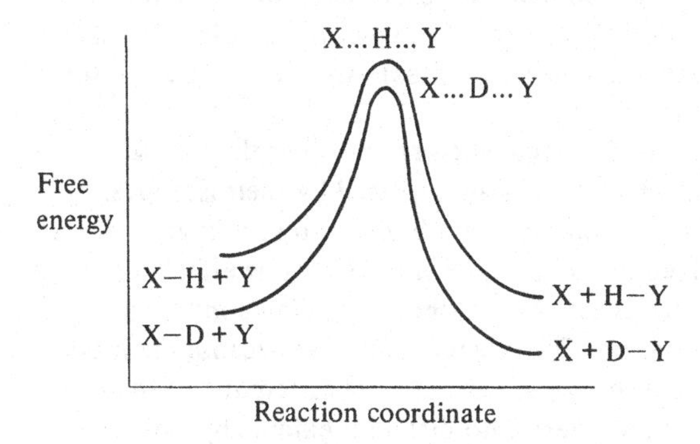

Fig. 1.9

The magnitude of the isotope effect k_H/k_D will vary with the structure of the transition state. In the limiting case where X = Y and the migrating H is equally bonded to both sites in the transition state, the vibrational energies of X. . .H. . .Y and X. . .D. . .Y will be almost equal since the major allowed vibrational mode is symmetrical about the central atom and therefore independent of its mass. In such a situation k_H/k_D will be at its maximum value. Hydrogen-deuterium isotope effects are potentially large as isotope effects go because of the large relative difference in mass between the isotopes (this applies even more so to hydrogen-tritium isotope effects). The value of the ratio k_H/k_D ranges from about one to eight or nine depending on the structure of the transition state. Isotopes other than those of hydrogen can be used although the effects are much smaller; for ^{12}C–^{13}C isotope effects the values range from 1.02 to 1.10. The magnitude of isotope effects depends on the temperature at which they are measured, the values decreasing with increasing temperature. Thus, for example $k_H/k_D = e^{\Delta\Delta G/RT}$ where $\Delta\Delta G$ is the difference in free energy of activation for the reaction of deuterated and undeuterated species. Application of primary hydrogen–deuterium isotope effects is illustrated by the reaction (1.5).

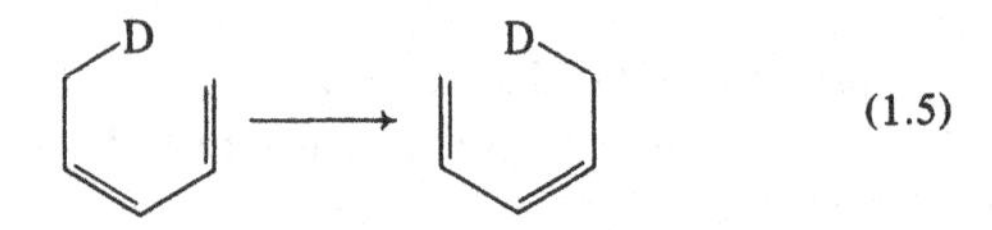

(1.5)

The deuterated diene rearranges more slowly than the undeuterated one (k_H/k_D = 5 at 470 K) indicating that the C—D bond is broken in the rate-determining step. A value of 5 for k_H/k_D at 470 K corresponds to a value of 12.2 at 298 K. This exceptionally large value for k_H/k_D indicates that the transition state is highly symmetrical.

Primary isotope effects are also observed in reactions involving simple cleavage or formation of one bond (as opposed to reactions involving transfer of the isotopically labelled atom between two sites). An example is given for the retro-Diels-Alder reaction (§5.2).

Secondary isotope effects are observed when atoms other than those directly involved in the bond which breaks are replaced by their isotopes. They give information basically similar to that obtained from primary isotope effects. They can be subdivided into $\alpha, \beta, \ldots$ secondary isotope effects depending on whether the isotopes are attached to the atom undergoing bond breaking, or the one adjacent to it, and so on. The effects again arise because the modes of vibration of the reactant and the transition state are different and their zero point vibrational energies are affected to different extents by isotopic substitution: they can be considered simply as reflecting changes in hybridisation. Secondary isotope effects are smaller than primary isotope effects and, depending on how the vibrational modes change on passing from reactants to transition state, k_H/k_D can be greater or less than one. When k_H/k_D is greater than one, the secondary isotope effects are normal; when k_H/k_D is less than one, they are referred to as inverse. An example of their application is to the Diels-Alder reaction, (1.6), which involves the formation of two new bonds.

Replacement of any of the hydrogen atoms *a, b, c* or *d* by deuterium results in an increase in the reaction rate ($k_H/k_D \triangleq 0.9$), indicating that

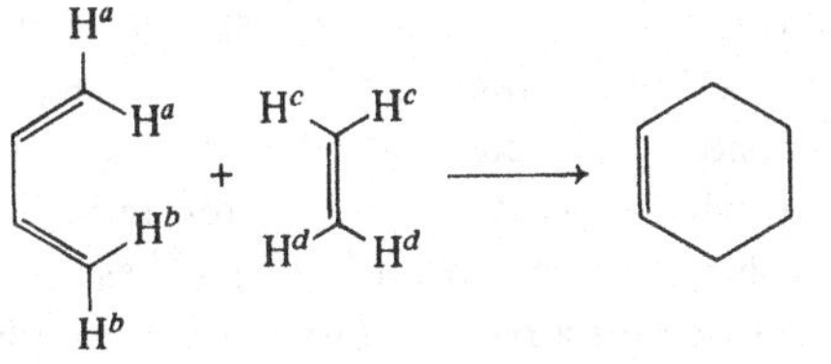

(1.6)

formation of both new σ bonds occurs simultaneously: there is a simultaneous change of hybridisation at all four termini. In a stepwise reaction, formation of one of the bonds only would be rate determining, so that only isotopic substitution on those atoms undergoing change of hybridisation would be expected to have any effect.

Measurement of activation parameters

The entropy of activation $\Delta S^\ddagger$ reflects the difference in ordering between reactants and transition state in the rate-determining step of a reaction. This includes not only reactant molecules but also solvent molecules. In a concerted cycloaddition such as the Diels–Alder reaction the reactants have to be highly orientated with respect to each other. A concerted intramolecular reaction such as the Cope rearrangement also requires a critical orientation of atoms. Such reactions show large negative $\Delta S^\ddagger$ values. In general, concerted cycloadditions and rearrangements show large negative entropies of activation. Stepwise processes in general do not require such critical alignment, and more positive values are to be expected. This tends to be the case, although this criterion must be used cautiously. For example, a concerted cycloaddition must have a large negative $\Delta S^\ddagger$, but a cycloaddition with a large negative $\Delta S^\ddagger$ need not necessarily be concerted. Concerted and radical reactions do not greatly perturb the ordering of the solvent so that $\Delta S^\ddagger$ is mainly determined by the orientational requirements of the reactants. However, solvent structure is likely to be greatly affected in a reaction proceeding through a highly solvated zwitterion intermediate. Stepwise ionic reactions therefore frequently show large negative $\Delta S^\ddagger$ values.

Not all types of concerted reactions show large negative $\Delta S^\ddagger$. In a concerted retro-cycloaddition, for example, the atoms are restricted similarly in both reactant and transition state. In the latter the σ bonds which break are slightly lengthened compared with those in the reactant. Such a process will therefore be expected to have a $\Delta S^\ddagger$ value of almost zero. In a stepwise retro-cycloaddition where only one bond breaks to give an open-chain intermediate the ordering of the system is relaxed on going to the transition state which leads to the intermediate, so that a positive entropy of activation is expected.

The degree of ordering in the transition state has also been related to the *volume of activation*[8]. This gives a measure of the change in volume of reactants (including solvent) in going to the transition state. In a cycloaddition it is often assumed that a concerted reaction in which two bonds are being formed will involve greater compression than a stepwise process in which only one new bond is formed. Concerted processes would therefore be expected to show larger negative volumes of activation ($\Delta V^{\ddagger}$). (These considerations would, of course, be reversed for a fragmentation.) This simple picture involves three important assumptions: that the first step in a stepwise reaction is rate determining, that the transition state of a concerted process is more product-like than that of a two-step process, and that solvent effects can be neglected. These assumptions are not always valid; for example, in some 2 + 2 cycloadditions the rate-determining step is closure of a zwitterionic intermediate. Also, in a process involving rate-determining formation of a zwitterionic intermediate considerable changes in solvent ordering and volume occur on solvation of the intermediate so that such processes can show the large negative values of $\Delta V^{\ddagger}$ similar to those found for the concerted Diels–Alder cycloaddition.

Thus activation volume studies can throw light on the nature of the transition state for a reaction but their interpretation must be undertaken with care and in conjunction with other data.

Concerted processes are usually characterised by having a small enthalpy of activation, $\Delta H^{\ddagger}$. Again, however, the converse is not necessarily true: a low $\Delta H^{\ddagger}$ does not prove a concerted mechanism.

In conclusion, to prove whether a reaction is concerted or stepwise is not a simple matter, since, with the exception of isotope effects, the evidence for a concerted reaction is of a negative rather than a positive nature. The total evidence has to be considered. Frequently, reactions are assumed to be concerted on the basis of one criterion alone, such as stereoselectivity. Although this may be justified in a series of closely related reactions where the concerted nature for some has been proved, such incomplete evidence should always be treated with reserve.

References

1. Woodward, R. B. and Hoffmann, R., *Angew. Chem., Int. End Engl.*, **8**, 781 (1969).
2. For a full discussion of the theory see (*a*) Frost, A. A. and Pearson, R. G., *Kinetics and mechanism*, 2nd edn, Wiley, New York, 1961; (*b*) Leffler, J. E. and Grunwald, E., *Rates and equilibria of organic reactions*, Wiley, New York, 1963. Experimental techniques for determining mechanism are given in Friess, S. L., Lewis, E. S. and Weissberger, A. (eds.), *Investigation of rates and mechanisms of reactions, Technique of organic chemistry*, vol. 8, 2nd edn, Interscience, New York, 1961.
3. Review: Salem, L. and Rowland, C., *Angew. Chem., Int. Edn Engl.*, **11**, 92 (1972).
4. An excellent introduction is Barltrop, J. A. and Coyle, J. D., *Excited states in organic chemistry*, Wiley, London, 1975.
5. For a review of the detection of intermediates on the basis of kinetic evidence, see Huisgen, R., *Angew. Chem., Int. Edn Engl.*, **9**, 751 (1970).
6. Lepley, A. R. and Closs, G. L., *Chemically induced magnetic polarization*, Wiley, New York, 1973; Richard, C. and Granger, P., *NMR*, vol. 8, ed. P. Diehl, E. Fluck and R. Kosfeld, Springer-Verlag, Berlin, 1974.
7. Melander, L., *Isotope effects on reaction rates*, Ronald Press, New York, 1960.
8. Asano, T. and le Noble, W. J., *Chem. Rev.*, **78**, 407 (1978).

2 Theory of concerted reactions

The development of a general theory of concerted reactions has been due chiefly to the work of R. B. Woodward and R. Hoffmann.[1] They have taken the basic ideas of molecular orbital theory and used them, mainly in a qualitative way, to derive selection rules which predict the stereochemical course of various types of concerted reactions. These rules are best understood in terms of symmetries of interacting molecular orbitals. However, it is important to appreciate that the rules do not depend on the use of any particular bonding theory for their validity. The same selection rules have been obtained by various other theoretical approaches, including a valence bond treatment,[2] and several alternative descriptions of the rules are commonly used in the chemical literature. In this chapter the aim is to describe the most important of these theoretical approaches and to show, as far as possible, how they are interrelated.

2.1. Molecular orbitals

Since extensive use is made of orbital pictures in describing the course of concerted reactions, the basis of these orbital representations will be outlined.

Molecular orbitals are made up of combinations of atomic orbitals. Each molecular orbital can contain a maximum of two electrons. The interaction of two equivalent atomic orbitals leads to the formation of two new molecular orbitals, one of lower energy than the atomic orbitals (the bonding orbital) and the other of higher energy (the antibonding orbital). The positive overlap between the atoms in the bonding orbital can be represented by shading: the shading in the larger lobes of the C—C σ bond illustrated (fig. 2.1) show this bonding interaction. For the σ^* bond, however, the lack of overlap is indicated by discontinuity of shading between the atoms. This is a formalised way of indicating a change of phase in the wave function.

Here the shapes of the interacting lobes represent those of the hybrid orbitals on the individual atoms, and are not drawn to represent areas of electron density in the resulting C—C bond. The important feature of this

Fig. 2.1

type of bond is that the electron density is concentrated along the axis joining the nuclei.

An important difference between the bonding and antibonding orbitals is their *symmetry*: the C—C σ orbital is symmetric with respect to a plane perpendicular to the axis of the bond, but the σ^* orbital is antisymmetric.

The same formalised system can be used to represent a C—H σ bond; here the interacting orbitals are a hybrid orbital on carbon and an s orbital on hydrogen (fig. 2.2).

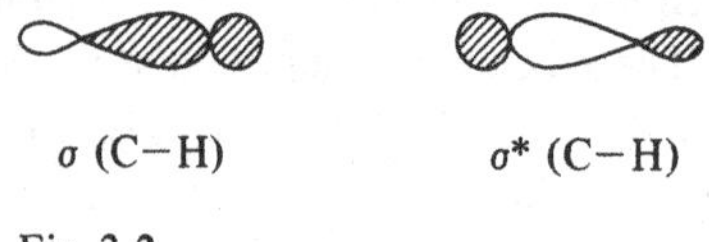

Fig. 2.2

In a similar way delocalised molecular orbitals can be constructed from a system of equivalent overlapping p atomic orbitals. The combination of *n* such p orbitals will lead to the formation of *n* molecular orbitals, symmetrically distributed in energy about the non-bonding level. Thus, for example, two such p orbitals can combine to give a π bonding orbital and a π^* antibonding orbital as in ethylene (fig. 2.3). As before, the shading indicates positive overlap and bonding interaction.

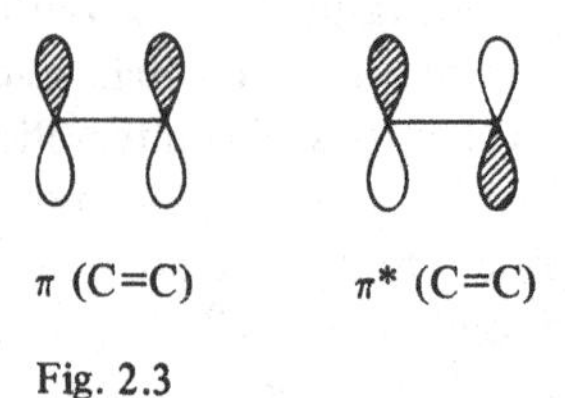

Fig. 2.3

Again, the different symmetries of the two orbitals distinguish them. All planar π systems have a plane of symmetry (m_1) bisecting the p orbitals (the nodal plane) about which they are antisymmetric. There are, however, other symmetry elements in the ethylene π systems, including another mirror plane, m_2, perpendicular to the C—C bond, and a twofold axis, C_2, running through the C—C bond and perpendicular to it (fig. 2.4).

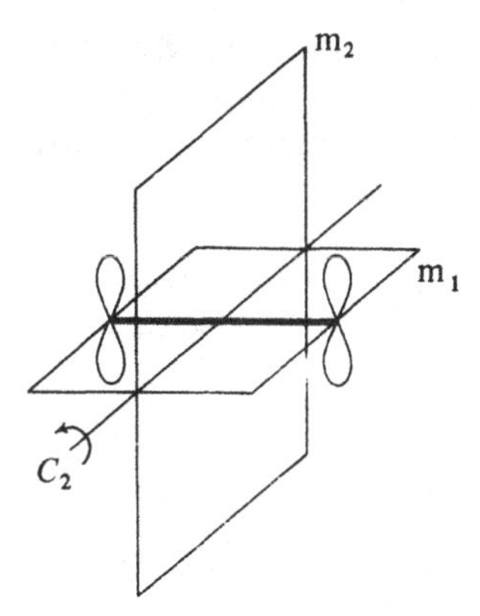

Fig. 2.4

The π orbital is symmetric with respect to m_2 and π^* is antisymmetric. On the other hand, if the twofold axis is taken as the element of symmetry, π is antisymmetric and π^* is symmetric. Thus, the element of symmetry must be specified when orbitals are classified as symmetric or antisymmetric.

In fig. 2.3 and in similar diagrams which follow, the π molecular orbitals are represented as combinations of p orbitals of the individual atoms. If the functions Φ_1 and Φ_2 describe the electron distribution in the two p orbitals, then the function describing the electron distribution in the two molecular orbitals is given by

$$\psi = c_1\Phi_1 + c_2\Phi_2$$

where c_1 and c_2 are numerical *coefficients* which indicate the relative contributions of Φ_1 and Φ_2 to the molecular orbital: c_1^2 and c_2^2 represent the probabilities of finding an electron in the neighbourhood of atoms 1 and 2. Because of the symmetry of ethylene, $c_1^2 = c_2^2$. If the total probability of finding an electron of a given spin state in the π orbital is unity, then $c_1^2 + c_2^2 = 1$ and $c_1^2 = c_2^2 = 1/2$; c_1 and c_2 have the numerical values of $\pm 1/\sqrt{2}$ or ± 0.707. For the π bonding orbital in which the atomic orbitals are in phase, the coefficients will have the same sign; thus for the π orbital

$$\psi = 0.707\Phi_1 + 0.707\Phi_2$$

For the π^* orbital, the coefficients have opposite signs; thus

$$\psi = 0.707\Phi_1 - 0.707\Phi_2$$

The allyl system (fig. 2.5) has three interacting p orbitals, and their combination gives three π molecular orbitals. The sum of the squares of the coefficients of each orbital is still unity, but the coefficients on each atom are now no longer equal. The size of the lobes in fig. 2.5 qualitatively represents the magnitude of the coefficients. The same three symmetry elements are present as were specified for the ethylene π system: the classification of the orbitals as symmetric (S) or antisymmetric (A) with respect to the mirror plane m_2 and the twofold axis is as shown.

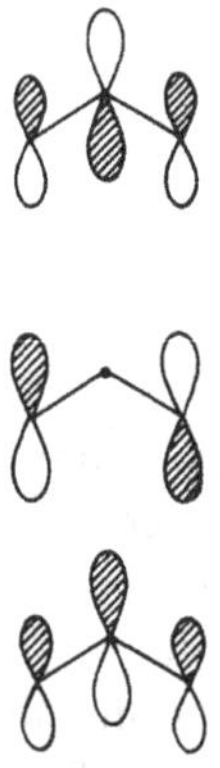

	Coefficients 1	2	3	Nodes	Symmetry m_2	C_2
ψ_3	0.500	−0.707	0.500	2	S	A
ψ_2	0.707	0	−0.707	1	A	S
ψ_1	0.500	0.707	0.500	0	S	A

1 2 3

Fig. 2.5

In the ground state, electrons occupy these orbitals, two in each, from the lowest upwards. Hence the highest occupied π orbital can be determined. For the allyl cation, with two π electrons, it is ψ_1; for the allyl radical and anion with respectively three and four π electrons, it is ψ_2. In general, odd electron systems correspond to the even electron systems which have one more electron.

The construction of the π orbitals of longer polyenes follows a similar pattern. The orbitals of butadiene and their symmetries in the *cisoid* conformation are shown in fig. 2.6. In general, for linear *cisoid* polyenes with n atoms, the n π orbitals can be drawn by putting in the signs with zero nodes for ψ_1, one node for ψ_2, and so on, up to $n - 1$ nodes for ψ_n. For polyenes in their ground states, the highest occupied molecular orbital (HOMO) will be symmetric with respect to m_2 for $2, 6, 10, \ldots \pi$ electron systems, and antisymmetric for $4, 8, 12, \ldots \pi$ electron systems. The lowest unoccupied orbital (LUMO) has the symmetry opposite to that of the HOMO. In the first excited state, the LUMO of the ground state will become singly occupied because of promotion of an electron, and it will thus become the new 'highest occupied' orbital. In the first excited state, therefore, the symmetry of the highest occupied orbital is opposite to that of the ground state.

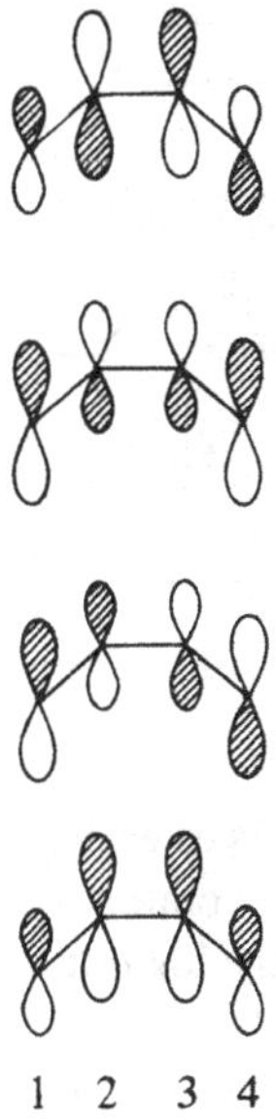

	Coefficients 1	2	3	4	Nodes	Symmetry m_2	C_2
ψ_4	0.371	−0.600	0.600	−0.371	3	A	S
ψ_3	0.600	−0.371	−0.371	0.600	2	S	A
ψ_2	0.600	0.371	−0.371	−0.600	1	A	S
ψ_1	0.371	0.600	0.600	0.371	0	S	A

1 2 3 4

Fig. 2.6

The coefficients for more complex molecules containing π systems are obtained by calculation using computer programmes based on approximate solutions of the Schrödinger wave equation of varying degrees of sophistication; by this means, for example, the effects of substituents on the orbital coefficients of polyenes have been calculated (chapter 4).

2.2. Frontier orbital approach[2]

In a bimolecular reaction, interaction between the two components is represented by interaction between suitable molecular orbitals of each. The extent of the interaction depends upon the geometry of approach of the components, since this affects the amount of possible overlap. It also depends upon the phase relationship of the orbitals and upon their energy separation, a small energy separation favouring a greater interaction.

When two orbitals interact two new orbitals are produced; one is lower in energy than either of the original orbitals and the other is higher. When both of the interacting orbitals are fully occupied (fig. 2.7*a*) there is no net stabilisation produced by the interaction because the lowering in energy of one is offset by the raising in energy of the other. Only when one orbital is not fully occupied does a net stabilisation result. This is illustrated in fig. 2.7 by the interaction diagrams for a low-lying filled orbital of one component and a higher energy orbital of the second component: a net lowering of energy results from the interaction with a part-filled orbital (*b*) or with a vacant orbital (*c*).

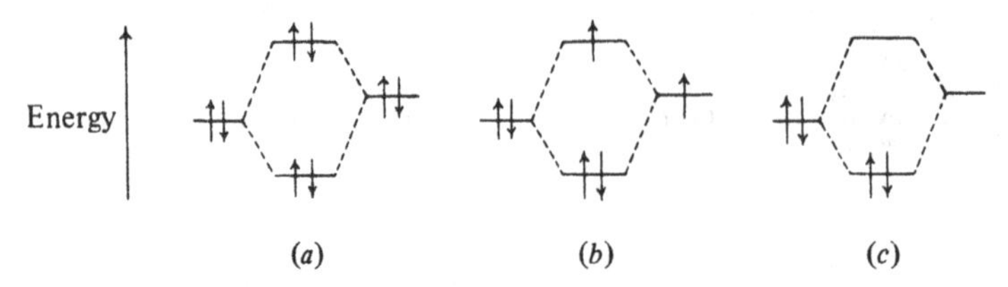

Fig. 2.7. Interaction of a filled orbital (*a*) with a filled orbital, (*b*) with a singly occupied orbital and (*c*) with a vacant orbital.

We can expect interactions of this type between all the orbitals of each component which are of the right phase and orientation, but the dominant stabilising interaction will be between those closest in energy - the highest occupied orbital of one component and the lowest vacant orbital of the other. By focussing attention on the phase relationships, or symmetries, of these orbitals in a particular transition state it should be possible to decide whether the transition state is likely to be energetically favourable or not. These HOMO-LUMO interactions are termed *frontier orbital* interactions: frontier orbital theory predicts the course of bimolecular pericyclic processes on the basis of the extent of overlap of the HOMO and LUMO of each component.

Consider two ethylene molecules approaching each other in such a way that the top of one π system interacts with the bottom of the other (fig. 2.8). The HOMO of one π system must then be matched with the LUMO of the other, as shown. In both cases the phase relationships at one end of the system are wrong for bond formation. On this basis the frontier orbital theory predicts that the cycloaddition of two ethylene units, *via the geometry shown in fig. 2.8*, cannot occur as a concerted process in which both new σ bonds are

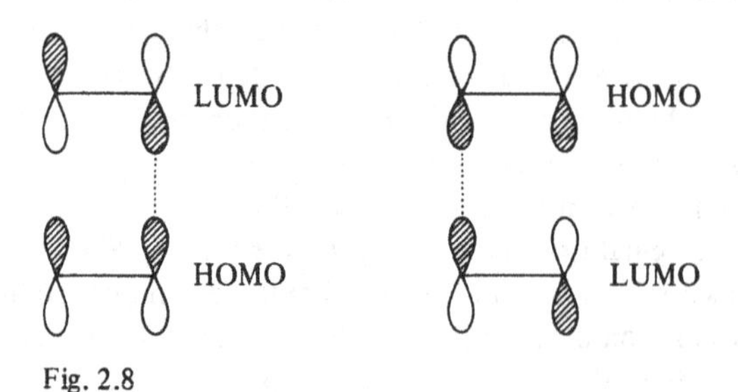

Fig. 2.8

formed simultaneously. The reaction of two ethylene units through the transition state of fig. 2.8 is termed a *symmetry-forbidden* reaction.

As an example of a reaction in which the frontier orbital interaction is a stabilising one, consider a similar interaction between ethylene and an allyl anion (fig. 2.9). In both cases the HOMO–LUMO interaction leads to bonding at both termini. The reaction shown in fig. 2.9 is therefore *symmetry-allowed*.

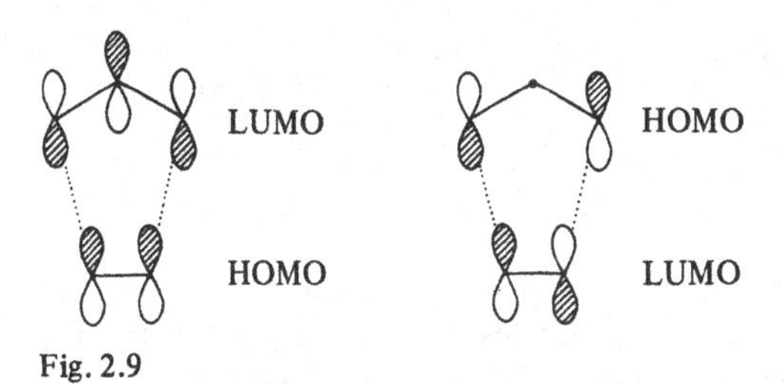

Fig. 2.9

It was pointed out in section 2.1 that the symmetries of the HOMO and LUMO of polyenes depend upon the number of π electrons which they contain, the symmetries alternating between systems containing $4n$ and $4n + 2$ electrons. An extension of the analysis shown in figs. 2.8 and 2.9 to other polyenes thus allows us to predict which types of intermolecular cycloadditions are symmetry-allowed, based on the number of π electrons in the polyene. The predictions are summarised in *selection rules*; the selection rules for cycloadditions are developed in chapter 4.

It is important to appreciate that these rules, and the terms 'symmetry-allowed' and 'symmetry-forbidden', only refer to the *concerted* processes. The geometry of approach is also crucial: by choosing an alternative transition-state geometry, the selection rules will be altered, and it is therefore possible to devise a symmetry-allowed concerted pathway for most reactions. Such pathways often involve rather severe distortion of the reactants and poor overlap of the interacting orbitals, and in these cases a stepwise mechanism is likely to provide an energetically attractive alternative. Thus a 'symmetry-allowed' reaction is not necessarily going to choose that pathway in preference to an alternative stepwise one.

Frontier orbital theory is able to provide information about a cycloaddition other than the prediction that it will be symmetry-allowed or symmetry-forbidden. The closer in energy the frontier orbitals are, the stronger the interaction between them, and therefore the more easily the reaction will occur. The orbital coefficients at the interacting centres can also influence the rate and the direction of addition. These points are developed in chapter 4.

The theory can also be used to obtain selection rules for intramolecular processes; indeed, the first ideas of symmetry control of concerted processes were developed to explain the stereochemical preferences observed in the

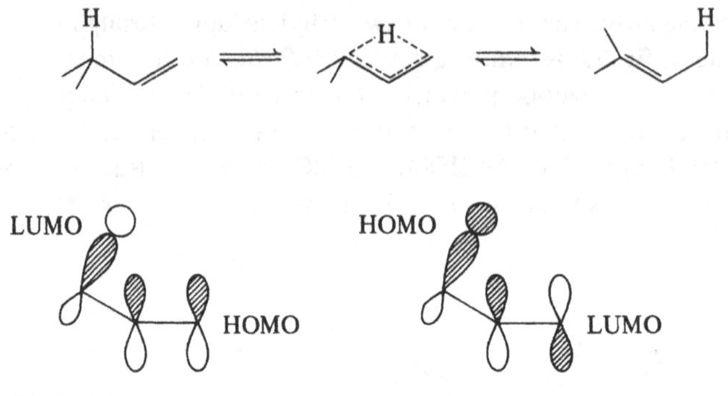

Fig. 2.10

intramolecular cyclisation of polyenes. In order to treat the intramolecular rearrangements as two component processes it is necessary to make a rather arbitrary division of the molecules into two interacting systems. If, for example, we wish to use the theory in a qualitative way to predict whether or not the migration of a hydrogen atom across the top face of an allyl system is symmetry-allowed, we can treat the system as an interacting C—H single bond and C=C double bond (fig. 2.10). By comparing the symmetries of the HOMO and LUMO of each component it is clear that the two components cannot interact in such a way as simultaneously to develop the new π bond and the new σ bond; the reaction is symmetry-forbidden. This can be contrasted with the analogous hydrogen migration across the top face of a pentadienyl system (fig. 2.11) where the interaction is now symmetry-allowed.

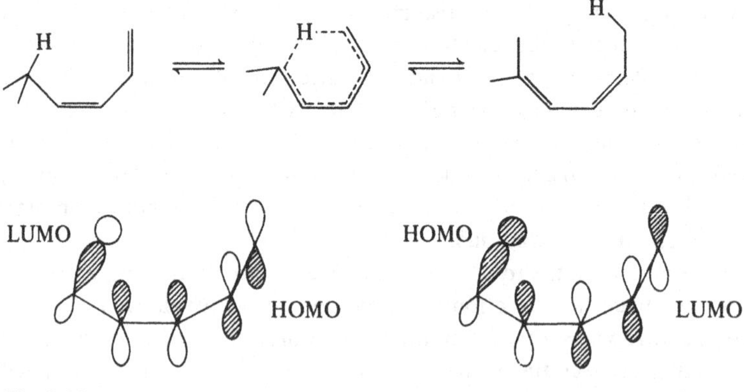

Fig. 2.11

The frontier orbital theory involves some drastic assumptions and oversimplifications, but it has proved particularly valuable in understanding the course of intermolecular cycloaddition reactions. Some of the other approaches to the theory of pericyclic processes are probably better for analysing the selection rules for intramolecular rearrangements, and these will be discussed in the following sections.

2.3. Correlation diagrams

Correlation diagrams, which have been used extensively by Woodward and Hoffmann in their development of the theory of concerted reactions, can be regarded as an extension of the frontier orbital approach. They provide a means of following the energy changes of all the participating electrons in a concerted reaction, and not only those in the HOMO. Relevant unoccupied orbitals are also included.

Correlation diagrams are not new; they were used in the early days of the development of molecular orbital theory to follow energy changes of the orbitals of separate atoms as they approach through space to form a molecule. Their use is discussed, for example, by Coulson.[3] Longuet-Higgins and Abrahamson showed that correlation diagrams could be applied to follow the energy changes in the orbitals of molecules as they approached through space to form a 'united molecule', and successfully explained the course of several concerted reactions by this method.[4]

Symmetry plays a vital role in the construction of these diagrams. The symmetry elements present in the transition state, starting material and product are defined, and the participating orbitals classified as symmetric or antisymmetric with respect to each of these elements. Each orbital is then transformed, with *conservation of symmetry*, into an orbital of the product. Substituents which do not fundamentally alter the energies of the participating orbitals, such as alkyl groups, are ignored for the purposes of determining the symmetry.

C_2 m (2.1)

Conrotatory Disrotatory

As an example of the use of correlation diagrams, consider the closure of butadiene to cyclobutene, and the reverse process (the conditions needed to bring about these reactions are discussed in chapter 3). The reaction involves rotation of the terminal CH_2 groups of the diene through 90°. The groups can rotate in the same direction (*conrotatory* movement) or in opposite directions (*disrotatory* movement); these are shown in equation 2.1.

Although the products of the two types of ring closure are not distinguishable in this case, they may be with substituted butadienes. The conrotatory mode maintains a twofold axis of symmetry (C_2) whereas in the disrotatory closure a mirror plane (m) is maintained. The symmetry classification of the participating orbitals depends on which mode of closure is used.

Having considered the various possible geometries for the transition state and the symmetry associated with each, the next step is to note which bonds are made and which are broken in the reaction, and to list the molecular orbitals associated with these bonds. In the closure of butadiene to cyclobutene, the π bonds of the diene are broken; those formed are the π bond between C-2 and C-3 and the σ bond between C-1 and C-4. The carbon σ framework is otherwise basically unchanged during the reaction and the orbitals associated with it are not included in the correlation diagram. Similarly the orbitals associated with the C—H bonds are not included because changes in their energies are small compared with the other bonding changes.

The orbitals associated with the bonds made and broken are:

butadiene: $\psi_1, \psi_2, \psi_3, \psi_4$

cyclobutene: $\sigma, \pi, \pi^*, \sigma^*$

Next, for each of the possible transition-state geometries, the orbitals are classified (S or A) with respect to the appropriate symmetry element. The symmetry elements chosen must bisect bonds made or broken during the reaction. For the conrotatory mode of closure, where a twofold axis is maintained, the symmetries are as shown (fig. 2.12).

The joining of orbitals of like symmetry completes the correlation diagram: thus, ψ_1 of butadiene correlates with π of cyclobutene, and ψ_2 with σ. The four electrons occupying ψ_1 and ψ_2 in the ground state of butadiene can therefore go into the π and σ bonding orbitals of cyclobutene without involving a high energy transition state: the reaction is symmetry allowed for the conrotatory mode. Similar correlations exist for the antibonding orbitals. The important characteristic of a correlation diagram of an allowed reaction is this linking of bonding with bonding orbitals, and antibonding with antibonding orbitals.

For the disrotatory mode of closure, the relationship of orbitals is different. The symmetries for the disrotatory closure are as shown in fig. 2.13.

The orbital ψ_2, occupied in the ground state of butadiene, cannot correlate with either of the bonding orbitals of the product, and ψ_2 increases sharply in energy as the transition state is approached; the reaction is therefore symmetry-disallowed.

We can see from figs. 2.12 and 2.13 that it is the effect on ψ_2, the HOMO of butadiene, which determines whether the reaction is symmetry-allowed or symmetry-forbidden. In the conrotatory closure shown in fig. 2.12 there is a

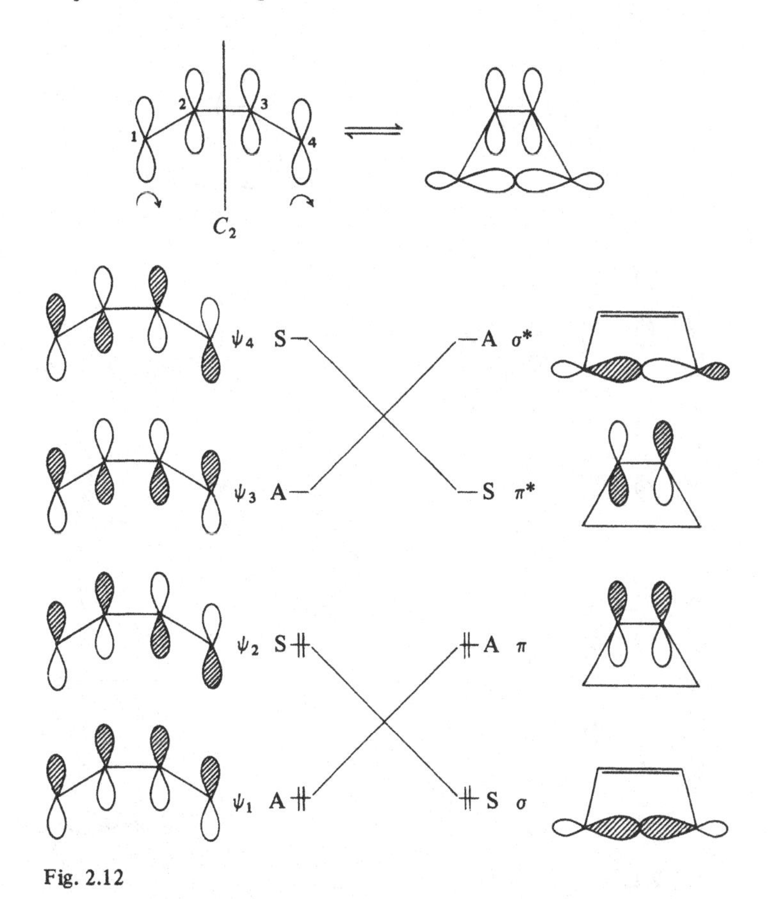

Fig. 2.12

bonding interaction between the terminal lobes of ψ_2, whereas the disrotatory closure shown in fig. 2.13 leads to an antibonding interaction, and an increase in energy. The preferred mode of electrocyclic ring closure of a polyene can thus be determined from the symmetry of the HOMO: further examples are given in chapter 3.

The same procedure is used in the construction of correlation diagrams for bimolecular reactions. In some cases it is not a simple matter to assign symmetries to all the participating orbitals. Consider, for example, a cycloaddition reaction in which two new σ bonds are formed, a plane of symmetry (m) being maintained throughout the reaction (fig. 2.14). The orbitals σ_1 and σ_2, localised on the new σ bonds, have no symmetry with respect to this plane.

However, by taking linear combinations of σ_1 and σ_2, two new σ bonding orbitals can be produced, having symmetry with respect to the mirror plane. The new orbitals are $(\sigma_1 + \sigma_2)$, which is symmetric with respect to m, and $(\sigma_1 - \sigma_2)$, which is antisymmetric (fig. 2.15). Each of these new σ bonding

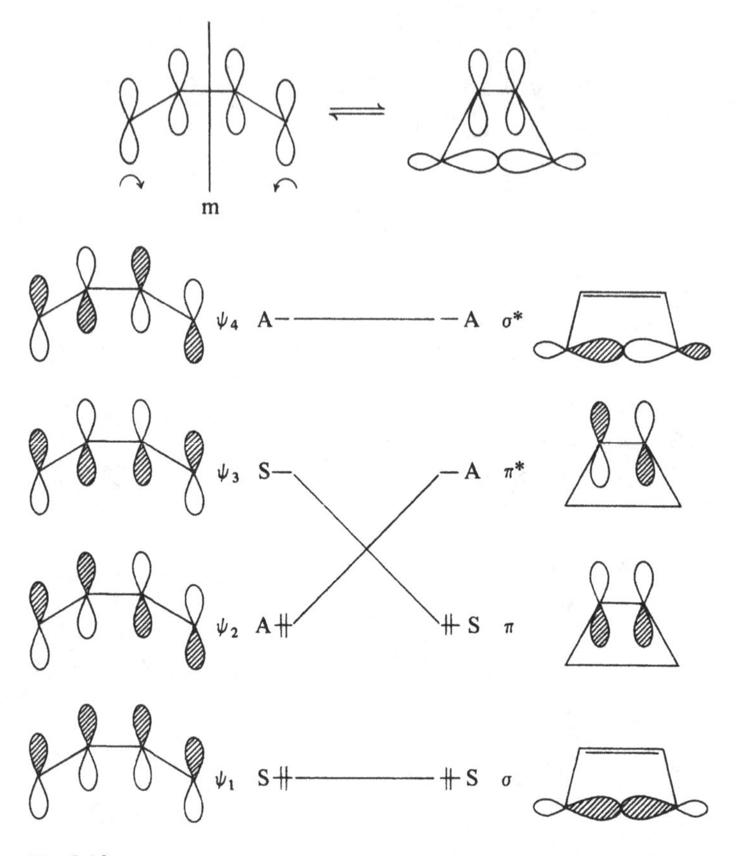

Fig. 2.13

orbitals, like any other molecular orbital, can contain a maximum of two electrons.

Combinations of the antibonding orbitals can be used in a similar way. This device enables correlation diagrams to be constructed for a wide range of reactions where there is symmetry in the transition state.

There are some difficulties in the general application of correlation diagrams to concerted reactions. One is that many transition states have no symmetry, even when non-reacting substituents are ignored. Another is that it is difficult to define the point at which the effect of substituents can no longer be neglected: for example, if one of the components in a Diels-Alder cycloaddition is highly polarised, because of the nature of its substituents or because its skeleton contains a heteroatom, it is probable that the transition state for the reaction will be asymmetric, the formation of one σ bond being much more advanced than of the other. The simple ethylene-butadiene

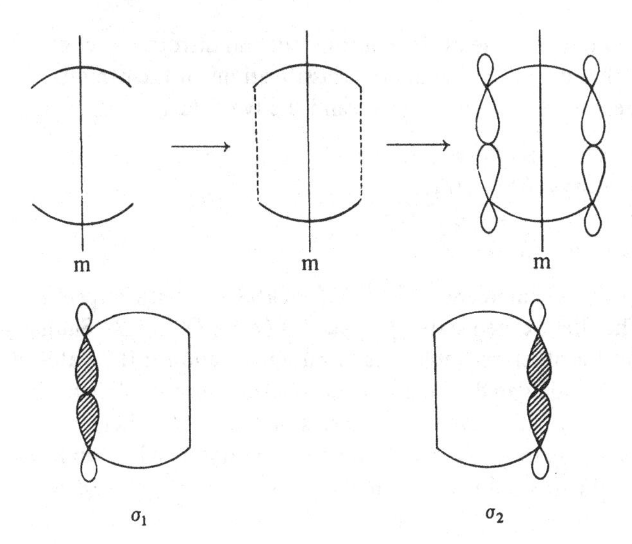

Fig. 2.14

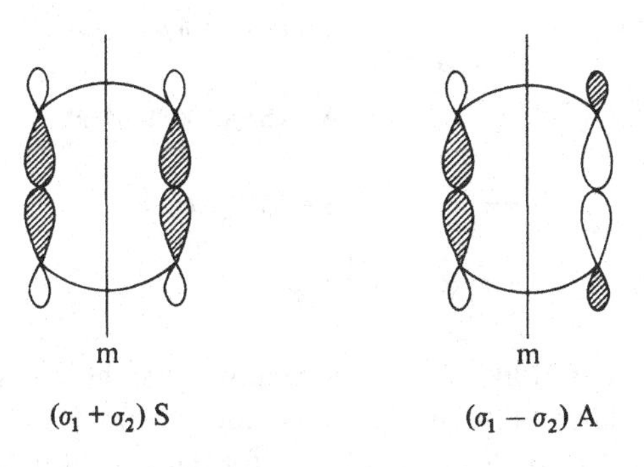

Fig. 2.15

correlation diagram might then be misleading. The application of correlation diagrams to explain the course of photochemical reactions also needs great caution (chapter 3). Correlation diagrams do, however, succeed in giving a good insight into the reasons why some reactions are thermally favoured and others are not.

The correlation diagrams that have been constructed so far involve relationships of orbitals; another type, *state correlation diagrams*, link states of like symmetry. The state symmetry for a molecule is obtained by multiplying together the symmetry labels for each electron, using the rules: S × S = S; A × A = S; S × A = A; A × S = A.

Consider again the example of the conrotatory and disrotatory closure of butadiene to cyclobutene. The symmetry classifications of the orbitals involved, with respect to the mirror plane and the twofold axis, are:

twofold axis S: $\psi_2, \psi_4, \sigma, \pi^*$
(conrotatory) A: $\psi_1, \psi_3, \pi, \sigma^*$
mirror plane S: $\psi_1, \psi_3, \sigma, \pi$
(disrotatory) A: $\psi_2, \psi_4, \pi^*, \sigma^*$

The ground state of butadiene ($\psi_1^2\psi_2^2$) is (AA)(SS) = S with respect to a twofold axis. The first excited state ($\psi_1^2\psi_2\psi_3$) is (AA)(S)(A) = A. Similarly, state symmetries for other excited states of butadiene and for the states of cyclobutene can be calculated, and a state correlation diagram drawn (fig. 2.16). The states which are linked have matching sets of symmetry labels for the electrons, as well as having the same overall state symmetry. The broken lines indicate the actual paths followed in the interconversion of states of A symmetry.

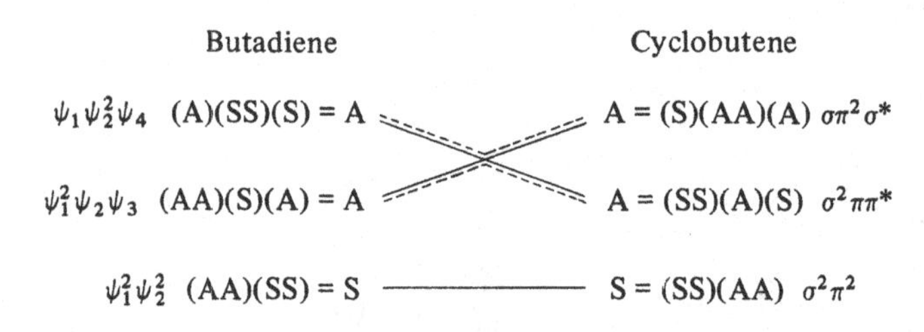

Fig. 2.16. State correlation diagram (conrotatory).

Because of the rule that states of like symmetry cannot cross, the actual correlations of the A states will be those shown by the dotted lines. Fig. 2.16 shows that the conrotatory mode of closure is allowed for the ground state but not for the first excited state. A similar state correlation diagram for the disrotatory closure (fig. 2.17) reveals the energy barrier to the ground state reaction.

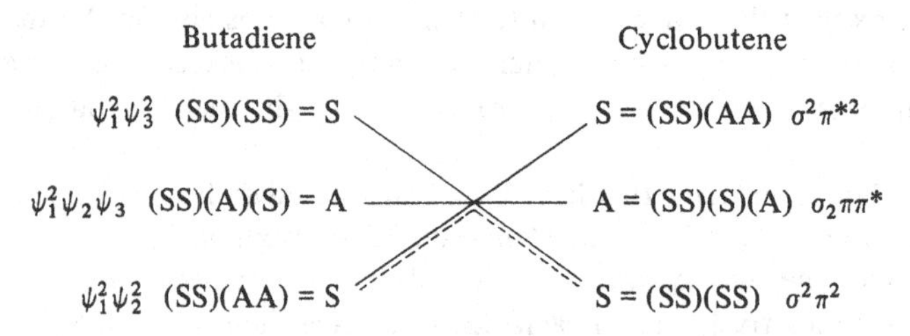

Fig. 2.17. State correlation diagram (disrotatory).

State correlation diagrams are particularly useful for predicting the course of reactions from excited states; an initial increase in energy, as indicated by the upward slope of the line from a given state, is a sign that the reaction from that state has a high energy barrier.

From the results obtained by the application of these concepts of symmetry to a wide range of concerted reactions, Woodward and Hoffmann have been able to formulate a generalised selection rule for predicting whether or not a reaction is symmetry-allowed. Equivalent forms of the same rule have been arrived at independently by Zimmerman[5] and by Dewar.[5] Since their approaches to the theory are different from that of Woodward and Hoffmann, these will be described first, and the Woodward–Hoffmann rule then compared with the alternative statements.

2.4. The 'aromatic transition state' concept[5]

One of the major concepts currently used in organic chemistry is that of aromaticity. The Hückel rule, that planar cyclic polyenes with $(4n + 2)\ \pi$ electrons (n being an integer) are more stable than their acyclic counterparts whereas those with $4n\ \pi$ electrons are less stable, is supported both by theory and by practical experience. This striking alternation in the stabilities of cyclic polyenes suggests a possible connection with the equally striking alternation in the stabilities of cyclic transition states of concerted reactions. This is the basis of the theories of both Dewar and Zimmerman; just as the $(4n + 2)\ \pi$ cyclic polyenes can be described as *aromatic* and the $4n\ \pi$ systems as *antiaromatic*, so the cyclic transition states of concerted reactions can be classified as aromatic or antiaromatic. To take a simple example, the thermal $2\pi + 2\pi$ cyclodimerisation of olefins prefers to go by a stepwise mechanism, whereas the $4\pi + 2\pi$ Diels–Alder cycloaddition is generally concerted. It may therefore be concluded that the four-electron cyclic transition state is destabilised relative to its acyclic counterpart whereas the six-electron cyclic transition state is more stable than its acyclic counterpart. To illustrate the point, the cyclic transition state (1) for the Diels–Alder reaction has an orbital arrangement equivalent to that of benzene and the acyclic transition

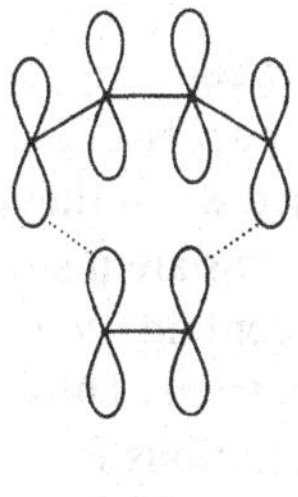

(1)

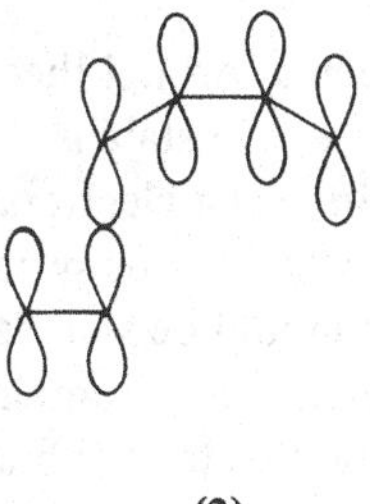

(2)

state (2), one like that of hexatriene. The cyclic transition state should be more stable for the same reasons that benzene is more stable than hexatriene.

This concept is extremely useful in practice. In order to make full use of the concept it is necessary to introduce another idea, that of Möbius systems. If a planar, linear polyene is twisted so that one end is turned through 180° relative to the other, and the ends of the polyene are then joined, the top portion of the π system will overlap with what was the bottom, and the molecule will be a so-called Möbius polyene. The topology of the π system is that of a Möbius strip: a simple model can be made by taking a narrow strip of paper, twisting one end through 180°, and then joining the ends together. The Möbius strip is remarkable in that it has a single continuous surface; similarly the p orbitals of the Möbius polyene form a single continuous ring (fig. 2.18) instead of the two separate rings of the normal cyclic π system.

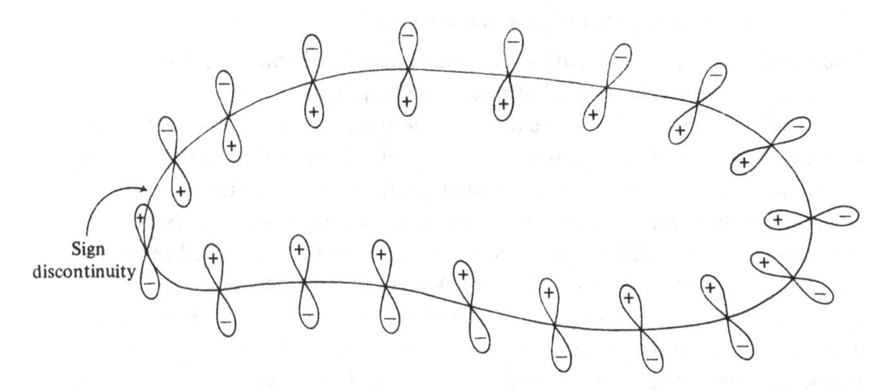

Fig. 2.18. Möbius π system.

In theory one can imagine similar cyclic polyenes with 2, 3, 4, ... twists. Those with zero or an even number of twists are classed as Hückel polyenes, and those with an odd number of twists are classed as Möbius polyenes. Twisted polyenes of this type are as yet unknown, and will presumably be highly strained, reactive species if they are ever made. On the other hand the possibility of having twisted Hückel or Möbius *transition states* in pericyclic reactions is a real one.

The predictions of molecular orbital theory for Hückel cyclic polyenes have already been stated: they are stabilised relative to their acyclic counterparts when they contain $(4n + 2)$ π electrons and destabilised when they contain $4n$ π electrons. For the Möbius systems the predictions are just the opposite. Möbius polyenes should be stabilised when they contain $4n$ π electrons and destabilised when they contain $(4n + 2)$ π electrons. Thus, if both types of cyclic transition state are obtainable, the predictions for thermal pericyclic reactions can be summarised as follows:

Number of electrons	*Hückel type*	*Möbius type*
0, 4, 8, . . . , $4n$	unfavourable	favourable
2, 6, 10, . . . , $(4n + 2)$	favourable	unfavourable

For photochemical reactions the rules are reversed.

As an example of a reaction which can be regarded as having a Möbius transition state, consider the conrotatory ring closure of butadiene described earlier. The signs on the p lobes of the butadiene are chosen to minimise the number of out of phase overlaps (fig. 2.19). This choice is an arbitrary one and is made for simplicity; the reversal of the sign on one of the lobes would just introduce two more sign inversions into the transition state, and leave its classification (as a Hückel or Möbius system) unaltered. The important point is that the method does not depend on knowing the symmetries of the various molecular orbitals of the polyene, and so can be applied to transition states with no symmetry.

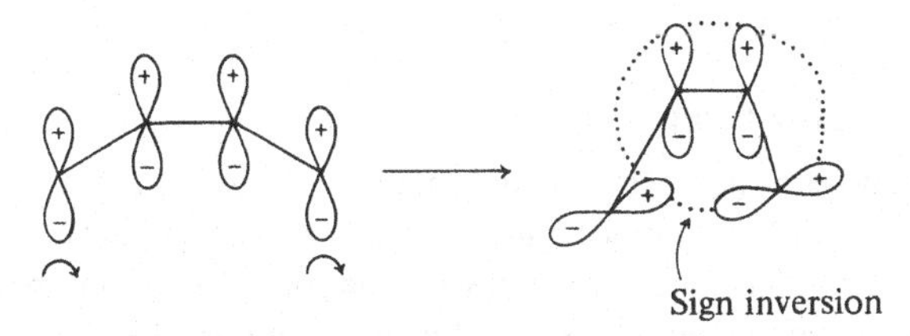

Fig. 2.19

In this reaction the transition state is a Möbius system because there is a sign inversion between the overlapping p lobes of C-1 and C-4, and since four electrons are involved, the reaction should be thermally favourable.

If the $4n$ π Möbius systems are included in the definition of aromatic compounds, the general rule for thermal pericyclic reactions can be succinctly stated as follows:

'Thermal pericyclic reactions take place via aromatic transition states.'

Although the basic idea of this rule was actually first proposed in 1939 by Evans,[6] its revival and justification is mainly due to Dewar. It is probably the simplest and clearest statement of the requirements for a concerted reaction to go thermally via a cyclic transition state. The procedure for using it is as follows:

1. Draw the transition state as a series of overlapping s and p orbitals, putting in + and – signs so that they minimise the number of sign changes in the participating orbitals.
2. Count the number of sign inversions in the cyclic array and the number of electrons involved. If the cycle includes both lobes of a single p orbital as in fig. 2.19, the necessary sign inversion across these two lobes is not counted.

3. Classify the transition state as being of the Hückel type (zero or an even number of sign inversions) or of the Möbius type (odd number of sign inversions). From the number of electrons determine whether or not it is aromatic.

Examples of the use of the rule are given extensively in later chapters, but a few illustrations here may help to clarify the procedure.

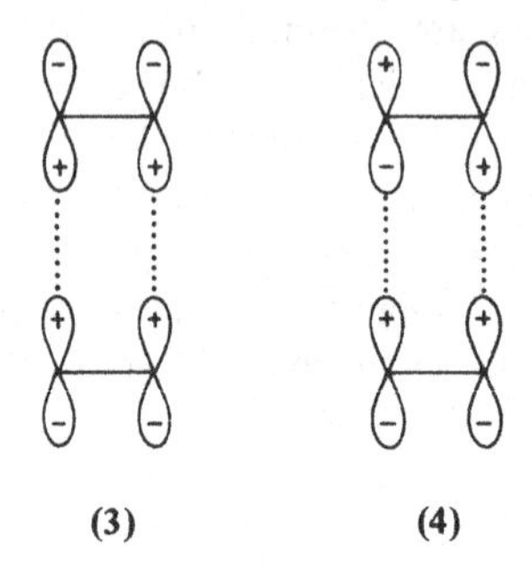

(3) (4)

Cyclodimerisation of ethylene

The simplest transition state (3) for the cyclodimerisation is shown, the signs chosen indicating minimum phase change in the participating components. The four carbon atoms are coplanar, with the top lobes of one π system interacting with the bottom lobes of the other. The transition state is of the Hückel type and as it contains four electrons it is antiaromatic. The reaction is therefore not favoured thermally. Note that the choice of signs for the lobes is arbitrary: if one is reversed, as in (4), the transition state now has two sign inversions but is still of the Hückel type.

Opening of a cyclopropyl cation

This reaction (fig. 2.20), like the cyclisation of butadiene, is an electrocyclic process, and as such, can be conrotatory or disrotatory. For the conrotatory opening the transition state (5) is of the Möbius type; for the disrotatory opening it is of the Hückel type (6). Since two electrons are involved, the opening should therefore be disrotatory.

1,3-Migration of a substituent through an allyl system

In this reaction (fig. 2.21) a group R migrates across the top face of an allyl system from C-1 to C-3.

Two transition states are possible, depending on whether R uses an s orbital (7) or a p orbital (8) for bridging.

Transition state (7) is of the Hückel type and (8) is of the Möbius type. Since four electrons participate, (7) is not allowed for a thermal reaction but (8) is. This prediction is borne out in practice; it has interesting stereochemical consequences, which are discussed in chapter 7.

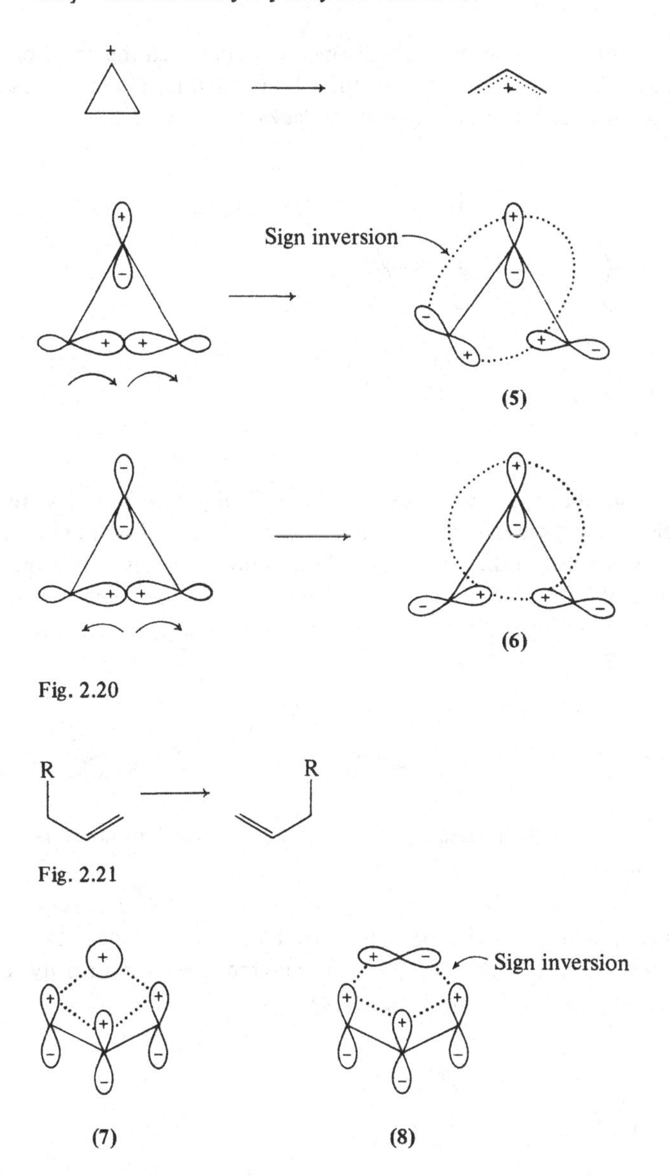

2.5. General rule for pericyclic reactions

A general rule for determining the stereochemical course of concerted reactions has been put forward by Woodward and Hoffmann. It requires some preliminary definitions and comments.

Consider a pericyclic reaction in which the electrons of a π bond are used in the transition state, and in which new bonds are being formed at the termini. There are two stereochemically distinct ways in which the new bonds can be formed: either to the same face of the π bond (that is, in a *suprafacial* way), or to opposite faces (that is, in an *antarafacial* way) (fig. 2.22).

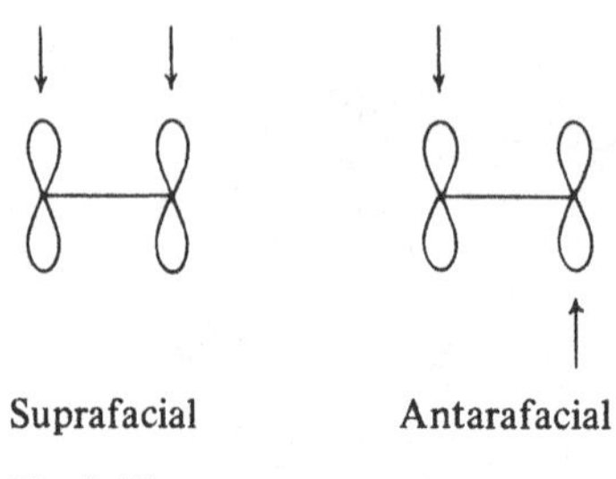

Suprafacial Antarafacial

Fig. 2.22

The same definitions apply to longer π systems. Similarly, there are stereochemically distinct ways in which a σ bond can be used. Suprafacial use of a σ bond involves bonding in the same way at both termini; that is, retention of configuration at both ends, or inversion at both ends. Antarafacial use of a σ bond involves retention at one end and inversion at the other. These are illustrated in fig. 2.23.

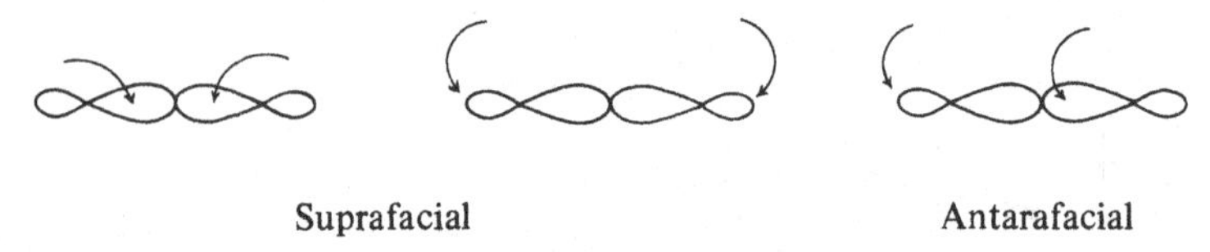

Suprafacial Antarafacial

Fig. 2.23

A non-bonding p orbital which participates in a pericyclic reaction can form bonds to both the flanking groups from the same lobe (suprafacially) or from opposite lobes (antarafacially) (fig. 2.24).

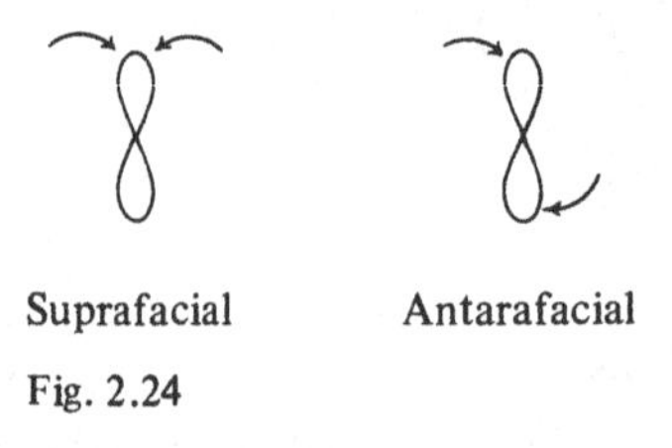

Suprafacial Antarafacial

Fig. 2.24

The symbols π, σ and ω are given respectively to the π systems, σ bonds and lone p orbitals which participate in the transition state, and the symbols (s) and (a) to indicate their suprafacial or antarafacial use. The notation is completed by the number of electrons supplied by each component. Thus,

${}_{\pi}2_s$ denotes a two-electron π system used in a suprafacial way, ${}_{\omega}0_a$ indicates a vacant p orbital used in an antarafacial way, and so on.

Woodward and Hoffmann state the general rule for pericyclic reactions as follows:

'*A ground state pericyclic change is symmetry-allowed when the total number of* $(4q + 2)$ *suprafacial and* $4r$ *antarafacial components is odd*' (q and r being zero or integers).

To apply the rule, an orbital picture of the reactants is constructed and the components used in a geometrically feasible way to achieve overlap. The $(4q + 2)$ electron suprafacial and $4r$ electron antarafacial components are counted, the others being ignored. If the total is an odd number, the reaction is predicted to be thermally allowed.

Although the rule may seem to bear little relationship to the aromatic transition-state concept described earlier, it is completely consistent. The use of the same examples as in §2.4 will illustrate the essential similarity.

Cyclodimerisation of ethylene

The reaction via a planar transition state is a ${}_{\pi}2_s + {}_{\pi}2_s$ reaction, **(9)**. There are two $(4q + 2)$ electron suprafacial components and no antarafacial components; the reaction is thermally disallowed, since the total number of 'counting' components is two, an even number.

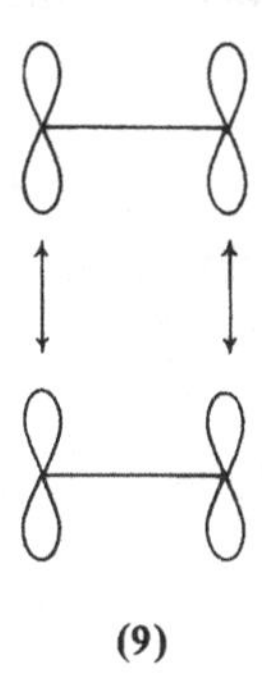

(9)

A comparison of diagrams **(9)** and **(3)** shows that the suprafacial, suprafacial interaction corresponds to a choice of signs for the ethylene components which gives the minimum phase change.

Opening of cyclopropyl cation

The conrotatory ring opening **(10)** is a thermally disallowed ${}_{\sigma}2_a + {}_{\omega}0_s$ reaction. Note the similarity between the overlap here and in diagram **(5)**. The geometrical constraints require that if the σ bond is used in an antarafacial way, then both bonds must be made to the same face of the lone p orbital. Alternatively, the σ bond could be used in a suprafacial way, but then the overlap must be to opposite faces of the p lobe, **(11)**; the reaction, now classed as ${}_{\sigma}2_s + {}_{\omega}0_a$, is still thermally disallowed. A disrotatory opening, for example the ${}_{\sigma}2_s + {}_{\omega}0_s$ process illustrated, **(12)**, is thermally allowed.

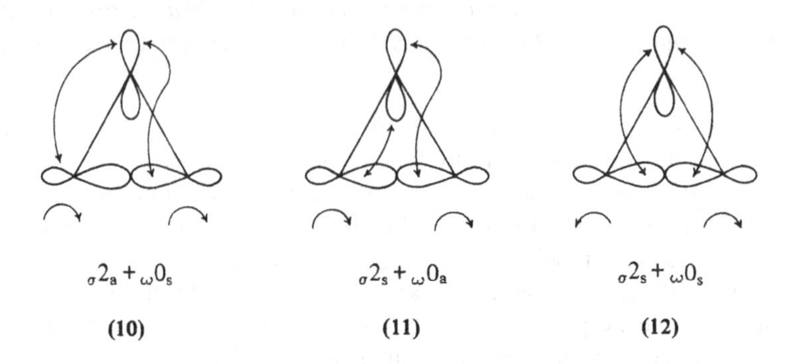

${}_{\sigma}2_a + {}_{\omega}0_s$ **(10)** ${}_{\sigma}2_s + {}_{\omega}0_a$ **(11)** ${}_{\sigma}2_s + {}_{\omega}0_s$ **(12)**

Migration of a substituent through an allyl system

The reaction could involve retention of configuration at the migrating group, **(13)**, or inversion, **(14)**. The former, illustrated as a ${}_{\sigma}2_s + {}_{\pi}2_s$ process, is thermally disallowed. The latter, shown as a ${}_{\sigma}2_a + {}_{\pi}2_s$ process, is allowed; in the transition state, rehybridisation of the sp^3 orbital to a p orbital will presumably take place, making this representation **(14)** equivalent to that of diagram **(8)**.

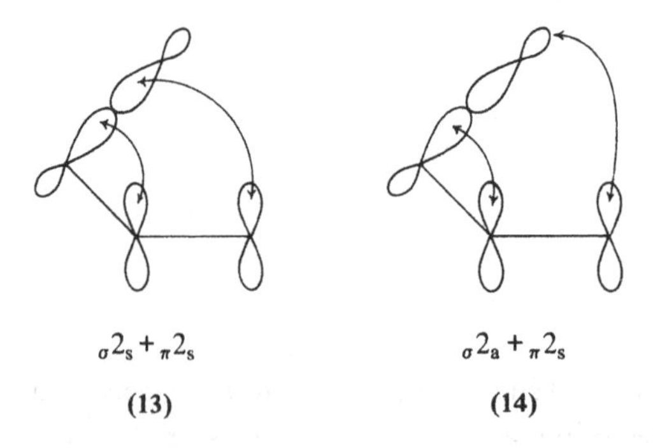

${}_{\sigma}2_s + {}_{\pi}2_s$ **(13)** ${}_{\sigma}2_a + {}_{\pi}2_s$ **(14)**

2.6. Conclusions

There are thus several approaches by which selection rules for pericyclic reactions can be developed.[7,8] Fortunately they almost invariably lead to the same predictions, and indeed some have been shown to be linked theoretically.[8] Some give more insight into particular types of process than others: for example, the advantages of frontier orbital theory in exploring the course of cycloaddition reactions have already been mentioned. In analysing the different classes of pericyclic reactions in the following chapters, we have tended to emphasise the theoretical approaches which give the best picture of those types under discussion.

There are a number of other important factors which the simple treatment given in this chapter has not dealt with, but which will be discussed, in the context of particular examples, in the chapters which follow. For example, the effect of substituents on the systems is clearly of major importance: not only do they have a steric effect but they also can considerably alter the energy levels of the π orbitals. Replacement of carbon atoms in the π systems by heteroatoms can have a similar effect on the orbital energies. Catalysts are capable of altering both the rates and the mechanisms of pericyclic processes. The simple rules do not give clear predictions of the behaviour of systems containing unpaired electrons, nor do they give an adequate picture of the likely course of many photochemical reactions. The applications and limitations of the simple theory are illustrated first for electrocyclic reactions, because reactions of this type provided the original basis for the development of the Woodward-Hoffmann rules.[9]

References

1. Reviews: Woodward, R. B. and Hoffmann, R., *Angew. Chem., Int. Edn Engl.*, **8**, 781 (1969); Hoffmann, R., *Science*, **167**, 825 (1970).
2. An excellent descriptive account of the theory is presented by Fleming, I., *Frontier orbitals and organic chemical reactions*, Wiley, London, 1976. See also Fukui, K., *Theory of orientation and stereoselection*, Springer-Verlag, Berlin, 1975.
3. Coulson, C. A., *Valence*, 2nd edn, Oxford University Press, London 1961.
4. Longuet-Higgins, H. C. and Abrahamson, E. W., *J. Am. chem. Soc.*, **87**, 2045 (1965).
5. Zimmerman, H. E., *Acc. chem. Res.*, **4**, 272 (1971); Dewar, M. J. S., *Angew. Chem., Int. Edn Engl.*, **10**, 761 (1971); Dewar, M. J. S. and Dougherty, R. C., *The PMO theory of organic chemistry*, Plenum, New York, 1975.
6. Evans, M. G., *Trans. Faraday Soc.*, **35**, 824 (1939).
7. Several other presentations of the selection rules and refinements of the theory have been proposed. These include a generalised method which applies to reactions not readily analysed by the Woodward-Hoffmann rules [Wieland, P., *Tetrahedron*, **31**, 1641 (1975)] and a refinement of the correlation diagram approach [Halevi, E. A., *Angew. Chem., Int. Edn Engl.*, **15**, 593 (1976)].
8. Day, A. C., *J. Am. chem. Soc.*, **97**, 2431 (1975).
9. Woodward has described the experiments which led to the original formulation of the Woodward-Hoffmann rules in *Aromaticity, Chem. Soc. Spec. Publ.*, No. 21, 217 (1967).

3 Electrocyclic reactions

An electrocyclic reaction is the formation of a σ bond between the termini of a fully conjugated linear π system, or the reverse process. It is therefore a type of intramolecular cycloaddition or retroaddition.

Electrocyclic reactions can be brought about by heat, by ultraviolet irradiation, and sometimes by the use of metal catalysts. They are nearly always stereospecific. In many cases, detection of their stereospecificity depends on distinguishing chemically similar stereoisomers, a problem which has been overcome mainly by the development of spectroscopic methods of structure determination, especially n.m.r. Thus, the recognition that stereospecific electrocyclic reactions form a coherent group extends only over the last few years; even so, the group includes some important synthetic reactions as well as some of the most clearcut examples of the successful predictive power of the orbital symmetry theory.

3.1. Thermal electrocyclic reactions

There are four stereochemically distinguishable ways in which an electrocyclic reaction can take place; two are disrotatory and two conrotatory. These are illustrated in fig. 3.1 for a ring opening.

Not all these modes of ring opening will be distinguishable in a particular case. If A, B, C and D are identical, for example, there is only one possible product. If A, B, C and D are all different groups, there are four possible products. Basically, the theory distinguishes only the disrotatory modes of ring opening from the conrotatory modes, and does not distinguish between the two possible disrotatory modes or conrotatory modes. Predictions as to which of these will be preferred can usually be made on the basis of steric effects, as subsequent examples will show.

The mode of electrocyclic ring opening or closure depends simply on the number of π electrons in the open polyene. A conjugated linear polyene, whether neutral, or positively or negatively charged, with $4n$ π electrons, has a HOMO with a twofold axis of symmetry, in which the terminal lobes are out

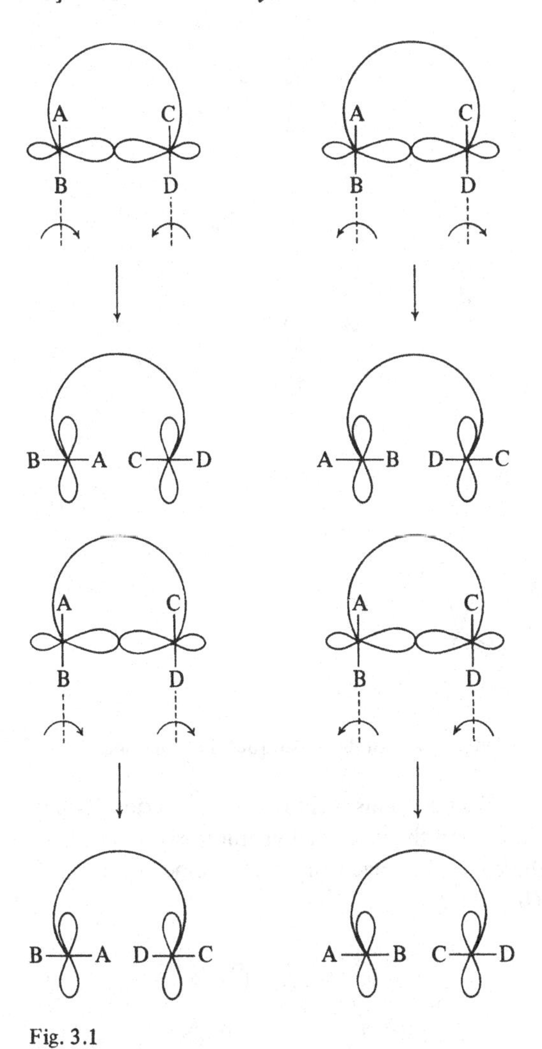

Fig. 3.1

of phase. Examples of such systems (where n = 1) are *cis*-butadiene, the allyl anion and the pentadienyl cation; the HOMOs are shown in fig. 3.2.

In these polyenes, thermal cyclisation must take place by a conrotatory closure. This was illustrated for butadiene in chapter 2; it is most simply pictured, in the general case, as involving a rotation of the termini of the HOMO in a direction necessary for overlap to form the new σ bond; or using the aromatic transition-state approach, as involving a Möbius-type transition state (fig. 3.3).

For linear π systems with $(4n + 2)$ π electrons, the thermal cyclisation is

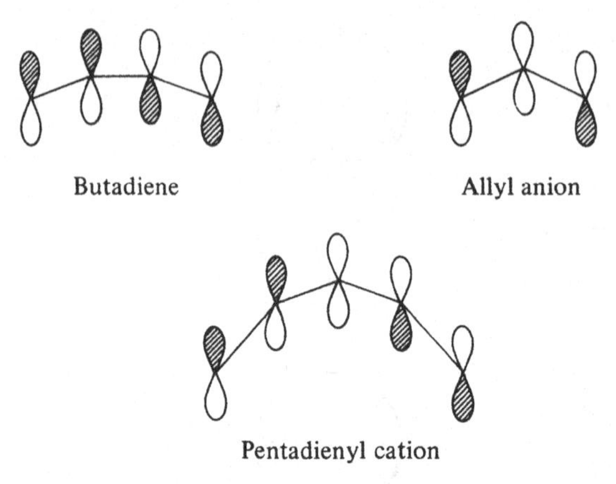

Fig. 3.2

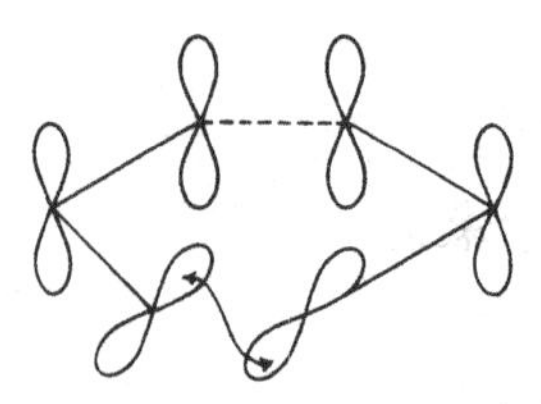

Fig. 3.3. Möbius transition state for $4n$ π electrocyclic reactions.

predicted to be disrotatory. Such systems include the allyl cation (two π electrons, $n = 0$), hexatrienes, and the pentadienyl anion (six π electrons, $n = 1$). The HOMOs of these systems have a plane of symmetry, and the terminal lobes in phase (fig. 3.4).

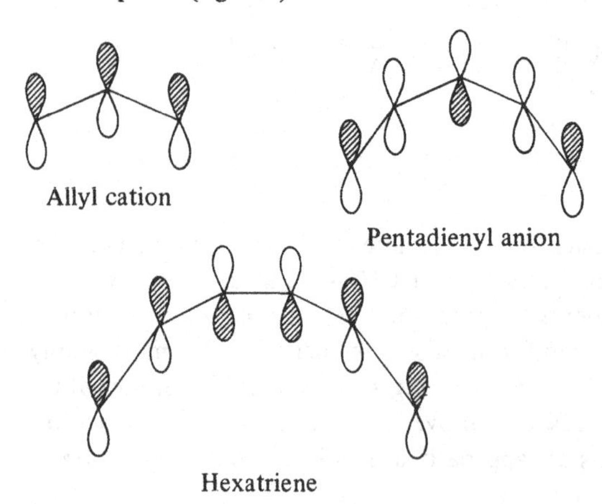

Fig. 3.4. HOMOs of $(4n + 2)$ π systems.

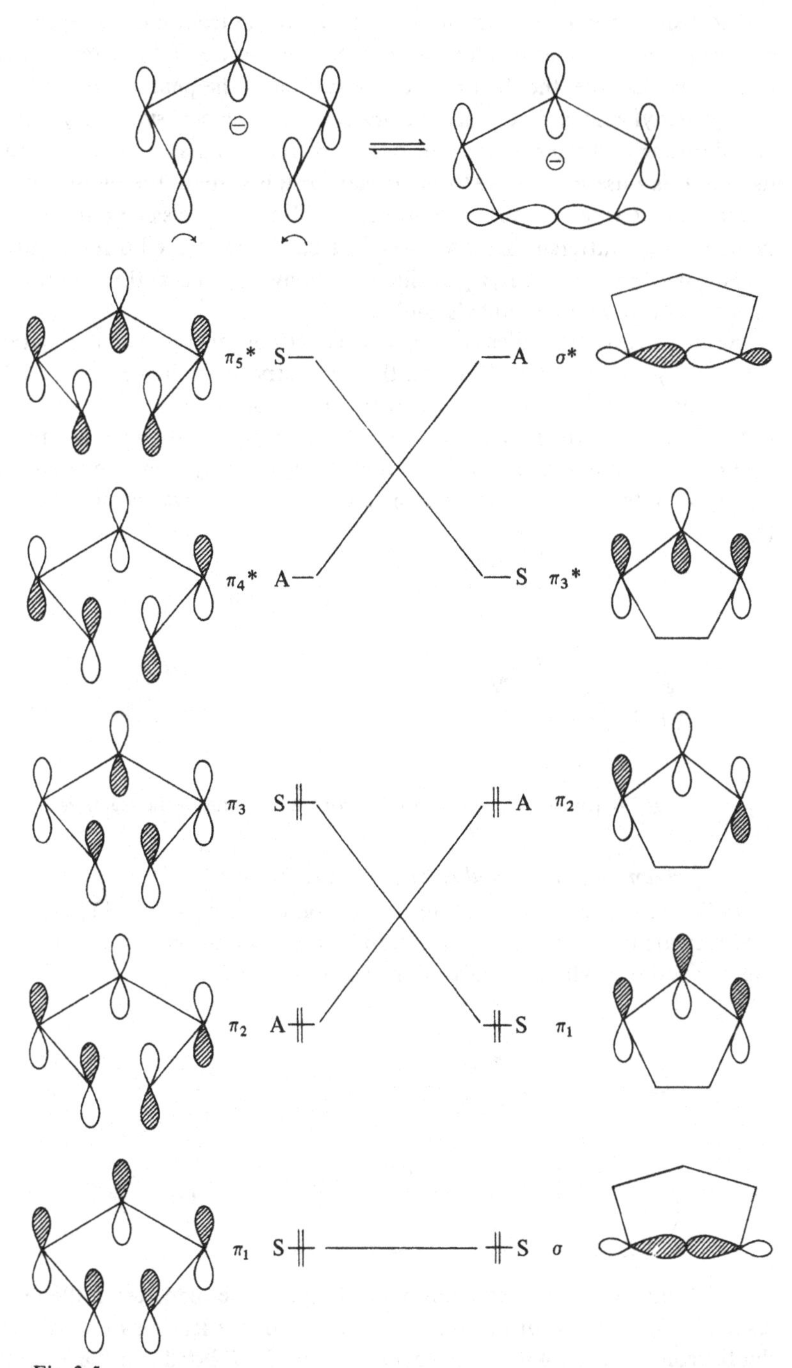

Fig. 3.5

The disrotatory mode of closure can be predicted from the appropriate correlation diagrams, just as for the conrotatory closure of 4π systems. Consider, as an example, the thermal interconversion of the pentadienyl and cyclopentenyl anions. In the disrotatory closure a plane of symmetry is maintained throughout the reaction; the orbital correlation diagram is shown in fig. 3.5. It is a useful exercise to show that for this system, the thermal conrotatory mode is forbidden. (The orbitals involved are the same but their symmetries are different because a twofold axis is maintained throughout, not a plane of symmetry. Thus, π_1 of the pentadienyl anion is antisymmetric with respect to the twofold axis, and so on.)

Again, the selection rules can be deduced without recourse to correlation diagrams by concentrating either on the symmetry of the highest occupied orbital, or on the 'aromatic' nature of the transition state. Thus, for a $(4n + 2)\,\pi$ system, the in-phase termini of the HOMO must move in a disrotatory fashion to overlap, or, as a Hückel $(4n + 2)\,\pi$ system, the aromatic transition state is achieved without sign inversion in the system of p orbitals (fig. 3.6).

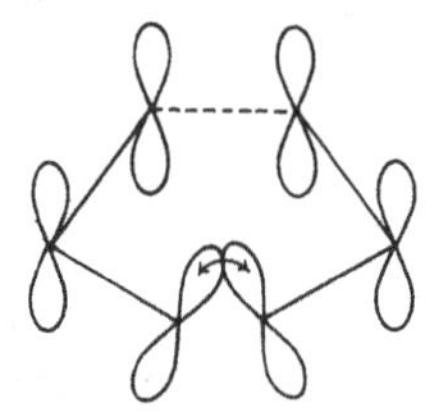

Fig. 3.6. Hückel transition state for $(4n + 2)\,\pi$ electrocyclic reactions.

Examples of thermal electrocyclic reactions

1. *2π Systems.* The cyclopropyl to allyl cation isomerisation (3.1) is a 2π electrocyclic ring opening. The reaction is an important one, especially in bicyclic systems, where it leads to ring expansion (3.2).

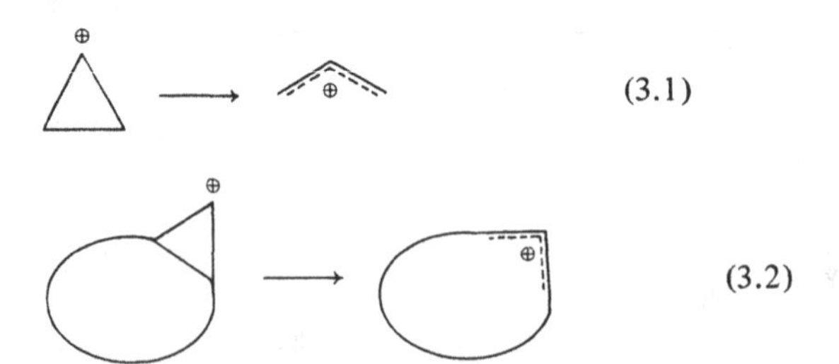

It is unlikely that free cyclopropyl cations are intermediates in the reported examples of reactions of this type; the removal of the leaving anion is assisted by the concerted breaking of the cyclopropane C—C bond and the generation of the allyl π system. This is elegantly shown by experiments with bicyclic cyclopropanes which reveal that the leaving group has to be *endo* for the smooth ring opening to take place. The effect is especially marked in cyclopropanes fused to five- or six-membered rings. This result was predicted by Woodward and Hoffmann in 1965, and has been confirmed in several systems since then. One example is provided by the isomeric 6-chlorobicyclo[3,1,0]-hexanes (1) and (2); the *endo*-isomer (1) is converted into 3-chlorocyclohexene by heating for three hours at 126 °C, but the *exo*-isomer (2) was found to be unchanged, even in much more vigorous conditions.[1]

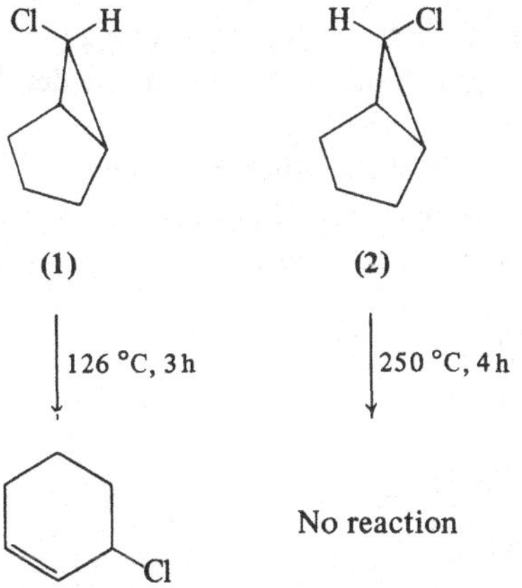

(1) —126 °C, 3 h→ 3-chlorocyclohexene

(2) —250 °C, 4 h→ No reaction (3.3)

Woodward and Hoffmann explain the preferred removal of the *endo*-anion as follows. The electrocyclic ring opening is a disrotatory one, and only the disrotatory mode which moves the bridgehead hydrogens outward is geometrically feasible in small, fused ring systems. In the *endo*-isomer, therefore, the C—C bond breaks in such a way as to increase the electron density behind the leaving group, and to bring about a type of S_N2 displacement (3.4).

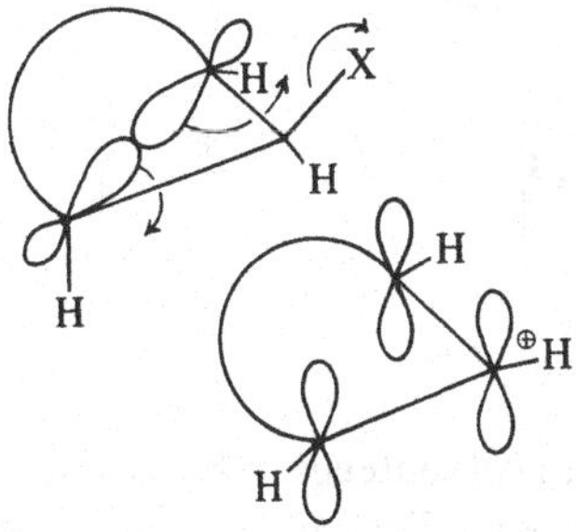

(3.4)

The effect is less marked in larger fused rings and in monocyclic cyclopropanes, but it still operates. For example, the thermal ring opening of cyclopropane (3) is faster than that of (4) because in (3) the required disrotatory mode allows an outward rotation of the methyl groups, but in **(4)**, the methyls must move towards each other in the transition state.[2]

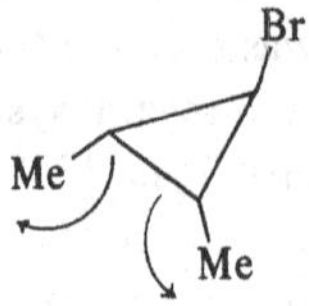

(3) 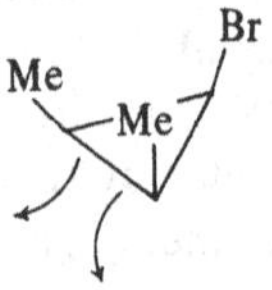 **(4)**

Other factors which affect the ease of electrocyclic ring opening include the nature of the substituents, which can stabilise or destabilise the developing positive charge, and the relief of strain in small bicyclic systems. Thus, the adjacent oxygen may assist the opening of the ether **(5)**, which goes smoothly at 60 °C; and the strained bicyclopentane **(6)** is unstable even at 0 °C (the ring opening in this case being facilitated by the relief of ring strain).

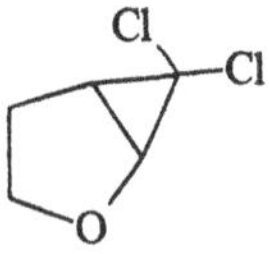

(5) 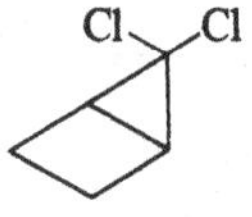 **(6)**

The cyclopropyl to allyl ring opening is one of the most useful of electrocyclic reactions, especially as a means of ring expansion, since the halogenocyclopropanes are readily available by carbene addition to the appropriate olefins.

An interesting extension of the reaction, which involves an iso-electronic heterocyclic ring expansion with the same orbital symmetry requirements, is the thermal conversion of the chloroaziridine (7) to isoquinoline (3.5).[3]

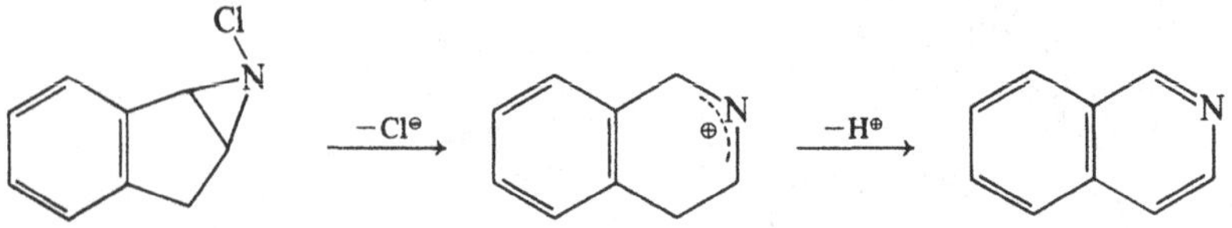

$-Cl^{\ominus}$ $\longrightarrow$ $-H^{\oplus}$ $\longrightarrow$ (3.5)

(7)

2. 4π *Systems.* The thermal ring opening of cyclobutenes to butadienes is stereospecific, and conrotatory, as the theory predicts. In most cases the ring opening goes to completion and there are very few examples of the reverse process, the thermal cyclisation of butadienes. The electrocyclic ring opening takes place smoothly in solution or in the gas phase at temperatures between

about 120 °C and 200 °C, the activation energy being about 146 kJ mol^{-1}. Typical examples occur with the *trans*- and *cis*-3,4-dichlorocyclobutenes, which open stereospecifically to *trans,trans*- and *cis,trans*-1,4-dichlorobutadiene, respectively (3.6).[4]

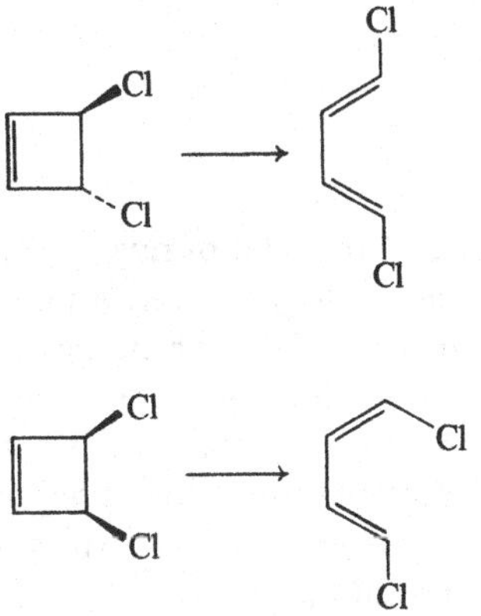

(3.6)

An example of the reverse process is provided by the conrotatory cyclisation of the strained *cis,trans*-1,3-cyclo-octadiene (**8**) which takes place at 80 °C.

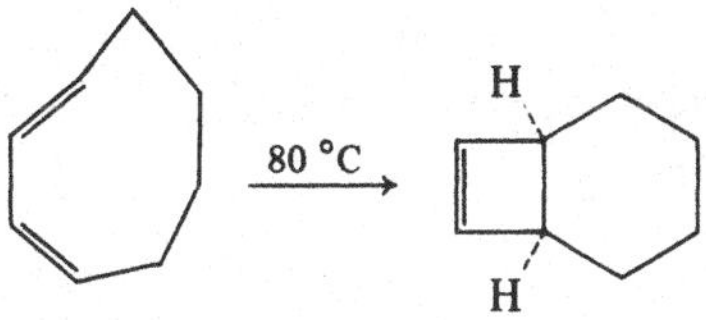

80 °C

(3.7)

(**8**) (**9**)

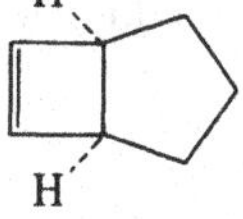

(**10**)

Fused cyclobutenes are thermally rather stable, especially those in which the second ring is five- or six-membered. For example, the cyclobutene (**10**) does not isomerise below about 380 °C.[4]

This might seem surprising, since the fusion of a second ring might be expected to increase the strain and make ring opening easier. However, the allowed ring opening is conrotatory, which would require the formation of dienes containing a *trans* fused double bond. The fused cyclobutenes prefer to take a path which leads to the disallowed *cis, cis* cyclic diene. For example, the bicyclo-octene (**9**) opens at 230–260 °C to give *cis,cis*-1,3-cyclo-octadiene

(3.8), and not the *cis,trans*-diene (8).[6] The activation energy, 180 kJ mol^{-1}, is much higher than for the opening of simple cyclobutenes.

H H (3.8)

(9)

Substituents can have a considerable effect on the rates of ring openings of this type which are constrained to go in a disrotatory manner. An interesting example is provided by the thermal rearrangement of Dewar benzene **(11)** and the mono- and dichloro derivatives **(12)** and **(13)**. All three compounds isomerise to the corresponding benzenes on heating, but the isomerisation of the mono-chloro compound **(12)** is faster than that of both **(11)** and **(13)**. It has been suggested that the unsymmetrical substitution pattern in **(12)** provides a 'push–pull' stabilisation of the antiaromatic transition state.[7]

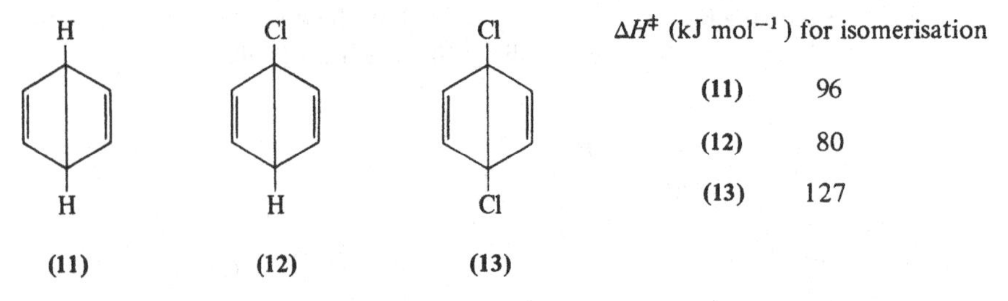

(11) **(12)** **(13)**

	$\Delta H^{\ddagger}$ (kJ mol^{-1}) for isomerisation
(11)	96
(12)	80
(13)	127

Conrotatory thermal ring opening is also predicted for heterocyclic compounds, e.g. **(14)**, in which the four electrons include a lone pair on the heteroatom X. Several examples of this type of ring opening are known,

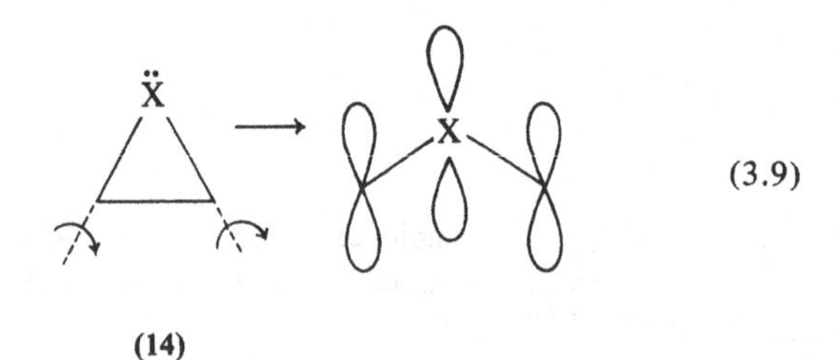

(3.9)

(14)

and where the stereochemistry has been determined, the opening is conrotatory. For example, many aziridines open when heated to about 100 °C, and form delocalised 4π intermediates which can sometimes be trapped in cycloaddition reactions. The ring opening is conrotatory, as Huisgen and his colleagues showed for the *cis*- and *trans*-diesters **(15)** and **(16)** (equation

3.10).[8] The stereochemistry was established in the cycloadducts of the 1,3-dipoles with dimethyl acetylenedicarboxylate.

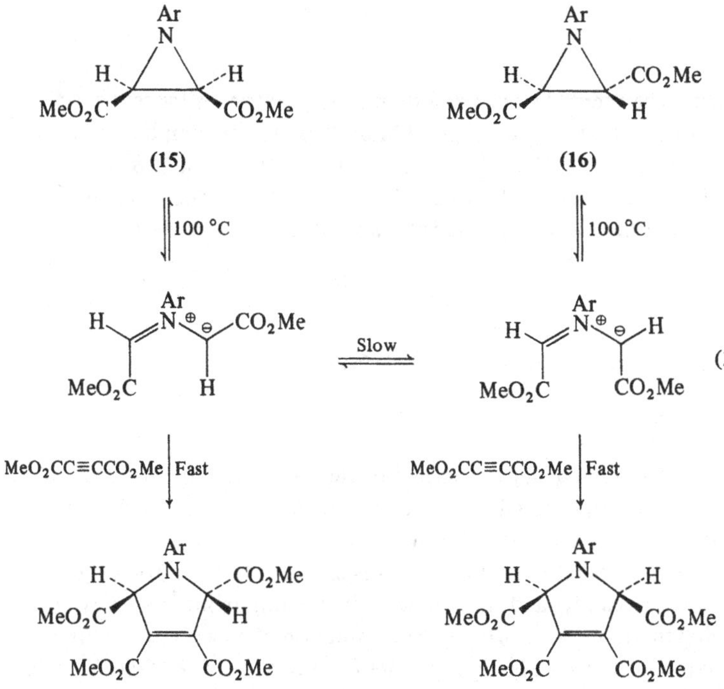

Ar
N
H
H
MeO2C
CO2Me
(15)
Ar
N
H
CO2Me
MeO2C
H
(16)
100 °C
100 °C
Ar
N
CO2Me
H
MeO2C
H
Slow
Ar
N
H
H
MeO2C
CO2Me
MeO2CC≡CCO2Me Fast
MeO2CC≡CCO2Me Fast
Ar
N
H
CO2Me
MeO2C
H
MeO2C
CO2Me
Ar
N
H
H
MeO2C
CO2Me
MeO2C
CO2Me

(3.10)

In the absence of a good dienophile, the two open structures are interconvertible, and this provides a mechanism for thermal isomerisation of the aziridines.

Just as with cyclobutenes, fusion of the aziridine in a structure which cannot open by the conrotatory mode inhibits the reaction. An example of such a system is the aziridine (**17**), which failed to give cycloaddition reactions even at 180 °C.[9] The opening of cyclopropyl to allyl anions is also slower in fused systems, as expected for the predicted conrotatory mode.[10]

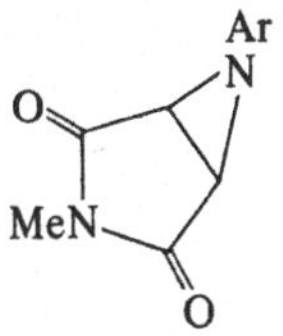

Ar
N
O
MeN
O

(**17**)

Another system which shows the same type of thermal ring opening is tetracyanoethylene oxide. The open form has been trapped in cycloaddition reactions: the addition to olefins at 130–150 °C is stereospecific (3.11).[8]

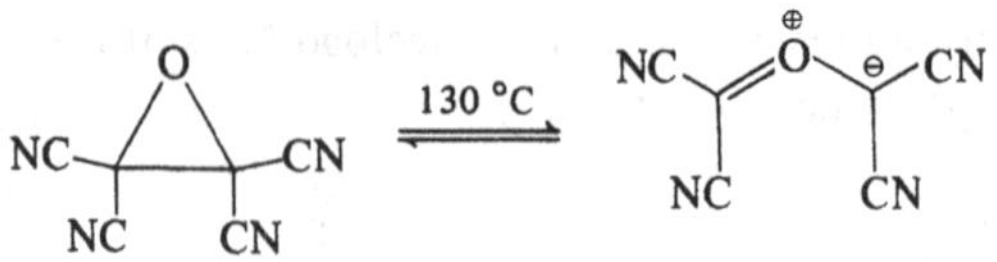

(3.11)

3. 6π *Systems.* The thermal cyclisation of acyclic hexatrienes takes the predicted disrotatory path, and the equilibrium lies almost completely in favour of the cyclic form. The thermal conversion of the triene **(18)** to the *cis*-dimethylcyclohexadiene **(19)** is a good example, the product contains less than 0.1 % of the thermodynamically more stable *trans*-isomer.[11]

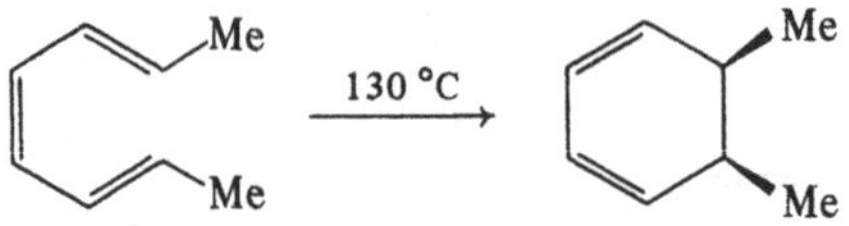

(3.12)

(18) **(19)**

Although a few other acyclic examples of stereospecific isomerisation of hexatrienes are known, especially in natural product work, the commonest reactions of this type are in cyclic hexatrienes. Cyclononatriene, cyclo-octatriene and cyclo-octatetraene are systems in which the electro-cyclic reaction goes very readily, and they show an interesting trend.[12] The cyclisation of cyclononatriene goes virtually to completion at room temperature, whereas the reverse reaction is slightly more favourable for the cyclo-octatriene equilibrium. For cyclo-octatetraene and bicyclo[4,2,0]octatriene **(20)**, the

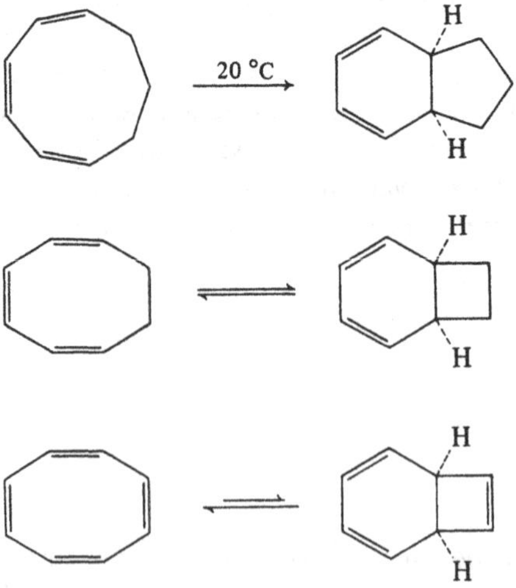

(3.13)

(20)

equilibrium strongly favours cyclo-octatetraene, to such an extent that the half-life of the bicyclo-isomer (**20**) is only a few minutes at 0 °C. Thus, as the bicyclic system becomes more strained, the equilibrium lies more in favour of the open structure.

Norcaradiene and cycloheptatriene structures are similarly inter-related through a disrotatory electrocyclic reaction (3.14). As with any rapidly equilibrating mixture, the structures cannot be distinguished by chemical means - catalytic reduction of a norcaradiene derivative gives a cycloheptane, for example - and the problem of whether a particular compound has the norcaradiene or the cycloheptatriene structure is best solved by the use of high resolution n.m.r.

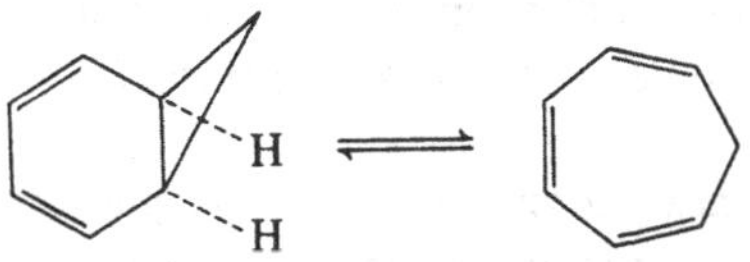

H H ⇌ (3.14)

N.m.r. can reveal a great deal of information in structural problems of this type. Not only can it distinguish the valence tautomers, but it can be used to study the rate of interconversion of the tautomers at temperatures where they are in equilibrium.

At temperatures below that required to establish the equilibrium, the n.m.r. spectrum represents only the separate isomeric structures, which can therefore be distinguished. Similarly, if the valence tautomers are in rapid equilibrium, but one tautomer is a very minor component (less than about 2 % of the total) in the equilibrium mixture, then the minor component is not detected and the n.m.r. spectrum represents only the structure of the major tautomer. These spectra are relatively simple to interpret. An example of the use of n.m.r. to distinguish a non-equilibrating system of this type is provided by the assignment of structure to the first simple norcaradiene to be prepared, the dicyano-derivative (**21**). N.m.r. established the norcaradiene structure rather than the cycloheptatriene one; the signals from the bridgehead hydrogen atoms appear at 3.47δ.[13]

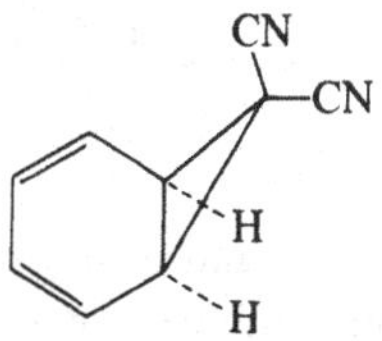

CN CN H H

(**21**)

A rapidly equilibrating mixture containing predominantly one tautomer (greater than about 98 % of the total) will also give a spectrum representing only the structure of the major tautomer.[14] In these cases the spectra may be temperature dependent. If Δt (s) is the lifetime of a particular tautomer and $\Delta \nu$ (Hz) the difference in chemical shifts of a proton in two different tautomers, then the spectra will be temperature dependent if the product $\Delta t \cdot \Delta \nu$ lies between 10^{-3} and 10. In such systems the spectra normally reflect three structural situations.

1. At low temperatures, the rate of interconversion of the tautomers may be so slow that the spectra of the individual tautomers can be seen, and are relatively invariant within this temperature range.
2. In an intermediate temperature range, the rate of interconversion of the tautomers is comparable with the n.m.r. time scale. The signals due to the individual tautomers broaden, become diffuse and then coalesce to form new signals at positions intermediate between those for the same protons in the separate tautomers.
3. At high temperatures, the tautomers are interconverted too rapidly to be detected individually, and the spectrum is an 'average', not equivalent to any single classical structure. When the sample is cooled, the low temperature spectrum should reappear, showing that no irreversible change has taken place.

Several examples of the use of this technique to detect valence tautomerism appear in later chapters. An application to the present problem is the interconversion of the two fused norcaradienes **(22)** and **(24)**, through the valence tautomer **(23)**.[15] In this system the equilibrium strongly favours the norcaradienes, because aromaticity is lost in the cycloheptatriene structure. At room temperature, the spectrum is that of the norcaradiene, with the signal from the *exo*-proton H^b at 1.30δ and that from the *endo*-proton H^a at -0.35δ.

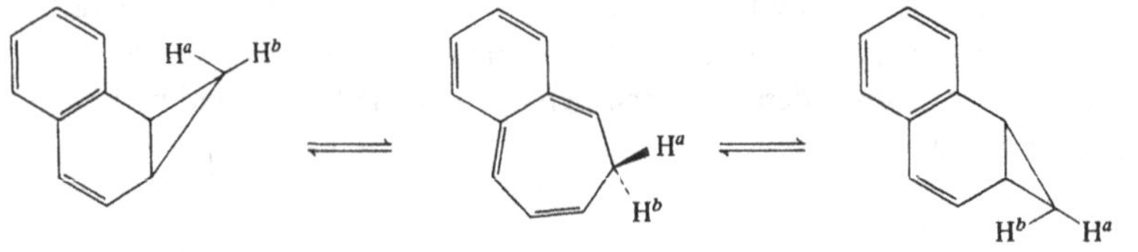

H^a H^b H^a H^b H^b H^a

(22) ⇌ (23) ⇌ (24) (3.15)

When the solution is warmed, the signals become diffuse and slowly merge, until at 180 °C they appear as a uniform flat absorption band centred at 0.46δ, midway between the original signals. Further increase in temperature sharpens the signal but does not alter its position. The other peaks in the spectrum are unchanged, and when the solution is cooled, the original signals reappear.

The explanation is that as the temperature is raised, the two norcaradienes are interconverted through the cycloheptatriene, though its concentration remains minute. In **(24)** the *exo*-proton of **(22)** has become *endo*, and vice versa, so that the signals in the two environments become indistinguishable when the interconversion is rapid.

In most derivatives of this system the equilibrium strongly favours one of the two forms, and the other is not detected. Most simple derivatives exist in the cycloheptatriene form, the dicyano derivative **(21)** being a rare exception.[16] In fused derivatives, like **(22)**, the norcaradiene structure may be favoured for special reasons, such as the loss of resonance energy in the cycloheptatriene.

Benzene oxide **(25)** and oxepin **(26)** are another example of a pair of valence tautomers interconvertible by disrotatory electrocyclic reactions. The system has been extensively studied by Vogel and his colleagues.[17] Benzene oxide and other arene oxides may be intermediates in the oxidative metabolism of aromatic substrates to phenols and catechols. 1,2-Naphthalene oxide **(27)** has actually been detected as an intermediate in the oxidation of naphthalene in a biological system.[18]

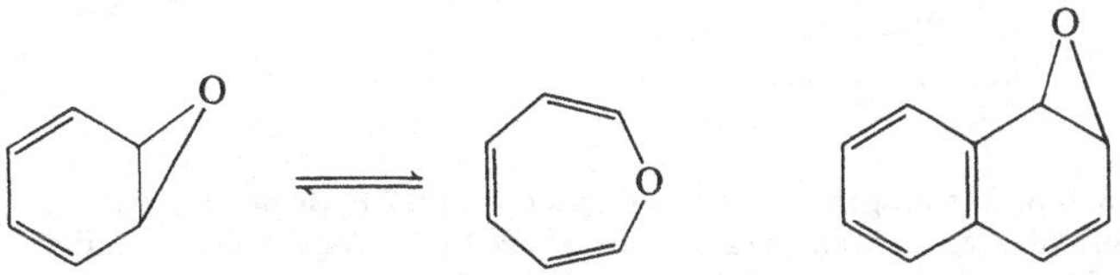

(25) **(26)** **(27)**

N.m.r. spectra of benzene oxide–oxepin give a great deal of information about the equilibrium. At room temperature, the tautomers are in equilibrium with appreciable amounts of both isomers in the equilibrium mixture (the exact proportions depend on the solvent and on the temperature). The n.m.r. spectrum is a complex one, showing 'weighted average' signals from the protons in the rapidly interconverting tautomers. Below about –80 °C the signals broaden and eventually separate into new signals which can be assigned to protons in the individual tautomers. The signal for the α-hydrogens at low temperatures is shown in fig. 3.7. The broad peak which appears at about 5δ in the spectrum run at –87 °C represents an 'average' for the α-hydrogens in the rapidly interconverting tautomers. As the temperature is lowered the signal becomes more diffuse, and below –113 °C it separates into two new signals at 5.7δ and 4δ. These can be assigned to the α-protons in oxepin and benzene oxide, respectively.

The activation energy barrier to the interconversion is much too low for the individual tautomers to be isolated at room temperature. As the temperature is lowered, the proportion of benzene oxide in the equilibrium

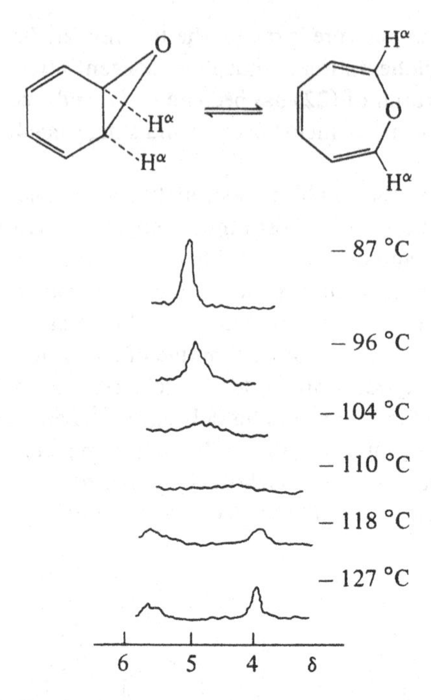

Fig. 3.7. Signals of α-hydrogens in [1]H n.m.r. spectrum (in CF_3Br/pentane) of benzene oxide–oxepin. Adapted from fig. 572.3 in the article by E. Vogel and H. Günther, *Angewandte Chemie International Edition*, **6**, 390 (1967).

mixture increases, since the unfavourable entropy term $T\Delta S$ becomes less significant.

Fused systems of this type can exist exclusively in the oxepin form or exclusively in the benzene oxide form, just like their all-carbon analogues. 1,2-Naphthalene oxide **(27)**, for example, has n.m.r. and ultraviolet spectra consistent only with this structure, not the oxepin structure. The spectra resemble those of the corresponding norcaradiene **(22)** very closely. On the other hand, the oxepin **(28)** exists in the open form because this contains a nearly planar, aromatic 10π carbon skeleton, whereas the tautomer **(29)** does not. A similar tautomeric equilibrium might be expected for the benzene imine–azepine system, **(30)** ⇌ **(31)**. However, the equilibria normally favour

O O

(28) **(29)**

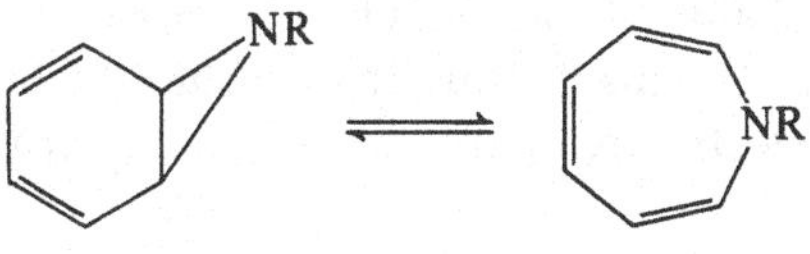

(3.16)

(30) (31)

(3.17)

(33) (32)

the azepine structures very strongly in the systems studied so far; the n.m.r. spectra are not temperature dependent.[19] An exception is the azepine (**32**), which does show a temperature-dependent spectrum;[20] the electron-withdrawing substituents may help to stabilise the benzene imine structure (**33**).[16]

Analogues of the hexatriene cyclisations can be expected in systems related to the pentadienyl anion, in which the six π electrons are distributed over five atoms. As the correlation diagram (fig. 3.5) showed, cyclisation of such systems is symmetry-allowed as a disrotatory process. An example is provided by the base-induced conversion of hydrobenzamide (**34**) into amarine (**35**) (3.18).[21]

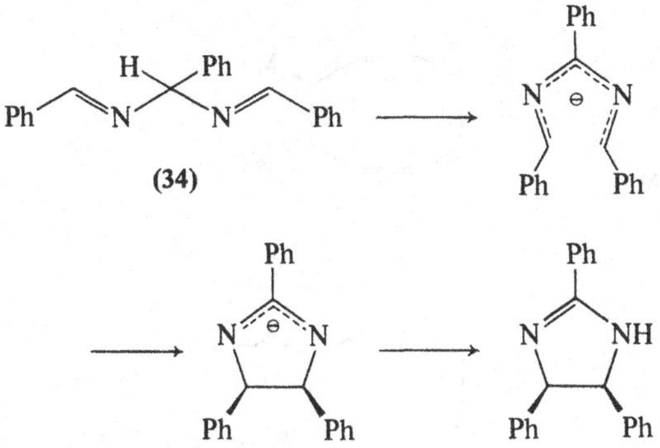

(34)

(3.18)

(35)

Related examples are the cyclisations of imidoyl azides, vinyl azide anions and acyl nitrile oxides shown in equations 3.19–3.21. Such reactions have been termed 1,5-*dipolar cyclisations.*[22] The movement of electrons in these cyclisations is not certain, since the compounds contain lone electron pairs at the termini which lie orthogonal to the π systems (that is, in the plane of the paper) and these allow the possibility of forming a bond between the termini without rotating the terminal atoms.[23] Thus, the transition states may not involve the same type of electron reorganisation as in the prototype pentadienyl anion.

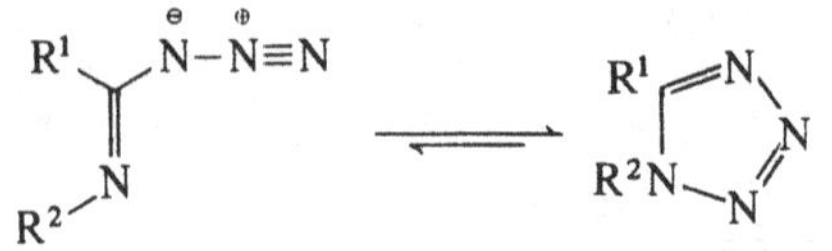

(3.19)

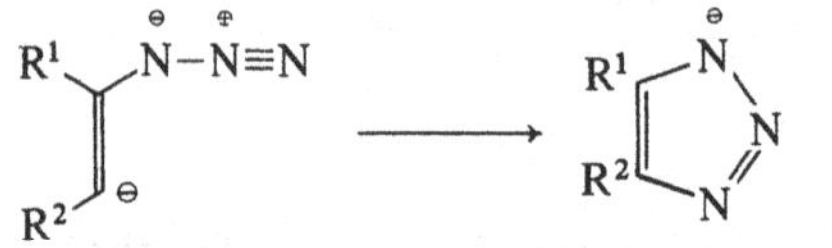

(3.20)

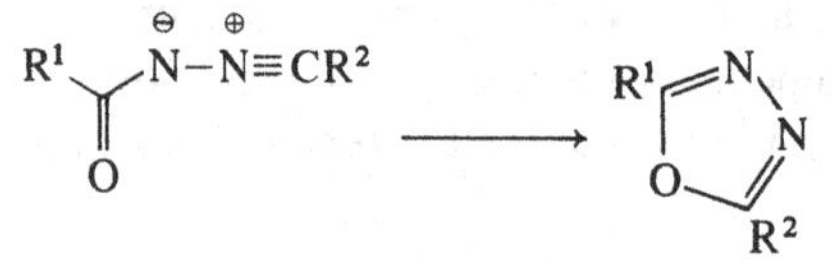

(3.21)

4. *Other systems.* The selection rules apply to 8π electron and higher systems. Thus, the thermal cyclisation of an 8π system should be conrotatory, for a 10π system it should be disrotatory, and so on. Examples of such reactions are few, but most conform to the expected pattern. One example is the cyclisation of the tetraenes **(36)** and **(37)** by conrotatory closure.[24]

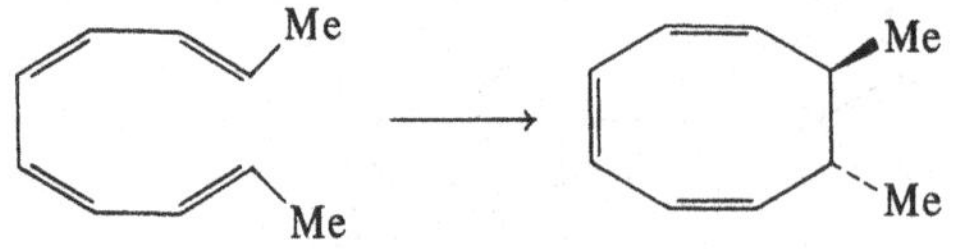

(3.22)

(36)

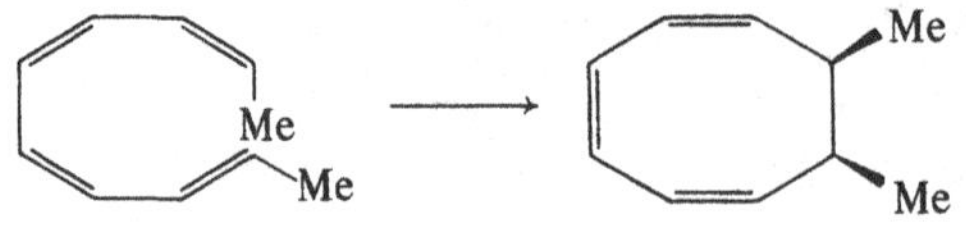

(3.23)

(37)

An example of a reaction which apparently fails to correspond to the expected pattern is the cyclisation of the all-*cis*-cyclodecatrienyl anion **(38)**.[25] This 8π system would be expected to cyclise by conrotatory closure, but the observed product **(39)** is the one expected for a disrotatory closure.

This apparent violation of the symmetry rules led the investigators to suggest that the strained anion **(38)** may first isomerise to the *cis–trans* isomer **(40)**, which would then give the observed product by an allowed conrotatory reaction. The power of the theory is nicely illustrated by this example: the theory suggests that a hitherto undetected intermediate may be involved, and this might then be amenable to experimental test.

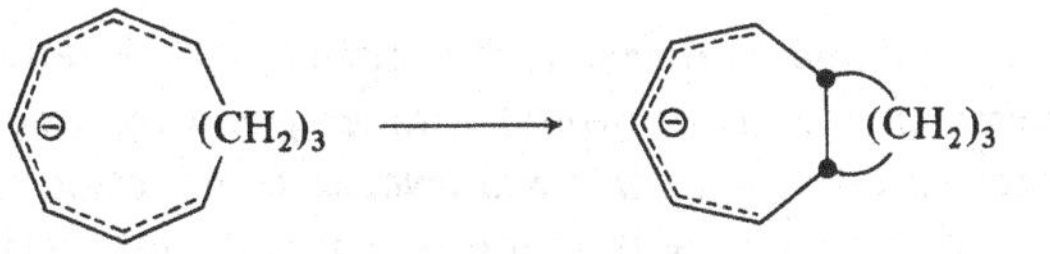

$(CH_2)_3$ → $(CH_2)_3$ (3.24)

(38) **(39)**

$(CH_2)_3$

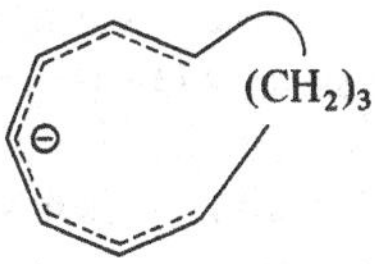

(40)

3.2. Electrocyclic reactions of radicals

The few electrocyclic reactions of π systems containing an unpaired electron are difficult to interpret using the simple qualitative theories. The symmetry of the HOMO of the radical system corresponds to that of the corresponding anion: thus, the allyl radical would be expected to cyclise in the same manner as the allyl anion, that is, in a conrotatory manner, on the basis of this theory. In fact, the interconversion of cyclopropyl and allyl radicals appears to go preferentially in a disrotatory manner: opening of the radicals **(41)** and **(42)** is preferentially disrotatory.[26] Correlation diagrams and simple calculations based on Hückel theory also tend to give ambiguous or incorrect predictions, and it is necessary to go to more sophisticated calculations to obtain reliable results. For radical cations and radical anions derived from butadiene and hexatriene, the preferred modes of reaction correspond to those of the neutral polyene. Thus, the butadiene radical cation and radical anion are predicted to cyclise preferentially in a conrotatory manner; the latter prediction has been verified experimentally.[27]

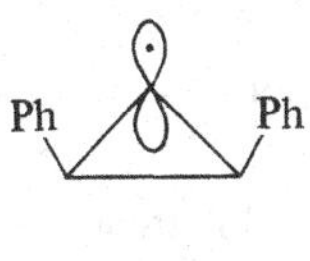

Ph Ph Ph Ph

(41) 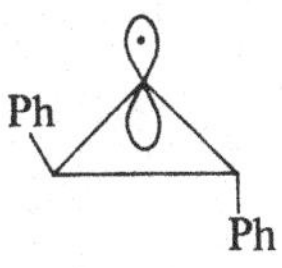 **(42)**

3.3. Photochemical electrocyclic reactions

Electrocyclic reactions brought about by irradiation with ultraviolet light are stereospecific, but in precisely the sense opposite to the thermal processes. If the thermal reaction is disrotatory, then the photochemical reaction is conrotatory, and vice versa. This can be important for synthetic applications, since by choosing the appropriate conditions, the stereochemistry of the product can be selected.

The LUMO of a conjugated linear polyene is of symmetry opposite to that of the HOMO. Thus the simple picture of orbital symmetry control predicts that excitation of an electron into the LUMO will reverse the direction of ring closure or ring opening, compared with that of the ground state system. For 2, 6, . . . , $(4n + 2)$ π systems, photochemically-induced ring closure should be conrotatory, and for 4, 8, . . . , $4n$ π systems, it should be disrotatory.

This simple picture agrees remarkably well with the observed reaction paths. However, it does raise problems, which have been discussed by Oosterhoff and others.[28] Consider the butadiene–cyclobutene ring closure, which can be brought about photochemically and which in substituted butadienes has been shown to be disrotatory. The energy levels of the ground states and the spectroscopic singlet states of butadiene and cyclobutene are shown in fig. 3.8; the diagram indicates that the spectroscopic singlet state of cyclobutene is about 210-250 kJ mol^{-1} higher in energy than that of butadiene. Simple orbital correlation diagrams predict a correlation between these two states in the 'photochemically allowed' disrotatory ring closure, but the energy difference between the excited states makes this extremely unlikely.

Van der Lugt and Oosterhoff have carried out valence bond calculations which provide a possible solution to the problem, but which indicate that the photochemical process is more complicated than the simple theory suggests. As the disrotatory closure of butadiene to cyclobutene proceeds, the first excited singlet S_1 increases in energy. However, a higher excited singlet state S_x *decreases* in energy and goes through a minimum before

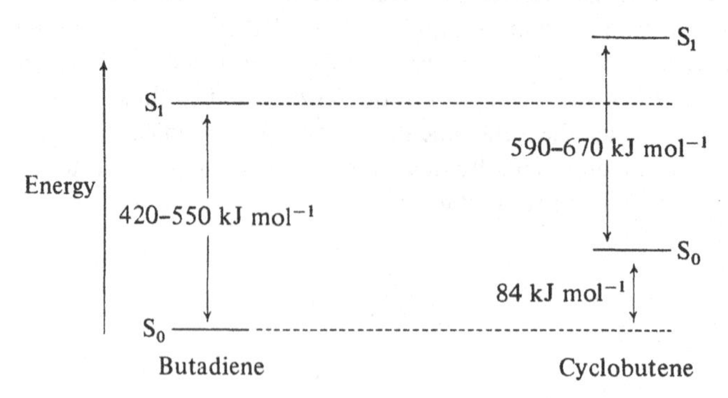

Fig. 3.8. Relative energy levels of butadiene and cyclobutene.

increasing in energy again. This energy minimum is at about the same point along the reaction path as the energy maximum in the ground state closure. The proposal is that nuclear vibrations cause the molecule to pass adiabatically into the second excited state at the point along the path where the excited states coincide, and then it drops down from the energy minimum to the ground state (S_0) contour, thus going on to give ground state cyclobutene (fig. 3.9).

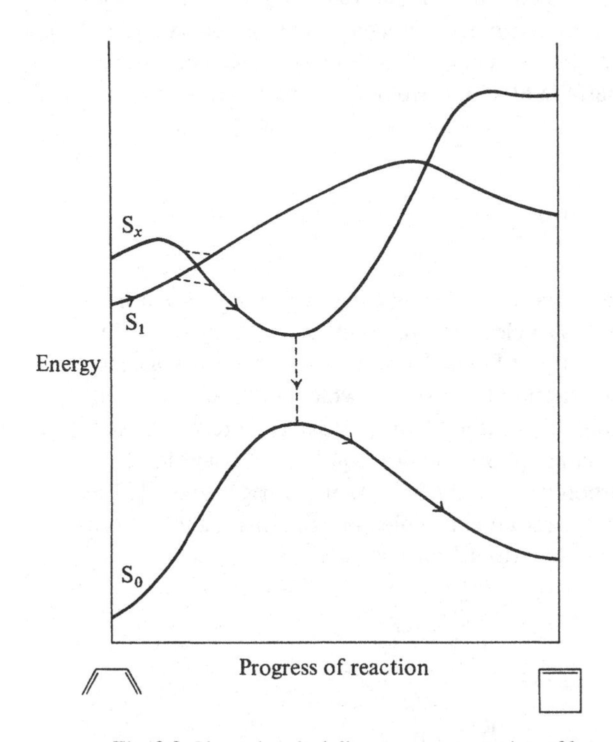

Fig. 3.9. Photochemical disrotatory conversion of butadiene into cyclobutene.

For the conrotatory closure there is no comparable energy minimum in the excited state, so the butadiene chooses the disrotatory path. The direction of closure is still as the simple theory predicts, therefore, but the cyclobutene is formed in the electronic ground state.

Photochemical butadiene-cyclobutene interconversions are disrotatory, and most involve conversions of butadienes into cyclobutenes, not the reverse. This is simply a result of the fact that the diene is a stronger absorber of light at the wavelengths used in the reactions, so it is the diene which is excited. The most instructive examples are those where the diene forms part of a ring so that a fused cyclobutene is formed (3.25).

H

(CH$_2$)$_n$ $\xrightarrow{h\nu}$ (CH$_2$)$_n$ (3.25)

H

Since the reverse reaction is thermally disallowed, the products may be unexpectedly stable. Thus, for example, cyclopentadiene can be partially converted photochemically into bicyclo[2,1,0] pentene, which despite its strained structure has a half-life of about two hours at room temperature (3.26).[29]

$\xrightarrow[\text{EtOH}]{h\nu}$ (3.26)

The photochemical equilibration of benzene derivatives and Dewar benzenes **(43)** can be regarded as an electrocyclic reaction of this type (3.27). The surprising thermal stability of Dewar benzenes can again be explained by the fact that their isomerisation to benzenes, which must be disrotatory because of geometrical constraints, is a thermally disallowed reaction. 2-Pyrone **(44)** undergoes a similar photochemical equilibration, and in the presence of iron pentacarbonyl, the bicyclic tautomer is decarboxylated to give a cyclobutadiene–iron tricarbonyl complex (3.28). This is a useful synthetic application of the photochemical ring closure.[30]

R R R R R R $\underset{}{\overset{h\nu}{\rightleftharpoons}}$ R R R R R R (3.27)

(43)

O O $\overset{h\nu}{\rightleftharpoons}$ O O $\xrightarrow{\text{Fe(CO)}_5}$ Fe (CO)$_3$ (3.28)

(44)

For 6π systems, the photochemical cyclisation is conrotatory, again in striking contrast to the thermal reactions. The first proposals of orbital symmetry control in organic reactions were put forward to explain the difference in the thermal and photochemical cyclisations of hexatrienes. Thus, Havinga and Schlatmann, in a study of the isomerisation of compounds related to calciferol (vitamin D, **45**) noted the difference in the stereochemistry of thermal and photochemical cyclisation.[31] Calciferol thermally tautomerises to precalciferol (**46**) (a reaction which we would now classify as a [1,7] sigmatropic hydrogen shift - see chapter 7). At 150-200 °C precalciferol then cyclises to give two *cis* fusion products, pyrocalciferol (**47**) and isopyrocalciferol (**48**); we recognise these as the products of the two possible modes of disrotatory closure. If precalciferol is irradiated, however, it equilibrates with a different ring-closed isomer, the *trans* fused lumisterol (**49**). This is a product of conrotatory closure.

HO (45) Me HO (46) (3.29)

Heat $h\nu$

Me H HO (47) + Me H HO (48) Me H HO (49)

Many other examples of this type of contrasting behaviour have since been discovered. The all-*cis*-cyclodecapentaene (**50**), for example, equilibrates photochemically at low temperatures with *trans*-9,10-dihydronaphthalene, by a conrotatory six-electron electrocyclic reaction, but it is converted thermally into *cis*-9,10-dihydronaphthalene by disrotatory closure.[32]

H H $h\nu$ (Conrotatory) (50) Heat (Disrotatory) H H (3.30)

The work of Heller and his colleagues on the cyclisation of sterically crowded molecules such as (**51**) provides many fine examples.[33] Thus, the yellow imides (**51**) and (**52**) are converted photochemically into the red isomers **(53)** and **(54)**; these when heated can open in a disrotatory manner to give the yellow isomers of opposite stereochemistry. The latter are eventually lost by irreversible hydrogen shifts, but similar systems have been designed which can undergo thermal and photochemical interconversion many millions of times without being destroyed.

H Ph O NPh O Ph **(51)** —— *hν* (Conrotatory) ——→ Ph H O NPh Ph O **(53)**

Heat (Disrotatory) (3.31)

Ph H O NPh O Ph **(52)** —— *hν* (Conrotatory) ——→ Ph H O NPh Ph O **(54)**

3.4. Metal-catalysed electrocyclic reactions[34]

Transition-metal catalysts can alter the normal thermal reactions of a polyene in several ways. First, the catalysed reaction can give products which are not related to any pericyclic process. Catalysis can also give a product resulting from a formally disallowed thermal reaction: the catalysed ring opening of strained cyclobutenes provide examples. Finally there are a few examples of reactions of pre-formed transition-metal complexes which follow the same course as the thermal reaction of the ligand, but at a different rate.

Silver ions and other metal catalysts can have a dramatic effect on the rate of ring opening of strained cyclobutene derivatives. These reactions are necessarily disrotatory because of the stereochemistry of the systems, and as such are thermally disallowed as concerted processes. The slow thermal conversion of bicyclo[4,2,0]octene **(9)** to *cis,cis*-1,3-cyclo-octadiene has already been referred to; the reaction is enormously speeded up in the presence of silver ions. Similarly, the disrotatory opening of the fused cyclobutenes **(55)** and **(56)** is catalysed by silver ions (3.32, 3.33). The conversion of **(55)** to dibenzocyclo-octatetraene is virtually instantaneous at room temperature in the presence of silver ions, and the activation energy is reduced from 96 kJ mol^{-1} for the uncatalysed reaction to about 33 kJ mol^{-1} for the catalysed process.

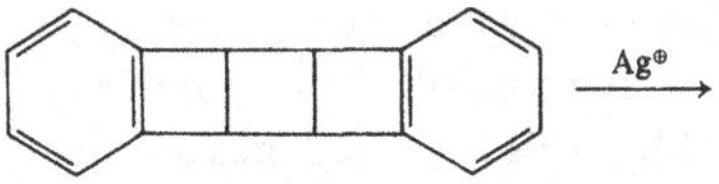

(55)

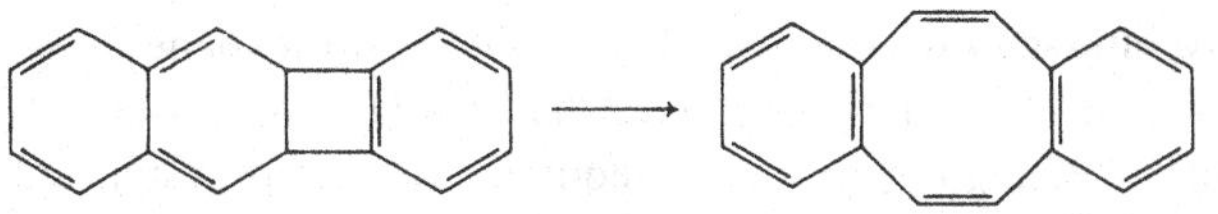

Ag$^{\oplus}$

(Not isolated)

(3.32)

Ag$^{\oplus}$

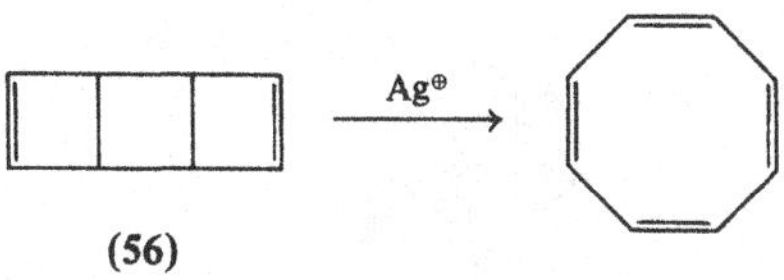

(56)

(3.33)

In a related reaction, the pre-formed iron tetracarbonyl complex **(57)** undergoes ring opening in boiling hexane to give the cyclohexadiene–iron tricarbonyl complex **(58)**. The reaction is inhibited by carbon monoxide, and it is suggested that complex **(59)** is an intermediate.[35] The rearrangement of **(59)** to **(58)** would then be a true electrocyclic ring opening of a complexed cyclobutene.

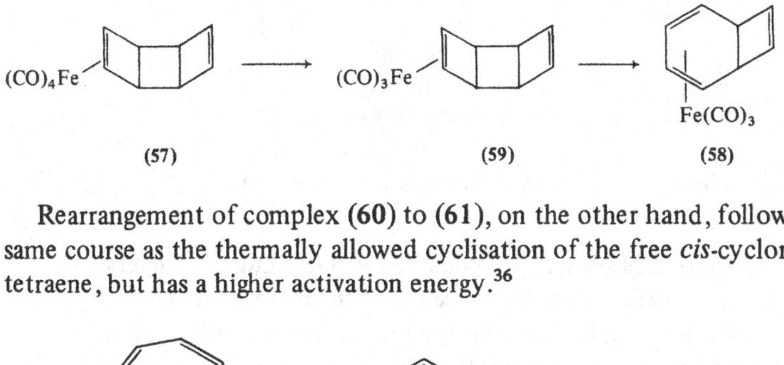

$(CO)_4Fe$ (57) → $(CO)_3Fe$ (59) → $Fe(CO)_3$ (58) (3.34)

Rearrangement of complex **(60)** to **(61)**, on the other hand, follows the same course as the thermally allowed cyclisation of the free *cis*-cyclononatetraene, but has a higher activation energy.[36]

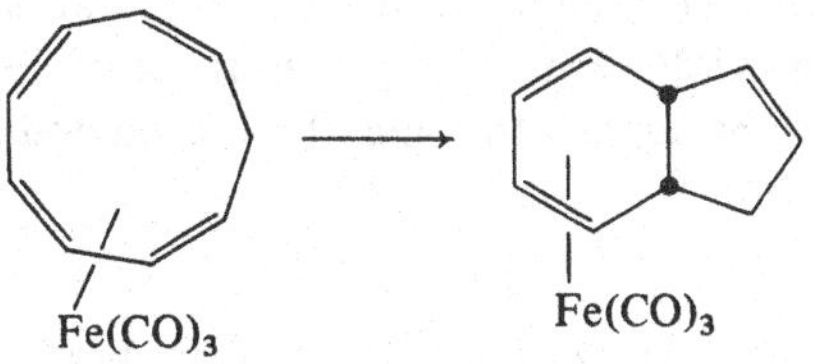

$Fe(CO)_3$ (60) → $Fe(CO)_3$ (61) (3.35)

For those reactions in which a thermally disallowed reaction occurs there have been several tentative suggestions of ways in which the catalysts might operate. First, it is possible that in the presence of the metal, a completely different mechanism operates, which is a route of low energy involving one or more intermediates. This seems to be so, for example, in the case of certain catalysed sigmatropic reactions (chapter 7), in which metal hydride intermediates are probably involved, and in some cycloadditions. An attractive alternative suggestion is that the metal supplies filled and empty orbitals of the right symmetry to interact with the LUMO and HOMO of the diene, thus effectively transferring electron density from the HOMO to the LUMO (fig. 3.10) and therefore reversing the symmetry requirements of the system; that is, making the forbidden disrotatory reaction an allowed one.

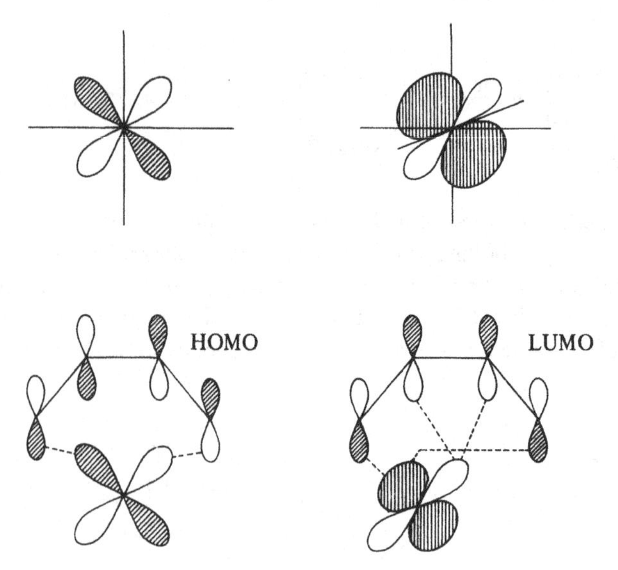

Fig. 3.10. Interaction of metal d orbitals with diene π orbitals.

Another explanation has been expounded by van der Lugt,[37] namely that the reaction remains forbidden in the presence of the metal, but that the activation energy is lowered because the excited electronic configurations of the metal-substrate complex are much lower in energy than those of the substrate alone. This principle can also be applied to some of the catalysed cyclodimerisations of olefins (chapter 6).

References

1. Baird, M. S., Lindsay, D. G. and Reese, C. B., *J. chem. Soc. (C)*, 1969, 1173.
2. Baird, M. S. and Reese, C. B., *Tetrahedron Lett.*, 1969, 2117.
3. Horwell, D. C. and Rees, C. W., *Chem. Communs*, 1969, 1428.
4. Criegee, R., Seebach, D., Winter, R. E., Börretzen, B. and Brune, H.-A., *Chem. Ber.*, 98, 2339 (1965).
5. Shumate, K. M., Neuman, P. N. and Fonken, G. J., *J. Am. chem. Soc.*, 87, 3996 (1965).

6. Branton, G. R., Frey, H. M. and Skinner, R. F., *Trans. Faraday Soc.*, **62**, 1546 (1966).
7. Breslow, R., Napierski, J. and Schmidt, A. H., *J. Am. chem. Soc.*, **94**, 5906 (1972).
8. Review: Huisgen, R., *Angew. Chem. Int. Edn Engl.*, **16**, 572 (1977).
9. Huisgen, R. and Mäder, H., *Angew. Chem., Int. Edn Engl.*, **8**, 604 (1969).
10. Newcomb, M. and Ford, W. T., *J. Am. chem. Soc.*, **96**, 2968 (1974).
11. Marvell, E. N., Caple, G. and Schatz, B., *Tetrahedron Lett.*, 1965, 385.
12. Glass, D. S., Watthey, J. W. H. and Winstein, S., *Tetrahedron Lett.*, 1965, 377.
13. Ciganek, E., *J. Am. chem. Soc.*, **89**, 1454 (1967).
14. Schröder, G., *Angew. Chem., Int. Edn Engl.*, **4**, 752 (1965).
15. Vogel, E., Wendisch, D. and Roth, W. R., *Angew. Chem., Int. Edn Engl.*, **3**, 443 (1964).
16. A simple explanation, based on frontier orbital theory, has been given for the effect of substituents on the equilibrium: Hoffmann, R., *Tetrahedron Lett.*, 1970, 2907.
17. Review: Vogel, E. and Günther, H., *Angew. Chem., Int. Edn Engl.*, **6**, 385 (1967).
18. Jerina, D. M., Ziffer, H. and Daly, J. W., *J. Am. chem. Soc.*, **92**, 1056 (1970). Review: Jerina, D. M., Yagi, H. and Daly, J. W., *Heterocycles*, **1**, 267 (1973).
19. Paquette, L. A., Barrett, J. H. and Kuhla, D. E., *J. Am. chem. Soc.*, **91**, 3616 (1969).
20. Prinzbach, H., Stusche, D. and Kitzing, R., *Angew. Chem., Int. Edn Engl.*, **9**, 377 (1970).
21. Hunter, D. H. and Sim, S. K., *J. Am. chem. Soc.*, **91**, 6203 (1969).
22. Reimlinger, H., *Chem. Ber.*, **103**, 1900 (1970).
23. Burke, L. A., Elguero, J., Leroy, G. and Sana, M., *J. Am. chem. Soc.*, **98**, 1685 (1976).
24. Huisgen, R., Dahmen, A. and Huber, H., *J. Am. chem. Soc.*, **89**, 7130 (1967).
25. Staley, S. W. and Heyn, A. S., *J. Am. chem. Soc.*, **97**, 3852 (1975).
26. Sustmann, S. and Rüchardt, C., *Chem. Ber.*, **108**, 3043 (1975).
27. Bauld, N. L. and Cessac, J., *J. Am. chem. Soc.*, **99**, 23 (1977); Bauld, N. L., Cessac, J., Chang, C.-S., Farr, F. R. and Holloway, R., *J. Am. chem. Soc.*, **98**, 4561 (1976).
28. van der Lugt, W. Th. A. M. and Oosterhoff, L. J., *J. Am. chem. Soc.*, **91**, 6042 (1969); Grimbert, D., Segal, G. and Devaquet, A., *J. Am. chem. Soc.*, **97**, 6629 (1975); Bruckmann, P. and Salem, L., *J. Am. chem. Soc.*, **98**, 5037 (1976); Michl, J. *Top. Curr. Chem.*, **46**, 1 (1974).
29. Brauman, J. I., Ellis, L. E. and van Tamelen, E. E., *J. Am. chem. Soc.*, **88**, 846 (1966).
30. Corey, E. J. and Streith, J., *J. Am. chem. Soc.*, **86**, 950 (1964).
31. Havinga, E. and Schlatmann, J. L. M. A., *Tetrahedron*, **16**, 146 (1961).
32. Masamune, S. and Seidner, R. T., *Chem. Communs*, 1969, 542.
33. Hart, R. J. and Heller, H. G., *J. chem. Soc., Perkin 1*, 1972, 1321; Hastings, J. S., Heller, H. G. and Salisbury, K., *J. chem. Soc. Perkin 1*, 1975, 1995, and earlier papers.
34. For a comprehensive review of these reactions and their mechanisms, see Mango, F. D., *Coord. chem. Rev.*, **15**, 109 (1975).
35. Slegeir, W., Case, R., McKennis, J. S. and Pettit, R., *J. Am. chem. Soc.*, **96**, 287 (1974).
36. Reardon, E. J. and Brookhart, M., *J. Am. chem. Soc.*, **95**, 4311 (1975).
37. van der Lugt, W. Th. A. M., *Tetrahedron Lett.*, 1970, 2281.

4 Cycloadditions: introduction

Cycloadditions form a very extensive and rapidly expanding area of chemistry, the synthetic potential and theoretical significance of which have only fairly recently been fully recognised. Indeed a desire to understand the subtleties of the effect of substituents on the rates and orientation of cycloadditions continues to stimulate refinements in the basic theory of pericyclic reactions.

A cycloaddition is a process in which two or more reactants combine to form a stable cyclic molecule during which no small fragments are eliminated and σ bonds are formed but not broken.[1,2] This definition covers reactions like the Diels–Alder reaction (4.1*a*) and the dimerisation of olefins (4.1*b*) but excludes such processes as the Dieckmann cyclisation (4.1*c*). Most cycloadditions involve the formation of two new σ bonds, as in the Diels–Alder reaction, but electrocyclic reactions like the cyclisation of butadiene in which only one new σ bond is formed (4.1*d*) (together with a π bond) can also be considered as intramolecular cycloadditions. In cyclotrimerisation three new σ bonds are formed.

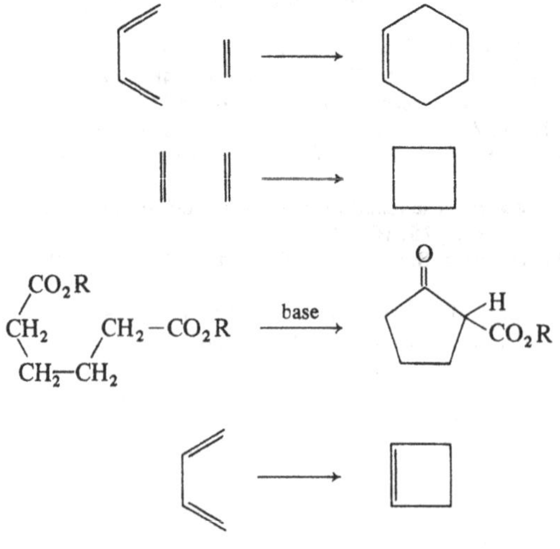

(*a*)

(*b*)

(4.1)

CO_2R

CH_2 CH_2-CO_2R base

CH_2-CH_2

O

H

CO_2R (*c*)

(*d*)

Since cycloaddition is such a varied process further classification is often desirable. Huisgen[2] has suggested that such a classification should be independent of mechanism and his system uses only the number of ring atoms provided by each component. Thus the Diels–Alder reaction (4.1*a*) is a 4 + 2 cycloaddition, and the olefin dimerisation (4.1*b*) a 2 + 2 cycloaddition. Another system recognises the fundamental importance of the number of electrons involved in the cycloaddition and classifies the reactions on this basis. The Diels–Alder reaction thus becomes a ($4\pi + 2\pi$) process and the olefin dimerisation a ($2\pi + 2\pi$) one.

In these two examples, the number of electrons happens to be the same as the number of atoms in each component. This is not always the case; in 1,3-dipolar cycloaddition, for example, the 1,3-dipole has four π electrons distributed over three atoms. The addition of a 1,3-dipole to an olefin is therefore a 3 + 2 addition according to the first classification but ($4\pi + 2\pi$) according to the second (4.2).

4π electrons

a b c ⊕ ⊖ → a b c (4.2)

4.1. Selection rules for thermal polyene cycloadditions

Many cycloadditions show all the characteristics of concerted processes but others are obviously stepwise and involve an intermediate which is either a zwitterion or a diradical. Why some cycloadditions should be concerted and others not has intrigued and stimulated chemists for many years. We are now in a position to rationalise these differences along the lines described in chapter 2.

Perhaps the simplest approach uses frontier orbital theory. In the concerted cycloaddition of two polyenes, bond formation at each terminus must be developed to some extent in the transition state. Thus orbital overlap must occur simultaneously at both termini. For a low energy concerted process (an allowed reaction) to be possible such simultaneous overlap must be geometrically feasible and must also be potentially bonding.

Suprafacial, suprafacial addition

Suprafacial, suprafacial (s,s) approach of two polyenes (fig. 4.1) is normally sterically suitable for efficient orbital overlap. The vast majority of concerted additions involves the s,s approach.

However, this type of overlap will only be energetically favourable when the HOMO of one component and the LUMO of the other can interact in a

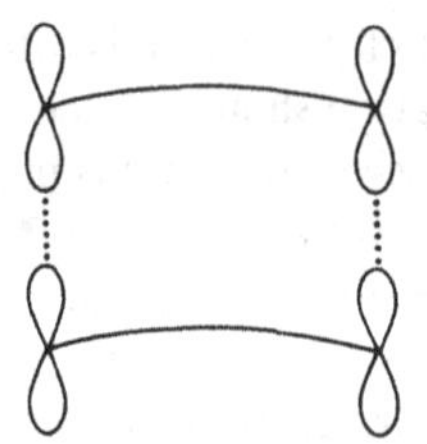

Fig. 4.1. Suprafacial, suprafacial approach of the two polyenes.

bonding fashion at both termini. Thus these orbitals must be of the correct phase or symmetry. In the Diels-Alder reaction of a diene with a monoene the HOMO and the LUMO of each reactant are of the appropriate symmetry so that mixing of these orbitals will result in simultaneous potential bonding character between the terminal atoms (fig. 4.2).

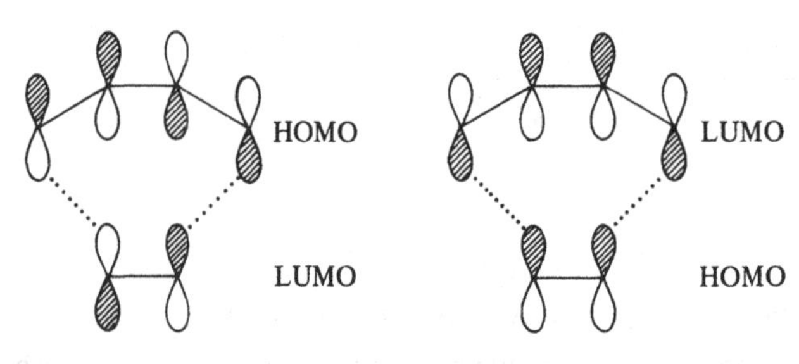

Fig. 4.2

In contrast, a similar concerted s,s approach of two olefins does not lead to a stabilising interaction since the HOMO and LUMO are of incompatible phase for simultaneous bonding interaction to occur at both termini (fig. 4.3). Thus the initial approach of reactants for a concerted s,s addition is favourable for the Diels-Alder reaction, which is therefore an allowed process, but not for olefin dimerisation, which is therefore disallowed.

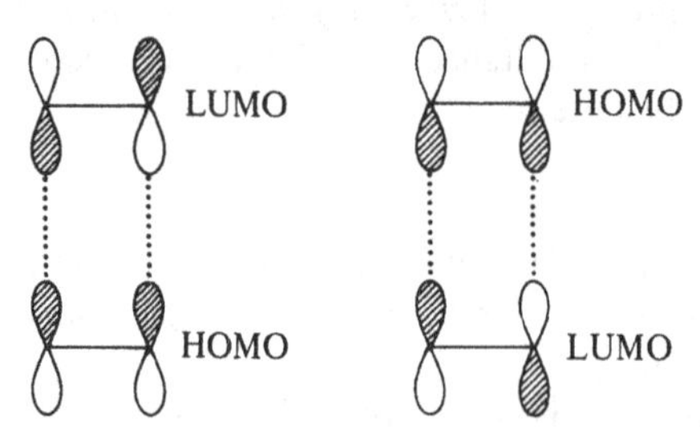

Fig. 4.3

The symmetries of the HOMO and LUMO of polyenes depend on the number of π electrons in the polyene. Thus the favourableness of initial interaction for concerted addition also depends on the number of π electrons provided by each component. The total number of π electrons is therefore fundamental to whether a concerted s,s cycloaddition is allowed. In general, s,s cycloaddition can be concerted for $(4n + 2)$ π electrons but not for $4n$. Exactly the same conclusion is reached using the aromatic transition state approach. Only those cycloadditions which have a 'quasi-planar' Hückel aromatic $(4n + 2)$ π electron transition state can be concerted (§2.4).

So far with the frontier molecular orbital approach, we have considered only whether the *initial* interaction between the reactants is bonding or antibonding in order to decide whether a concerted cycloaddition is possible. It has been assumed that if this initial interaction is favourable the trend will

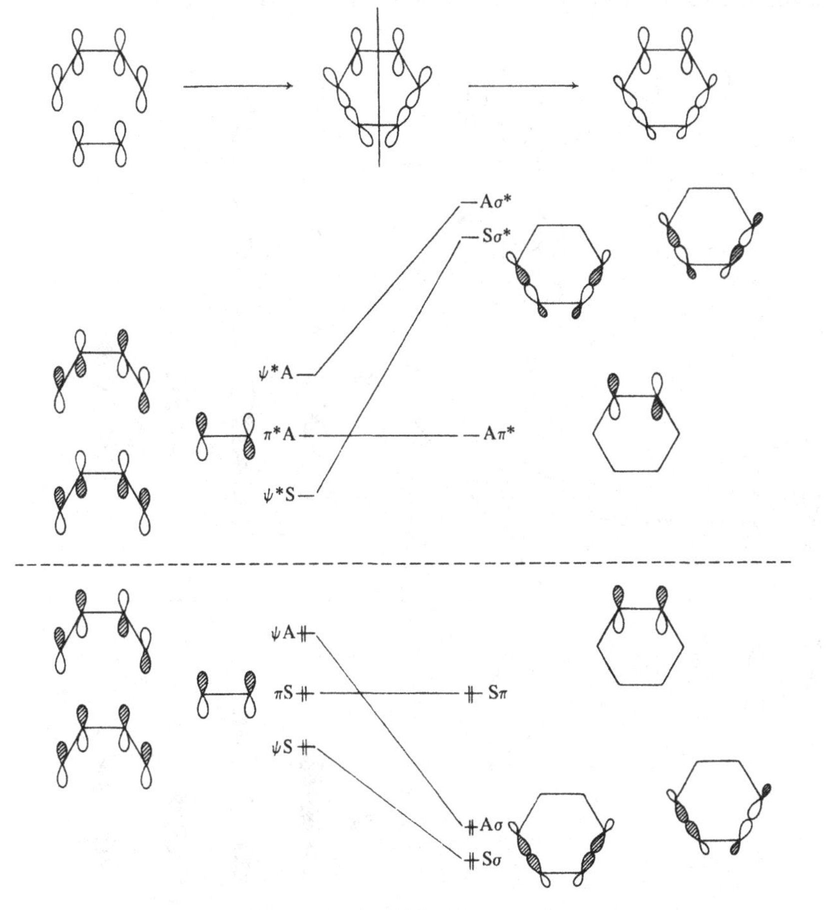

Fig. 4.4. Correlation diagram for the Diels–Alder reaction.

continue throughout the process and lead to a low activation energy. Correlation diagrams enable the interaction of the reactant orbitals to be followed right through to product orbitals. For allowed, concerted processes reactant bonding orbitals must be able to be transformed into product bonding orbitals without any change of symmetry. Correlation diagrams lead to exactly the same conclusions as frontier orbital theory. This is illustrated for a typical allowed reaction, the Diels-Alder reaction, and for a typical disallowed reaction, the ${}_{\pi}2_s + {}_{\pi}2_s$ dimerisation of ethylene.

The correlation diagram for the ${}_{\pi}4_s + {}_{\pi}2_s$ Diels-Alder reaction is shown in fig. 4.4. The symmetry of the orbitals is labelled with respect to the plane bisecting both reactants. The bonding reactant orbitals are all transformed into bonding product orbitals without a change in symmetry. There is thus no unfavourable rise in energy as the reaction proceeds and a concerted transition is favoured. Alternative stepwise mechanisms would involve diradical or

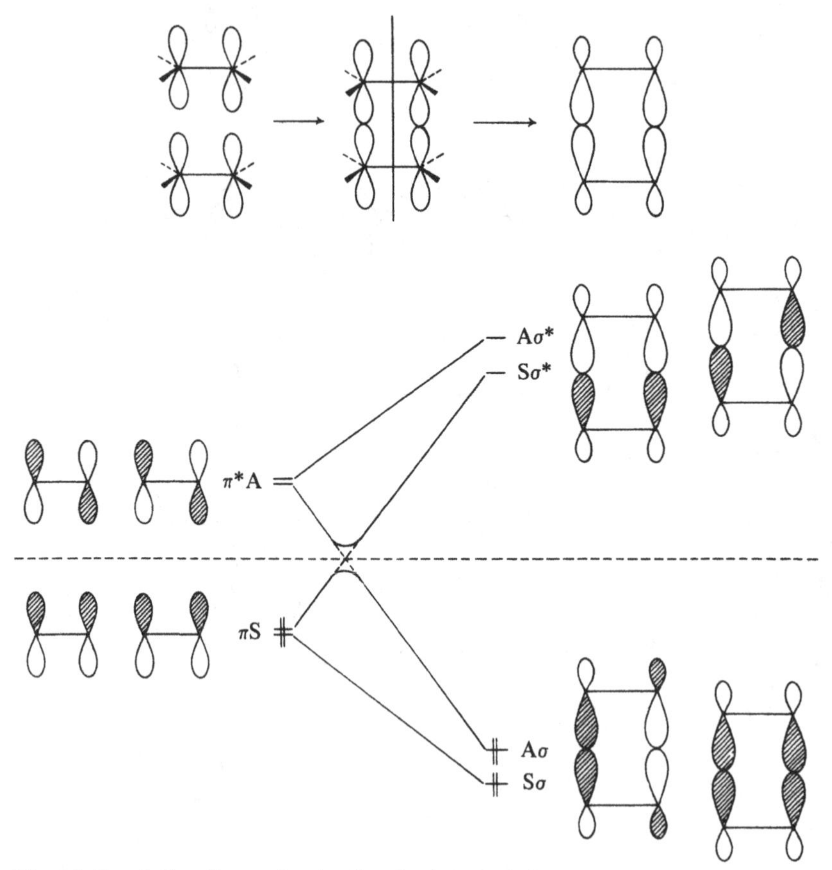

Fig. 4.5. Correlation diagram for the dimerisation of ethylene.

zwitterion formation and so, except in cases where such intermediates are particularly stabilised, these are unlikely to compete with the concerted process.

Note that the monoene π orbital correlates with the cyclohexene π orbital. This is because levels of like symmetry cannot cross.

A similar correlation diagram for the ${}_{\pi}2_s + {}_{\pi}2_s$ dimerisation of olefins is shown in fig. 4.5. Here, the orbitals are classed as symmetric (S) or antisymmetric (A) with respect to a plane through the π systems, just as in fig. 4.4. The diagram shows the typical characteristic of a disallowed process - the correlation of a bonding reactant orbital with an antibonding product orbital.

A more sophisticated version of this correlation diagram is shown in fig. 4.6. The degeneracy of the reactant orbitals is removed by their classifica-

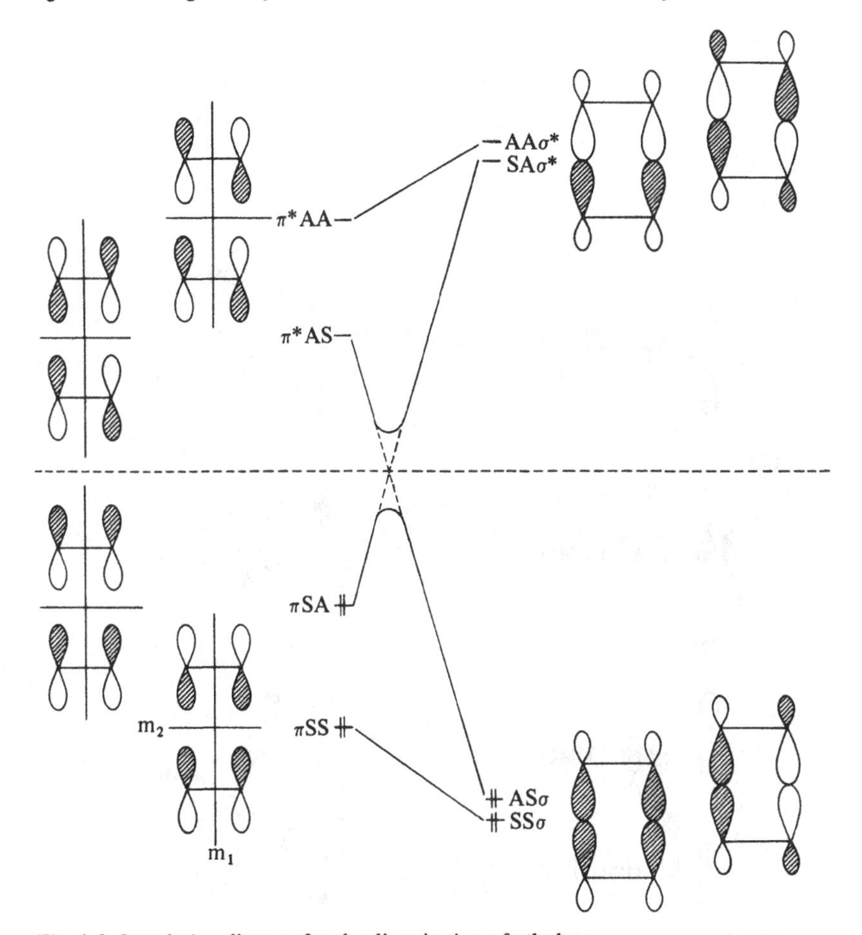

Fig. 4.6. Correlation diagram for the dimerisation of ethylene.

tion with respect to the two mirror planes m_1 and m_2. The SA combination increases in energy as the transition state is approached and the electrons from this orbital should ultimately enter the bonding AS cyclobutane orbital because a crossing of energy levels occurs at high energy. However, this energy barrier is sufficiently high to make it likely that alternative stepwise processes are more favourable.

Other geometries of approach

Two polyenes can also overlap in the suprafacial, antarafacial (s,a) sense as shown in fig. 4.7.

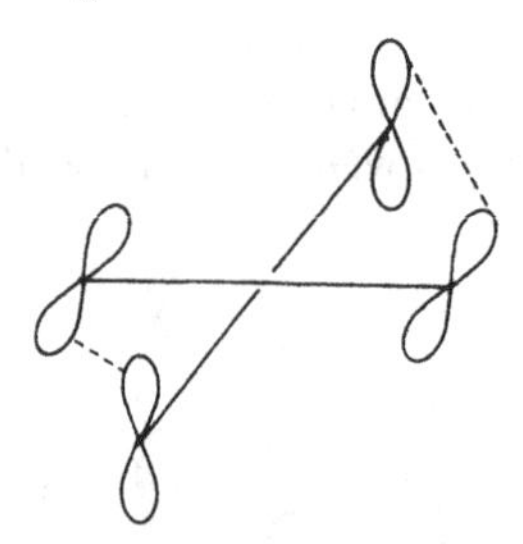

Fig. 4.7

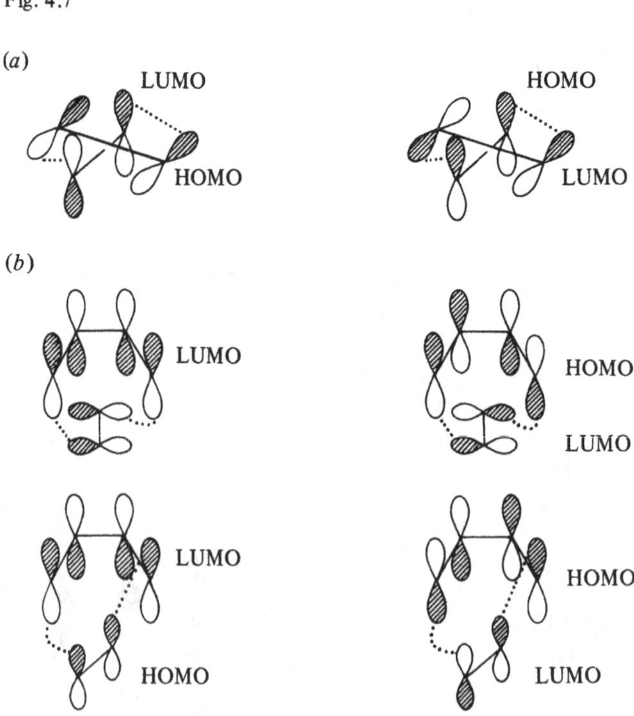

Fig. 4.8

This has different stereochemical consequences: addition to one component is *cis*, to the other *trans*. The efficiency of this type of orbital overlap will depend greatly on the geometry of the reactants, but it is normally considerably less than for the s,s mode. This s,a approach of two olefin molecules leads to a bonding interaction (fig. 4.8*a*) but the similar approach of a diene and a monoene, regardless of which is the antarafacial or suprafacial component, gives an initial antibonding interaction (fig. 4.8*b*). The former is therefore allowed, although geometrically improbable, and the latter disallowed.

A correlation diagram for the $_{\pi}2_s + _{\pi}2_a$ dimerisation of ethylene is shown in fig. 4.9. The most reasonable transition state for this process is the 'crossed' one illustrated; the methylene groups twist as indicated as the reactants proceed to the product. A C_2 axis of symmetry (at right angles to the plane of the page) is therefore the appropriate symmetry element for this process. The correlation of bonding with bonding and antibonding with antibonding orbitals, typical of an allowed process, is apparent.

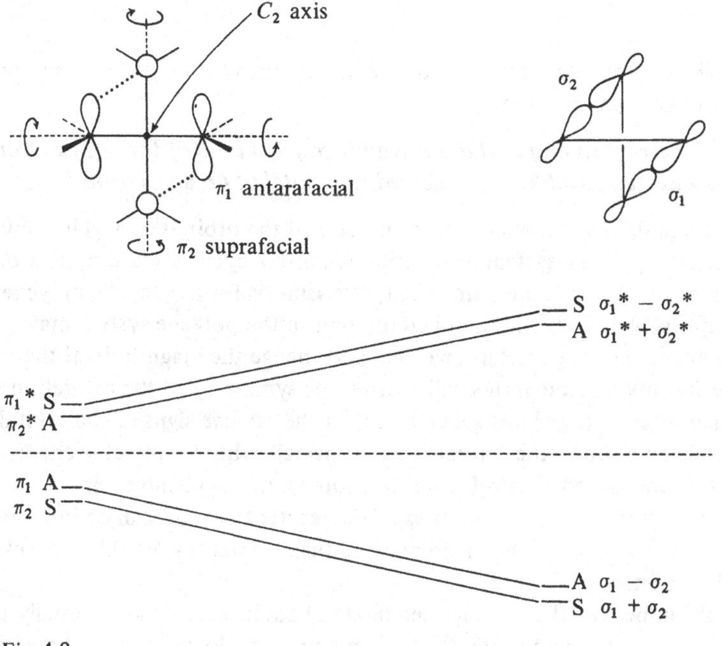

Fig. 4.9

In general, concerted s,a addition is allowed when the total number of π electrons is $4n$ but disallowed for $(4n + 2)$ π electrons; that is, a concerted cycloaddition involving $4n$ electrons must have one antarafacial component. Such reactions will involve twisted, Möbius type aromatic transition states (§2.4).

Antarafacial, antarafacial (a,a) addition (*trans* to each component) would be allowed for the reaction of a monoene with a diene (fig. 4.10) and for any $(4n + 2)\ \pi$ electron cycloaddition, but disallowed for olefin–olefin or any other $4n\ \pi$ electron cycloaddition. It is important to emphasise again, however, that some allowed modes of concerted addition such as these may be very unfavourable or impossible because for steric reasons the reactant orbitals cannot overlap effectively.

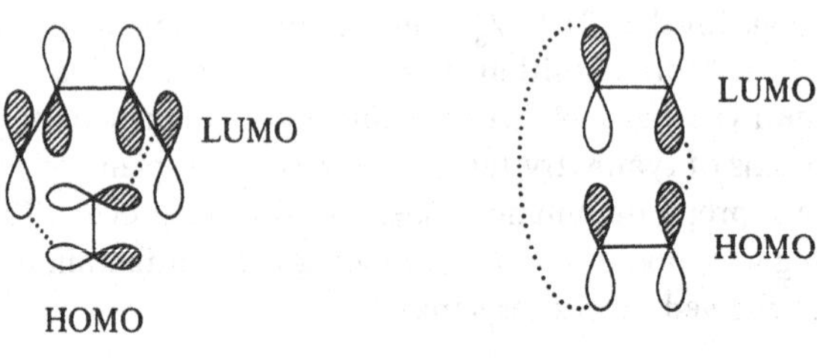

Fig. 4.10

The general selection rules for thermal cycloadditions of two components are summarised as follows:

> '*Concerted s,s or a,a addition is allowed for a total of* $(4n + 2)\ \pi$ *electrons. Concerted s,a addition is allowed for a total of* $4n\ \pi$ *electrons.*'

These rules are based on the symmetry of the orbitals of acyclic polyenes. When the polyene system is incorporated into a cyclic structure, its orbital symmetry characteristics are usually the same as for the acyclic polyene. Simple substituents and even heteroatoms in the polyene system may alter the energies of the orbitals involved, and change the magnitude of the orbital coefficients. Although this will destroy the symmetry of the orbitals in the strictest sense, it will not generally affect the relative signs of the orbital coefficients and so the orbital phase or overall orbital symmetry characteristics will be unchanged. The rules are therefore widely applicable. However, if there is any doubt that the symmetries of the reactant orbitals can be inferred from those of the simple open-chain polyene the symmetries should be checked by calculation.

It can be seen that a concerted mode of addition is always formally possible, if not by an s,s route, then by an a,s route. However, not all cycloadditions are concerted because, as already pointed out, the geometry necessary for efficient orbital overlap in the allowed mode may be impossible to attain for steric reasons. Thus, for example, concerted ${}_{\pi}2_s + {}_{\pi}2_a$ cycloaddition of two olefins is geometrically unfavourable, and since the geometrically favourable ${}_{\pi}2_s + {}_{\pi}2_s$ addition is forbidden as a concerted process, 2 + 2 additions are almost always stepwise. On the other hand, the allowed

$_{\pi}4_s + _{\pi}2_s$ Diels-Alder reaction involves geometrically favourable orbital overlap, and Diels-Alder reactions are mostly concerted and almost invariably occur through this sterically most favourable mode rather than via the allowed but geometrically unfavourable $_{\pi}4_a + _{\pi}2_a$ mode. Even when the concerted pathway is both allowed and geometrically favourable, substituents in the reactants can so stabilise an intermediate that an alternative stepwise route may be competitive, and occasionally predominant.

In some cycloadditions more than one relative orientation of reactants is possible for a particular mode of addition. For example, *N*-substituted azepines undergo allowed $_{\pi}4_s + _{\pi}6_s$ dimerisation through the transition state (**1**) rather than (**2**).

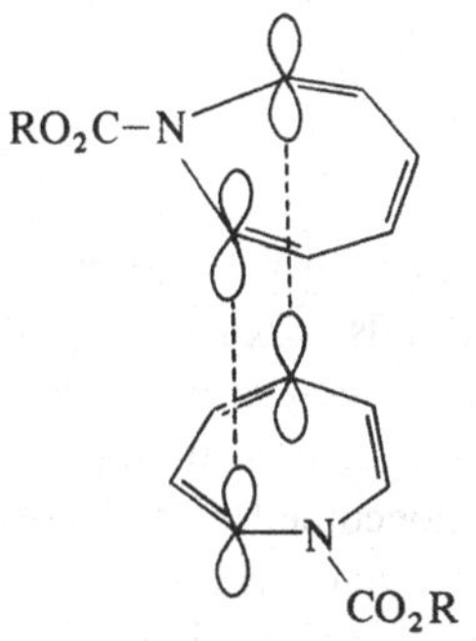

RO_2C-N CO_2R RO_2C-N RO_2C N

(**1**) 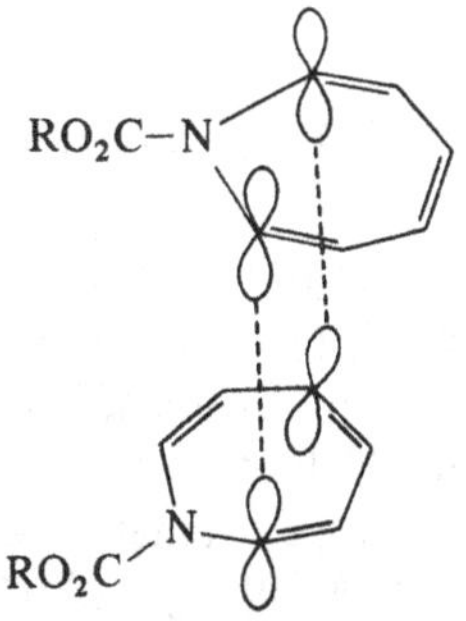 (**2**)

This can be explained by secondary orbital interactions favouring or disfavouring one particular transition state relative to the other (see §4.5) and this particular example is discussed in detail on p. 100. Other examples appear in §§5.1 and 6.9.

Cycloadditions with more than two components

The above types of argument apply to cycloadditions of more than two components but the geometrical considerations are more complicated. For a $2\pi + 2\pi + 2\pi$ concerted cycloaddition the s,s,s and the three modes of a,a,s (a,a,s; a,s,a; s,a,a) are allowed but the a,a,a and three modes of a,s,s are not. However, concerted intermolecular cycloadditions of more than two components are most unlikely to occur because of the unlikelihood of three-body collisions. Those that are known have at least two of the components constrained in one molecule (§5.6).

The application of simple frontier orbital theory to a three-component system requires some further comment since one has to consider interaction between one HOMO and two LUMOs or alternatively interaction between two HOMOs and one LUMO. In the first case all the orbital phases must be compatible for the reaction to be symmetry-allowed, but in the second case a phase discontinuity at one of the new bonds is associated with the allowed mode.[3] This is illustrated for the allowed ${}_{\pi}2_s + {}_{\pi}2_s + {}_{\pi}2_s$ cycloaddition (fig. 4.11).

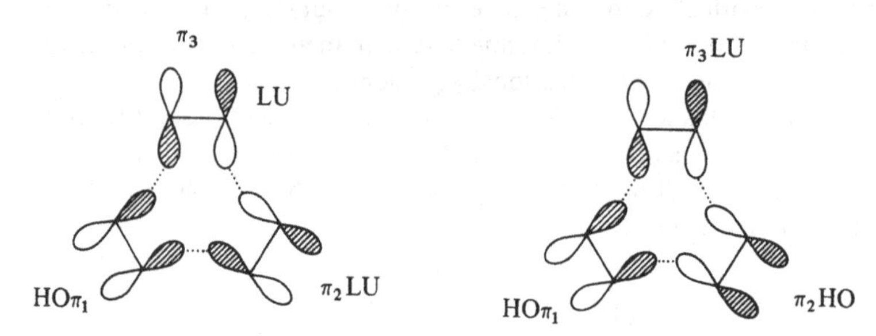

Fig. 4.11. ${}_{\pi}2_s + {}_{\pi}2_s + {}_{\pi}2_s$ cycloaddition.

The reason for this pattern of orbital phase relationships can be seen by considering successive mixing of the interacting orbitals. Mixing of π_1 HOMO with π_2 LUMO in a bonding fashion gives an occupied new combination HO, the terminal phases of which are compatible for simultaneous bonding to the π_3 LUMO (the antibonding combination will be unoccupied). On the other hand, mixing of π_1 HOMO with π_2 HOMO gives two new orbitals, the higher energy combination being the one in which π_1 and π_2 are combined in an antibonding or out of phase sense. This new combination HOMO then has terminal phases which are compatible for a stabilising mixing with the vacant π_3 LUMO.

4.2. Other types of cycloaddition

According to frontier orbital theory, the overriding factor in determining the selection rules for polyene cycloadditions is the symmetry of the HOMO and LUMO of the reactants, this in turn is related to the number of electrons involved in each of the reactants. Any component other than a neutral polyene which can supply the same number of electrons in an orbital of the same symmetry is potentially able to participate in a cycloaddition in place of the polyene. Some of the more important reactions of this type are briefly outlined here; they are discussed in more detail in chapters 5 and 6.

Three-atom components. Cycloadditions involving allylic systems

An allyl cation has two π electrons and a HOMO of the symmetry shown in fig. 4.12*a*. It is thus equivalent to an olefin in terms of orbital symmetry and should be capable of participating in cycloadditions as the two-electron component. Such reactions are known (§5.7).

Three-atom components with four electrons and the symmetry of the allyl anion, as shown in fig. 4.12*b*, should similarly be capable of the same types of cycloadditions as dienes. These reactions are well known; they are commonly referred to as 1,3-*dipolar cycloadditions* (§5.3).

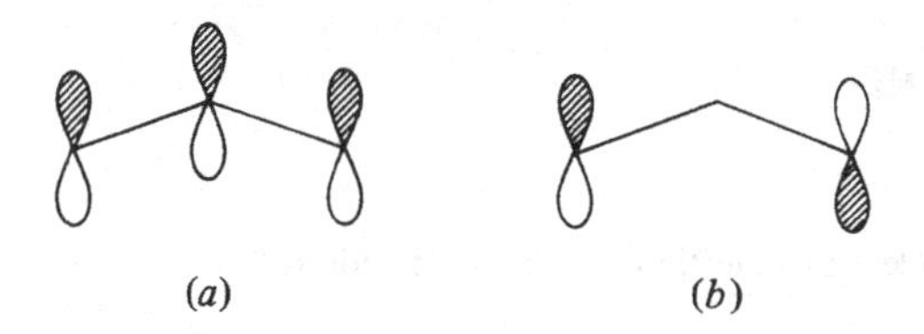

Fig. 4.12. HOMO of (*a*) allyl cation; (*b*) allyl anion.

One-atom components. Cheletropic reactions

Some small molecules have a filled and vacant orbital available on the same atom for bonding to other atoms. Sulphur dioxide, carbon monoxide and singlet carbenes are examples: in each case one atom (sulphur or carbon) has a lone pair of electrons in the plane of the molecule and a vacant p orbital orthogonal to it.

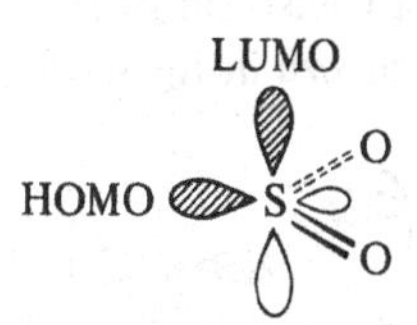

In cycloadditions and eliminations involving such molecules, the σ bonds which are formed or broken are to the same atom; for example, the addition and elimination of sulphur dioxide shown in equation 4.3. Woodward has proposed that these should be called *cheletropic reactions*[4] (Greek, *chele* - claw).

$$SO_2 \rightleftharpoons SO_2 \qquad (4.3)$$

Approach of a sulphur dioxide molecule to a diene from below and in the plane bisecting the diene leads to favourable overlap of HOMO and LUMO (fig. 4.13), so that a concerted suprafacial, suprafacial addition is allowed. Such reactions are called *linear* cheletropic processes.

Concerted suprafacial, suprafacial addition to an olefin (fig. 4.14) is a disallowed process.

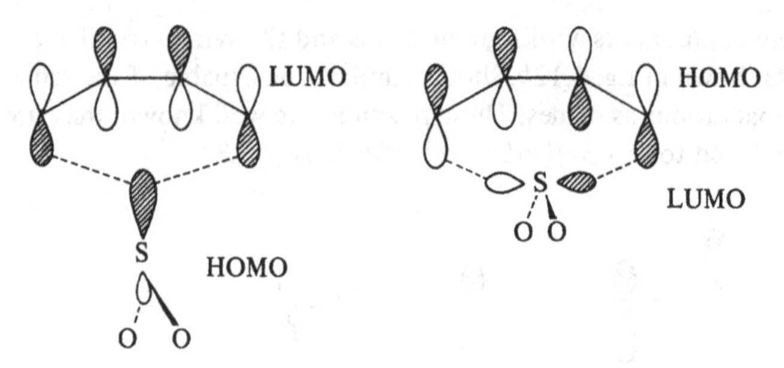

Fig. 4.13. Linear cheletropic addition of sulphur dioxide to butadiene.

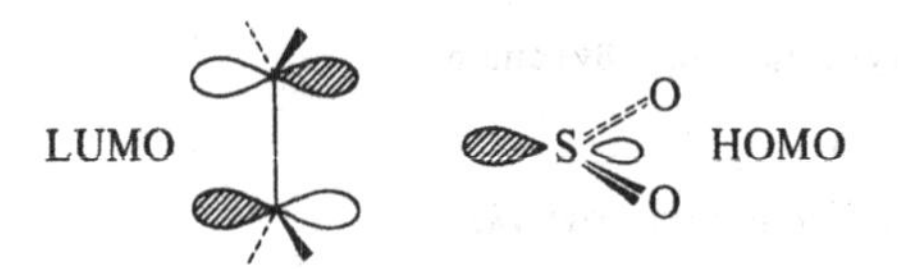

Fig. 4.14

If such a reaction is to be concerted the addition must be suprafacial, antarafacial. This means that in the approach or departure, the atom bearing the lone pair must tip up, so that its orientation with respect to the olefin is different from that in the linear cheletropic process. A transition-state geometry which meets the symmetry requirements is shown in fig. 4.15.

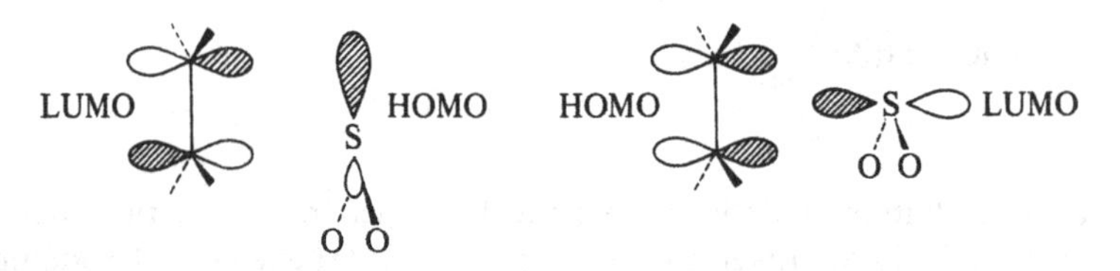

Fig. 4.15

Here, the axis of the lone pair is perpendicular to the axis of approach or departure. The addition is ${}_{\pi}2_s + {}_{\omega}2_a$; it is equivalent to the ${}_{\pi}2_s + {}_{\pi}2_a$ olefin dimerisation described in §4.1. Reactions which could involve such *non-linear* addition or extrusion are discussed in §6.6.

The concerted addition of sulphur dioxide to a triene could formally be either a linear or a non-linear cheletropic reaction. The $6\pi + 2\pi$ addition must be suprafacial, antarafacial. If the sulphur dioxide is the suprafacial component it is a linear cheletropic process and the terminal groups of the triene must rotate in a conrotatory sense, as shown in fig. 4.16*a*. In the non-linear cheletropic reaction the sulphur dioxide acts as the antarafacial component and the terminal groups must rotate in a disrotatory sense, (fig. 4.16*b*). The experimental evidence is that the rotation is conrotatory, so the linear cheletropic pathway is more favourable (§6.9).

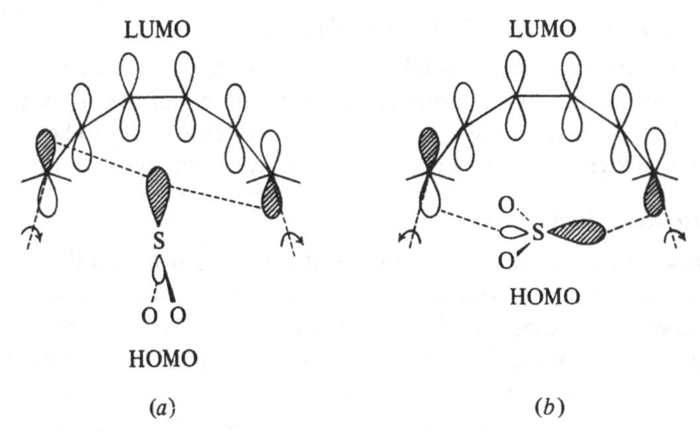

Fig. 4.16

The retro-additions or extrusions are the microscopic reverse of the forward reactions and follow the same selection rules. These rules are summarised as follows: linear cheletropic reactions in which the polyene is a suprafacial component (that is, involving disrotatory motion of the termini) are allowed for a total of $(4n + 2)$ electrons. Linear cheletropic reactions in which the polyene is an antarafacial component (involving conrotatory movement of the termini) are allowed for a total of $4n$ electrons. The rules are reversed for a non-linear cheletropic change.

Photochemical cycloadditions

Photochemical excitation of an electron to the next highest orbital produces an excited state species of which the HOMO is of opposite symmetry to that of the ground state molecule. It might be expected, therefore, that cycloadditions which are thermally disallowed should be photochemically allowed (one excited reactant molecule adding to an unexcited reactant molecule). In practice, most photochemical cycloadditions appear to be stepwise processes. Many photochemical additions are carried out in conditions where the reacting species is likely to be in a triplet state, either because photosensitisers are used or because the excited singlet undergoes intersystem crossing to the triplet before intermolecular collision with other reactants. It is generally assumed that concerted addition of a triplet molecule is not possible because a reversal of electron spin would be required. On the other hand, intramolecular cycloadditions involving singlet excited species could well be concerted.

With the above points in mind it is nevertheless a useful practical guide that a reaction which cannot be achieved thermally, for example the formation of cyclobutanes from cycloaddition of two olefins, can often be brought about photochemically.

Transition-metal-catalysed cycloadditions

Certain disallowed cycloadditions take place very much more readily in the presence of transition-metal complexes (§6.8). The origin of the catalytic effect is still a matter for debate (see chapter 3). Evidence seems to be hardening in favour of stepwise mechanisms for these catalysed reactions.

Retro-cycloadditions

Since retro-cycloadditions are the reverse of the forward reactions the same selection rules apply. It is artificial to separate the forward and reverse reactions because the direction of the reactions depends merely on the relative free energies of reactants and products. Since cycloadditions involve an unfavourable entropy change (from the greater degree of ordering) but a favourable enthalpy change (from the formation of new σ bonds), the reactions can often be reversed by heating.

Retro-cycloadditions can be classified in terms of the σ bonds which are broken. The retro-Diels–Alder reaction shown in fig. 4.17*a* is an allowed concerted ${}_{\pi}2_s + {}_{\sigma}2_s + {}_{\sigma}2_s$ process; the disallowed mode of cyclobutane cleavage shown in fig. 4.17*b* is ${}_{\sigma}2_s + {}_{\sigma}2_s$. The latter reaction would have to be ${}_{\sigma}2_s + {}_{\sigma}2_a$, as shown in fig. 4.17*c*, in order to be concerted.

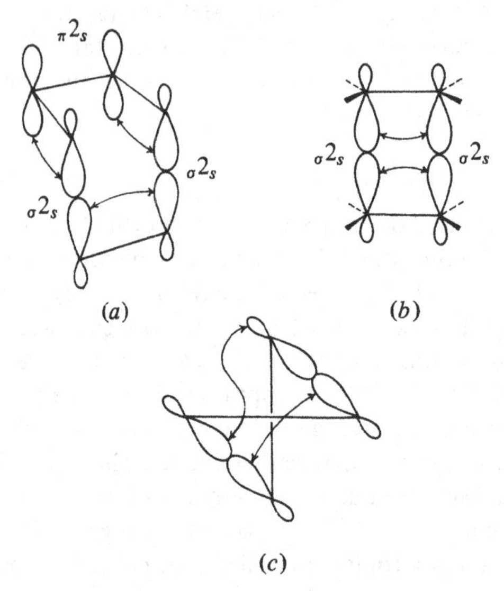

Fig. 4.17

Because of the fundamental importance of the total number of electrons in determining whether a cycloaddition or its reverse can be concerted or not, cycloadditions are discussed in detail in the next two chapters under headings based on the number of electrons involved.

4.3. Generalised perturbation molecular orbital theory in cycloadditions

So far considerations of orbital symmetry have allowed us to develop selection rules for cycloadditions. These rules emphasise that there are two types of situation, one in which a particular concerted mode of reaction is allowed and one in which it is disallowed (energetically unfavourable). If the allowed mode is associated with a transition state of feasible geometry then it is usually preferred over an alternative stepwise sequence leading to the same product. On the other hand if it is not, then a stepwise mechanism is to be expected and such simple considerations allow us to rationalise the concerted or stepwise nature and hence the stereoselectivity observed in a wide range of cycloadditions.

Thus it is reasonable that the allowed Diels–Alder reaction is very general, and is concerted and hence stereoselective. 2 + 2 Cycloadditions are likely to occur only when substituents capable of stabilising diradical or zwitterionic intermediates are present; moreover, such cycloadditions are not usually stereoselective because the σ bonds can rotate in these intermediates. However, several very important questions concerning cycloadditions remain which consideration of orbital symmetry alone cannot answer.

For example, there is considerable variation in the ease with which the allowed Diels–Alder reaction can take place. The prototype reaction of *cisoid* butadiene with ethylene is very slow and requires forcing conditions. However, Diels–Alder reactions in which the diene bears electron-releasing substituents and the dienophile electron-withdrawing substituents, or vice versa, proceed much more readily. Thus there is the question of ***reactivity***.

Also in the Diels–Alder reaction of an unsymmetrically substituted diene and dienophile it is found that one particular orientation is favoured; for example, in the case of a 2-substituted diene and dienophile the major product is as shown, regardless of the nature of R and X (4.4). We therefore need an explanation for this phenomenon of ***regioselectivity***.

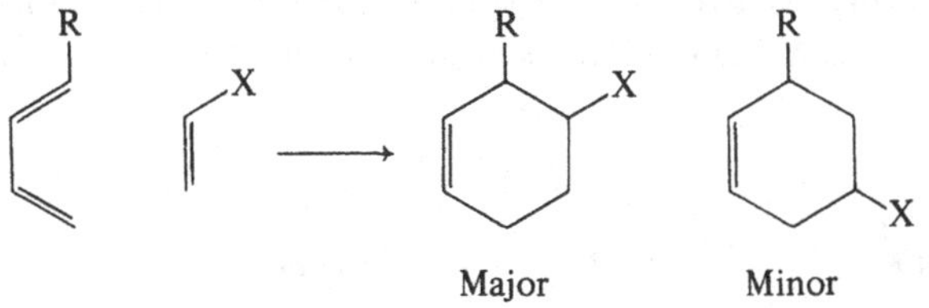

R X → R X (Major) R X (Minor) (4.4)

A final question concerns which is the preferred site of reaction if more than one is possible. For example, why does the Diels–Alder dimerisation of (3) lead to (**4a**) rather than (**4b**) and why in the case of the azepine (**5**) is the 6 + 4 cycloaddition mode preferred over the alternative allowed 4 + 2 modes? The former example illustrates a special case (site selectivity) of the more general phenomenon of *periselectivity* (preference for one pericyclic mode over others) illustrated by the latter.

NC → NC CN NC CN

(3) (4a) (4b)

RO_2CN → RO_2C-N N CO_2R

(5)

An approach to these problems is to be found in perturbation molecular orbital (PMO) theory.[5] This leads to a general perturbation (GP) equation for the energy of interaction between two molecules. This has been applied extensively in cycloadditions and in this case the GP equation can be simplified to the point where reactivity, regioselectivity and periselectivity can be rationalised very simply in terms of the energies and atomic orbital coefficients of the reactant HOMOs and LUMOs. This section describes the GP equation and its simplifying approximations. A qualitative general guide to the effect of substituents on the energies and coefficients of polyene orbitals is given in section 4.4, followed, in section 4.5, by illustration of how the approach answers the questions posed above.

Perturbation theory leads to the expression (4.5) for the energy of interaction of two polyenes A and D at atoms d and a and d′ and a′, as shown in fig. 4.18.

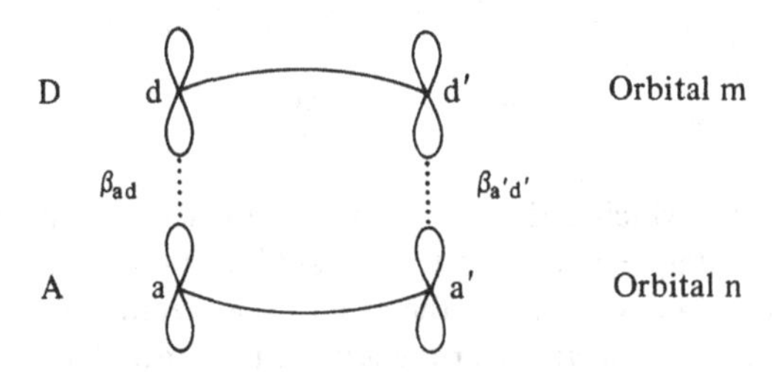

Fig. 4.18

$$\Delta E = -\frac{q_d q_a}{R_{ad}\epsilon} - \frac{q_{d'} q_{a'}}{R_{a'd'}\epsilon} + 2\sum_{m}^{occ}\sum_{n}^{unocc}\frac{(c_d^m c_a^n \Delta\beta_{ad} + c_{d'}^m c_{a'}^n \Delta\beta_{a'd'})^2}{E_m - E_n}$$

(electrostatic) (covalent)

$$+2\sum_{m}^{unocc}\sum_{n}^{occ}\frac{(c_d^m c_a^n \Delta\beta_{ad} + c_{d'}^m c_{a'}^n \Delta\beta_{a'd'})^2}{E_n - E_m} \qquad (4.5)$$

(covalent)

q_a and q_d	are the charges on the atoms a and d;
$q_{a'}$ and $q_{d'}$	are the charges on the atoms a′ and d′;
c_a^n and c_d^m	are the atomic orbital coefficients at atoms a and d in molecular orbitals n and m on molecules A and D respectively;
$c_{a'}^n$ and $c_{d'}^m$	are the atomic orbital coefficients at atoms a′ and d′ in molecular orbitals n and m on molecules A and D, respectively;
$\Delta\beta_{ad}$	is the change in resonance integral for overlap of orbitals m and n at atoms a and d in going from reactants to transition state;
$\Delta\beta_{a'd'}$	is the change in resonance integral for overlap of orbitals m and n at atoms a′ and d′ in going from reactants to transition state;
R_{ad}	is the distance between atoms a and d in a medium with dielectric constant ϵ;
$R_{a'd'}$	is the distance between atoms a′ and d′ in a medium with dielectric constant ϵ and
E_m and E_n	are the appropriate orbital energy levels.

Equation 4.5 involves an electrostatic part and a covalent part, the latter arising from second-order mixing of all occupied orbitals on D with all vacant orbitals on A and of all vacant orbitals on D with all occupied orbitals on A. The covalent part is normally the only one considered in a treatment of pericyclic processes; the electrostatic part is usually neglected and this is reasonable provided the reactants are not charged or particularly polar.

A key factor in determining the magnitude of the covalent terms is the energy separation between the interacting occupied and vacant orbitals. Since this appears in the denominator of equation 4.5, the closer these are in energy the greater the stabilisation due to their interaction. This leads to a further simplification of equation 4.5 because the orbitals closest in energy are bound to be the HOMO and LUMO pairs. The contribution of orbital interactions other than those of the HOMO and LUMO to $\Delta E_{covalent}$ will be smaller and can often be neglected in comparison with the frontier orbital interactions.

If only the frontier HOMO and LUMO of A and D are taken into account, equation 4.5 simplifies to equation 4.6.

$$\Delta E_{\text{covalent}} = 2\,\frac{(c_{\text{d}}^{\text{HO}} c_{\text{a}}^{\text{LU}} \Delta\beta_{\text{ad}} + c_{\text{d}'}^{\text{HO}} c_{\text{a}'}^{\text{LU}} \Delta\beta_{\text{a}'\text{d}'})^2}{E_{\text{D}}^{\text{HO}} - E_{\text{A}}^{\text{LU}}} + 2\,\frac{(c_{\text{d}}^{\text{LU}} c_{\text{a}}^{\text{HO}} \Delta\beta_{\text{ad}} + c_{\text{d}'}^{\text{LU}} c_{\text{a}'}^{\text{HO}} \Delta\beta_{\text{a}'\text{d}'})^2}{E_{\text{A}}^{\text{HO}} - E_{\text{D}}^{\text{LU}}} \qquad (4.6)$$

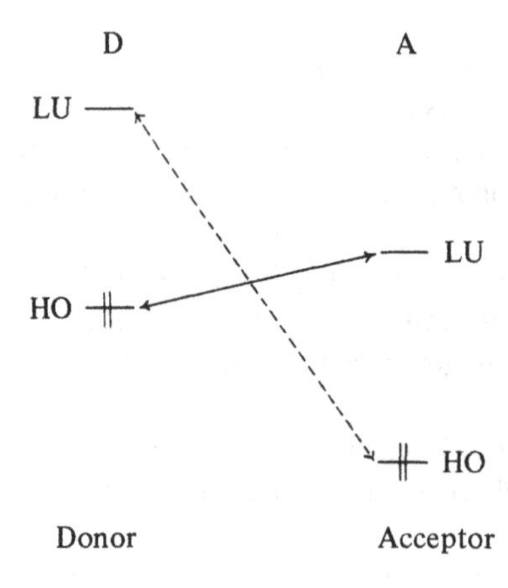

Fig. 4.19

When the orbital energies of D and A are as shown in fig. 4.19, the first of these two terms in equation 4.6 will dominate since $E_{\text{D}}^{\text{HO}} - E_{\text{A}}^{\text{LU}}$ is smaller than $E_{\text{A}}^{\text{HO}} - E_{\text{D}}^{\text{LU}}$ and in many cases the second term can be ignored. This is the situation where one reactant D is clearly an electron donor and the other, A, an electron acceptor. Such a reaction could be described as $\text{D}_{\text{HOMO}}\text{-A}_{\text{LUMO}}$ controlled and in this case equation 4.6 reduces to equation 4.7.

$$\Delta E_{\text{covalent}} = 2\,\frac{(c_{\text{d}}^{\text{HO}} c_{\text{a}}^{\text{LU}} \Delta\beta_{\text{ad}} + c_{\text{d}'}^{\text{HO}} c_{\text{a}'}^{\text{LU}} \Delta\beta_{\text{a}'\text{d}'})^2}{E_{\text{D}}^{\text{HO}} - E_{\text{A}}^{\text{LU}}} \qquad (4.7)$$

Equations 4.5–4.7 can be used to give semiquantitative estimates of the stabilisation of transition states in cycloadditions, given the necessary quantum mechanical data. Absolute values obtained by this method will not be very good but since they are based on a perturbational method, comparative values for similar systems will be much better and this is the great value of the treatment. Thus the equation is useful for rationalising reactivity trends in, for example, a series of Diels–Alder reactions. When different modes of addition are possible, the preferred mode (that is, the regioselectivity or periselectivity of the process) can be deduced from the relative stabilisation of the various transition states. This requires a knowledge of orbital energies and coefficients.

Before concluding this general discussion of the GP equations 4.5-4.7, it should be pointed out that the covalent terms naturally lead to the simple FMO symmetry selection rules outlined in chapter 2. This is because the signs of the atomic orbital coefficients reflect the symmetry of the orbitals. A quick inspection reveals that for the ${}_{\pi}2_s + {}_{\pi}2_s$ interaction of two ethylene molecules ($\psi_{HOMO} = 0.71\ \phi_1 + 0.71\ \phi_2$, $\psi_{LUMO} = 0.71\ \phi_1 - 0.71\ \phi_2$) the numerator reduces to zero.† On the other hand, for the ${}_{\pi}4_s + {}_{\pi}2_s$ interaction of ethylene with butadiene ($\psi_{HOMO} = 0.61\ \phi_1 + 0.37\ \phi_2 - 0.37\ \phi_3 - 0.61\ \phi_4$, $\psi_{LUMO} = 0.61\ \phi_1 - 0.37\ \phi_2 - 0.37\ \phi_3 + 0.61\ \phi_4$) the numerator is positive and the interaction is stabilising.

The GP equation thus predicts no covalent stabilisation for the disallowed ${}_{\pi}2_s + {}_{\pi}2_s$ and ${}_{\pi}2_a + {}_{\pi}2_a$ processes. (see, however, §4.5.) The ${}_{\pi}2_s + {}_{\pi}2_a$ mode will only be weakly stabilised because of low $\Delta\beta_{nm}$ values resulting from geometrically ineffective overlap. Equations 4.5-4.7 do, however, indicate that a 1s + 1s interaction (fig. 4.20) of two alkenes would be stabilised by both pairs of frontier orbital interactions. This one-centre interaction is the type which would lead to a dipolar or diradical intermediate.

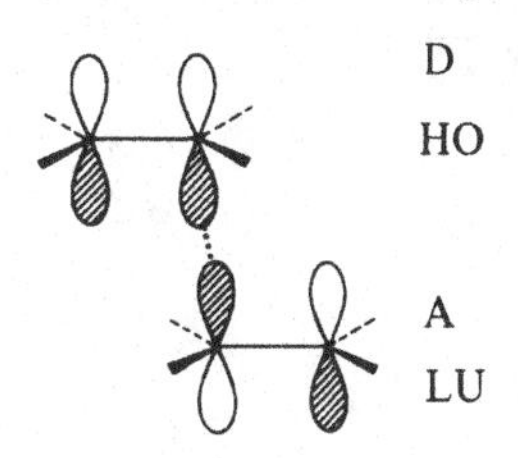

Fig. 4.20

4.4. Molecular orbitals and substituents

Equation 4.5 provides the basis for a rationalisation of reactivity, periselectivity and regioselectivity in cycloadditions given a good qualitative picture of the effect of substituents on the energies and orbital coefficients of polyenes. Houk[6] has collected this information which is presented in figs. 4.21, 4.22 and 4.23.

† $c_d^{HO} = 0.71$, $c_a^{LU} = 0.71$, $c_{d'}^{HO} = 0.71$, and $c_{a'}^{LU} = -0.71$; $\Delta\beta_{ad} = \Delta\beta_{a'd'}$ for a symmetrical transition state; thus the numerator of equation 4.7 becomes $2\Delta\beta^2\ (0.71 \times 0.71 + 0.71 \times -0.71)^2 = 0$. The signs of the orbital coefficients refer to one face of a polyene and are therefore appropriate when suprafacial additions of the polyene are considered. When the polyene is used in an antarafacial mode the sign of one terminal coefficient must be changed. Thus, for the ethylene HOMO in a ${}_{\pi}2_a$ process the appropriate coefficients are +0.71 and −0.71.

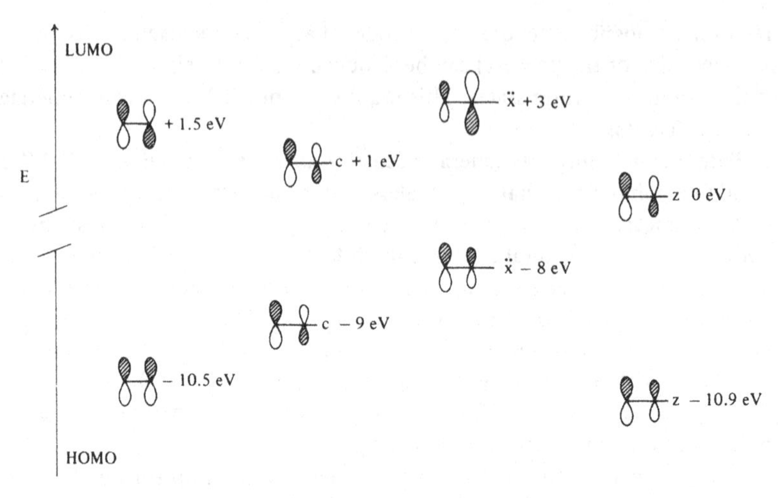

Fig. 4.21. Estimated frontier orbital energies and coefficients for alkenes. c = conjugating substituent: vinyl, aryl, etc.; ẍ = Me, OR, NR_2, etc.; z = CHO, CN, CO_2R, NO_2, etc.

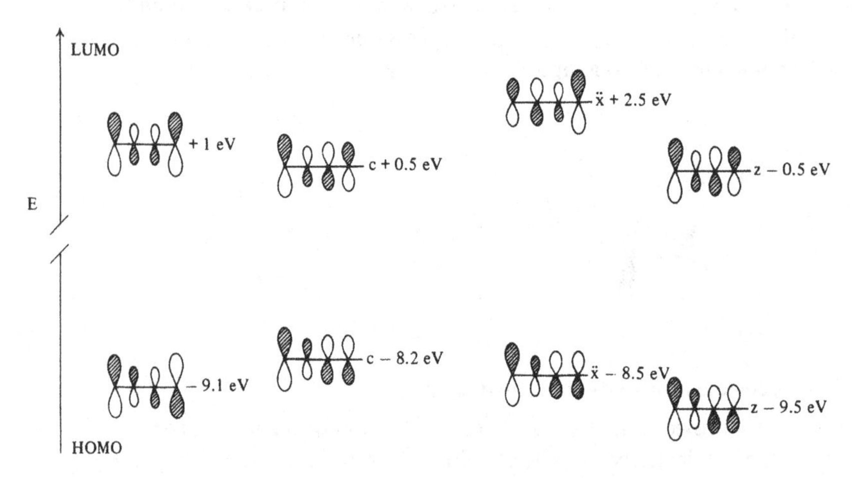

Fig. 4.22. Estimated frontier orbital energies and coefficients for 1-substituted dienes.

Relative orbital energies can be estimated by calculation or can be derived from experimental data. HOMO energies are related to ionisation potentials (IPs) ($E_{HOMO} = -IP$) and LUMO energies to electron affinities (EAs) ($E_{LUMO} = -EA$). There are plenty of data available for alkenes and the orbital energy trends summarised in fig. 4.21 correspond to experimentally determined average or typical values for different types of substituent. The orbital coefficients are derived by calculation; however, several methods all

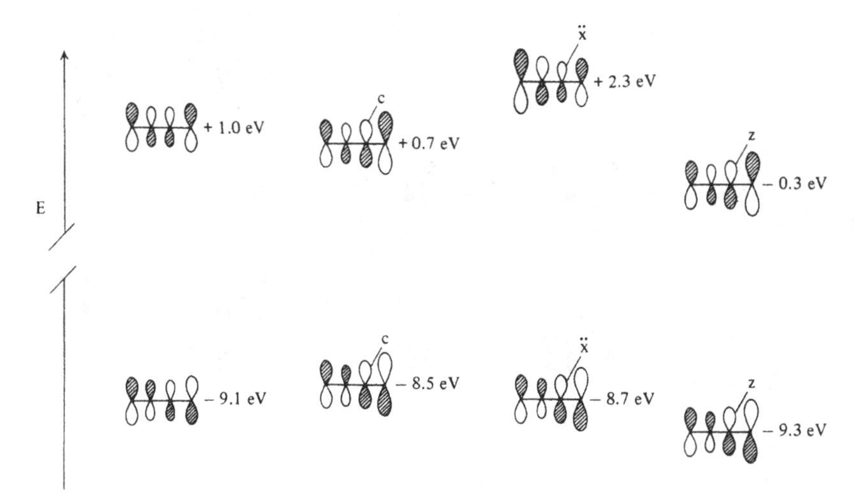

Fig. 4.23. Estimated frontier orbital energies and coefficients for 2-substituted dienes.

give the same trends in relative magnitudes except for the electron-deficient alkene HOMO which is discussed below. For convenience the magnitude of the coefficient is represented by the size of the lobe shown in fig. 4.21, the larger lobe corresponding to the larger coefficient. Figs. 4.22 and 4.23 for dienes were obtained similarly but are based on more limited experimental data.

Alkenes. The patterns which emerge for the alkenes in fig. 4.21 can be understood relatively easily by reference to the HOMOs and LUMOs of ethylene, butadiene and the allyl anion. These, as derived by extended Hückel or CNDO/2 calculations, are shown in fig. 4.24.

Butadiene is a good model for ethylene bearing a typical neutral conjugating substituent. The HOMO energy of butadiene is higher than that of ethylene and the LUMO energy is lower. The molecular orbital coefficients at C-1 and C-2 in both the HOMO and the LUMO of butadiene are unequal, the larger values being at C-1. Thus, in a conjugatively stabilised ethylene the larger coefficient is at the carbon remote from the substituent in both the HOMO and the LUMO.

In the case of an ethylene bearing an electron-releasing group which acts only inductively, the substituent will lower the electronegativity of the carbon to which it is attached, and hence it will lower the coefficient at this point in the HOMO and raise it in the LUMO. In the more common situation where electron release is by a mesomeric effect (NH_2, OR, etc.) the allyl anion system provides an extreme model. The HOMO will resemble the allyl anion HOMO and the substituent-bearing carbon will have a much reduced coefficient (corresponding to the node in the allyl anion HOMO).

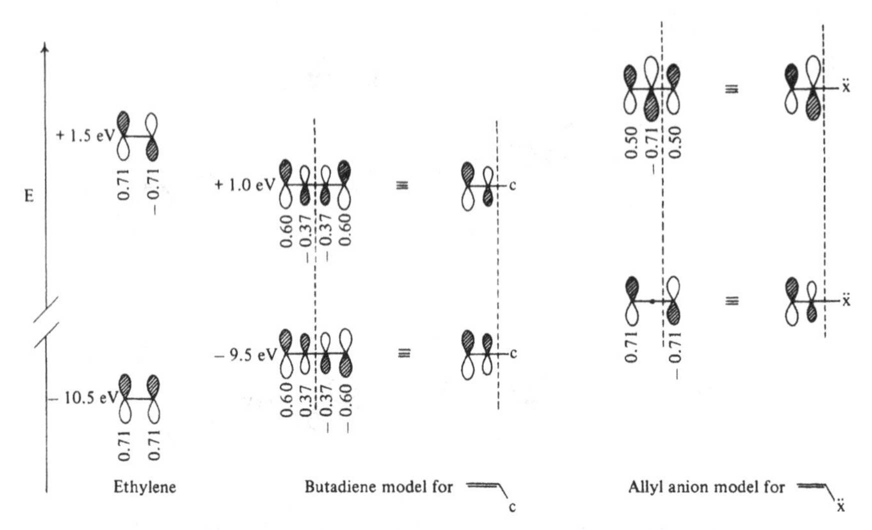

Fig. 4.24

The LUMO will have the larger coefficient at the substituent-bearing carbon atom. Both the HOMO and the LUMO will be raised in energy relative to those in ethylene, the former more than the latter.

For an electron-deficient alkene, a purely inductive electron-withdrawing substituent (e.g. CF_3) would raise the coefficient on the substituent-bearing carbon (more electronegative) in the HOMO and reduce it in the LUMO. However, most electron-withdrawing substituents are also conjugating groups (e.g. CO_2R, CHO) and, as in butadiene, the conjugating effect tends to diminish the coefficient at the substituted site in both HOMO and LUMO. Because of these opposing trends there is some doubt as to their relative magnitudes in the HOMO. The conclusion arrived at by Houk is that the coefficient of the substituted carbon in the HOMO is slightly smaller than the terminal coefficient. In the LUMO the trends reinforce each other and the terminal carbon has a much larger coefficient than the substituted carbon. The energy of the HOMO is reduced slightly (opposing inductive and conjugating trends) and the LUMO somewhat more relative to ethylene.

Dienes. The trends for substituted butadienes shown in figs. 4.22 and 4.23 can be rationalised similarly.

Polysubstituted alkenes. For polysubstituted alkenes the effect of substituents seems to be additive for both energies and coefficients. This is illustrated by the data for acrylonitrile (**6**) and 1,1-dicyanoethylene (**7**).

In the case of α-methylacrylonitrile (**8**) the larger coefficient is on C-2 in the HOMO (the effects of both Me and CN reinforce each other) but the

	0.66 −0.54		0.66 − 0.49	
LUMO	CN	− 0.02 eV	CN, CN	− 1.54 eV
	0.60 0.49		0.61 0.45	
HOMO	CN	− 10.92 eV	CN, CN	− 11.38 eV
	(6)		**(7)**	

coefficients at C-1 and C-2 are roughly equal in the LUMO where the effects of Me and CN oppose each other. For a cinnamic ester **(9)** the larger HOMO coefficient is on C-1 because phenyl (conjugating) has a larger effect on the HOMO than CO_2R (electron withdrawing and conjugating). In the LUMO, C-1 and C-2 have approximately equal coefficients as a result of opposing and nearly equal substituent effects.

$\overset{2}{C}H_2=\overset{1}{C}(Me)(CN)$ **(8)**

$Ph-\overset{2}{C}H=\overset{1}{C}H-CO_2R$ **(9)**

Alkynes. In the case of alkynes, the HOMO is lower in energy (0.4–0.9 eV) than that of a similar alkene HOMO but the LUMO energies are more nearly equal. The orbital coefficients vary in much the same way as for the alkene but it must be borne in mind that only one of the two orthogonal acetylene π bonds can conjugate with a substituent; the other can only be affected inductively.

Heteromultiple bonds. π Bonds other than those of alkenes and alkynes are also important in cycloadditions. Here the trends are as expected, electron density in the HOMO being higher at the more electronegative end of the bond. Thus systems of the type C=X where X=O or N will have a HOMO **(10)** and LUMO **(11)** as shown. Since X is more electronegative than carbon the energies of the HOMO and LUMO will be similar to those of electron-deficient alkenes.

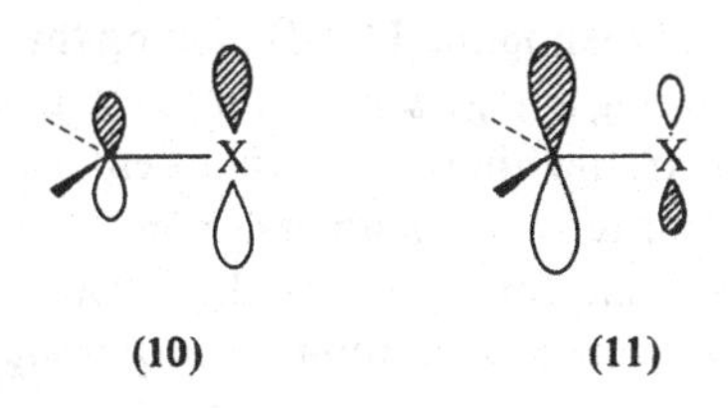

(10) **(11)**

4.5. Reactivity, regioselectivity and periselectivity

Reactivity

In equations 4.6 and 4.7, the HOMO–LUMO energy separations largely determine the magnitudes of the covalent term and it is not surprising that reactivity in cycloadditions is related to the HOMO–LUMO separation, higher reactivity being associated with a small HOMO–LUMO energy gap. We can now account for the reactivity trends observed in the Diels–Alder reaction (p. 85). An orbital interaction diagram for butadiene–ethylene is shown in fig. 4.25.

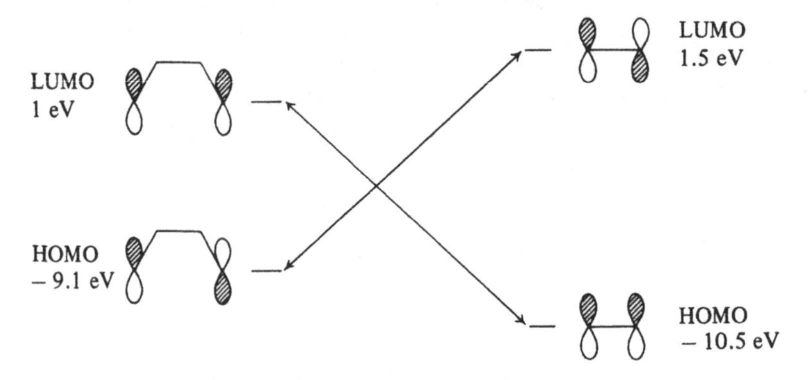

ΔE (diene HOMO–dienophile LUMO) = 10.6 eV
ΔE (diene LUMO–dienophile HOMO) = 11.5 eV

Fig. 4.25

The most important interaction in the covalent term of equation 4.5 involves the diene HOMO and dienophile LUMO although with an estimated difference of approximately only 1 eV both HOMO–LUMO interactions are probably important. Both energy separations are relatively large so the energy of interaction will be correspondingly small.

Electron-releasing groups in the diene will raise the diene HOMO and LUMO energies, decreasing the important diene HOMO–dienophile LUMO energy separation and increasing the less important diene HOMO– dienophile LUMO separation. Introduction of electron-withdrawing groups in the dienophile will lower both the HOMO and LUMO energies. This further closes the energy gap between the diene HOMO and dienophile LUMO making this interaction clearly dominant. The decreasing energy separation as electron-releasing groups enter the diene and electron-withdrawing groups enter the dienophile tends to maximise the covalent term in equation 4.5 and so explains the normal substituent effects found for the Diels–Alder reaction. The actual orbital energies for a diene bearing an electron-releasing group on

C-1 and a dienophile with an electron-withdrawing group, using the data from figs. 4.21 and 4.22, are shown in fig. 4.26.

In contrast, for dienes bearing electron-withdrawing groups and dienophiles with electron-releasing substituents the diene HOMO–dienophile LUMO interaction (energy separation 12.5 eV) becomes less than that between the diene LUMO–dienophile HOMO (8.5 eV) (fig. 4.27). Such Diels–Alder reactions with inverse electron demand should be, and indeed are, facile.

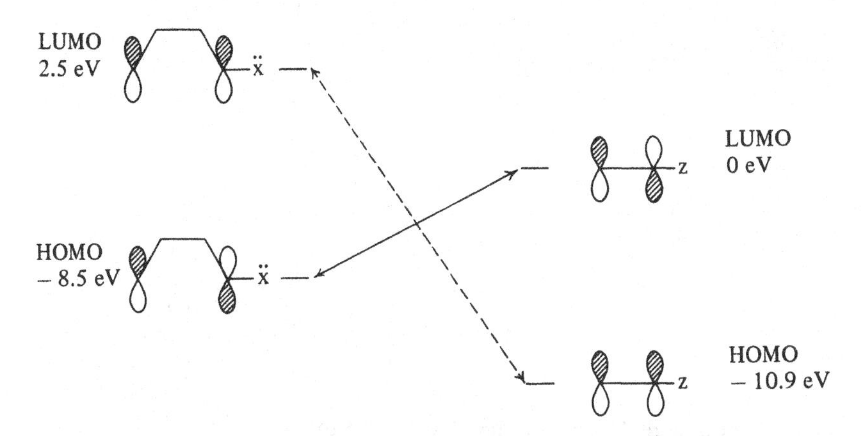

ΔE (diene HOMO–dienophile LUMO) = 8.5 eV
ΔE (diene LUMO–dienophile HOMO) = 13.4 eV

Fig. 4.26

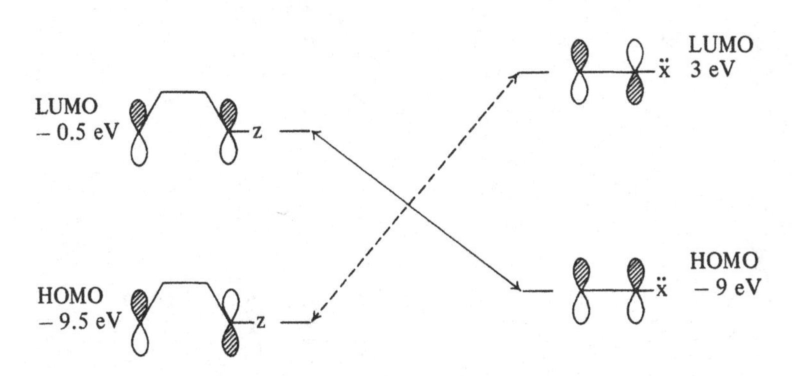

ΔE (dienophile HOMO–diene LUMO) = 8.5 eV
ΔE (dienophile LUMO–diene HOMO) = 12.5 eV

Fig. 4.27

Thus for the Diels–Alder reaction to occur readily complementary substituent effects are required. Reaction between a diene with electron-withdrawing groups and a dienophile also with electron-withdrawing groups (fig. 4.28) or between dienes and dienophiles both bearing electron-releasing groups (fig. 4.29) will be less favourable.

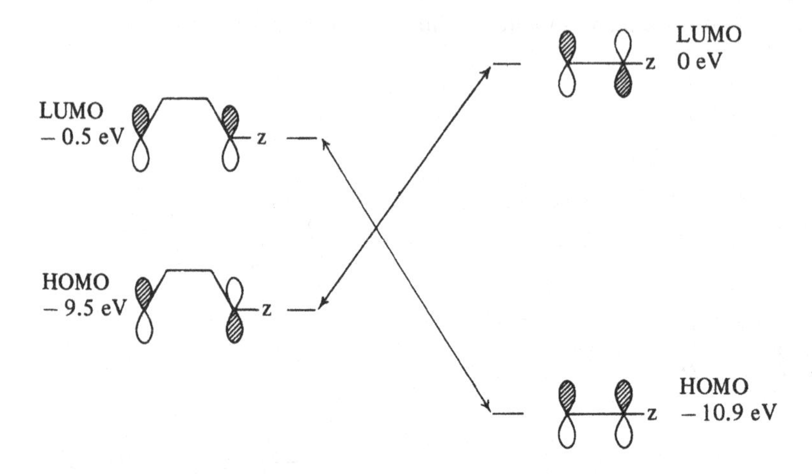

ΔE (diene HOMO–dienophile LUMO) = 9.5 eV
ΔE (diene LUMO–dienophile HOMO) = 10.4 eV

Fig. 4.28

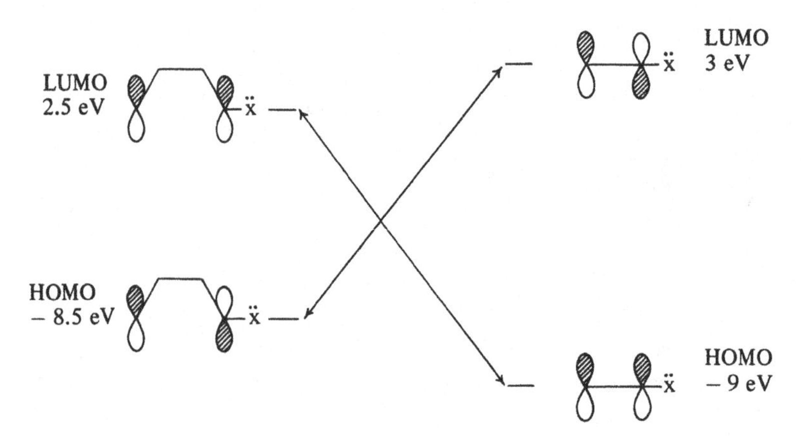

ΔE (diene HOMO–dienophile LUMO) = 11.5 eV
ΔE (diene LUMO–dienophile HOMO) = 11.5 eV

Fig. 4.29

The above approach allows us to rationalise qualitatively the reactivity ordering observed in Diels–Alder reactions. More quantitative correlations have been observed. The rate constant for a reaction is given by

$$k = A\mathrm{e}^{-E_a/RT}, \text{ therefore } \ln k = \ln A - E_a/RT$$

The energy of activation E_a can be considered as being made up of two parts, the stabilisation energy ΔE_{FMO} as calculated from equations 4.5–4.7 and the destabilisation energy resulting from van der Waals repulsions involving filled orbitals, ΔE_{VDW}. Thus,

$$\ln k = \ln A - \Delta E_{VDW}/RT - \Delta E_{FMO}/RT$$

Ln A is related to the probability of a collision between reactants having a favourable orientation and for a particular type of cycloaddition is likely to be fairly constant. So too is the ΔE_{VDW} term if we assume that the substituents on the reactants are of constant size. In these conditions

$$\ln k = \text{constant} - \Delta E_{FMO}/RT$$

and a plot of ln k against ΔE_{FMO} should be a straight line. This has been tested for a series of Diels–Alder reactions of cyclopentadiene with dienophiles, using experimental rate data and orbital energies and coefficients derived from Hückel molecular orbital calculations.[7] The best correlation was found for plots of ln k against $(E_{\text{HOMO diene}} - E_{\text{LUMO dienophile}})^{-1}$.

Inclusion of the orbital coefficient values in the numerator of equation 4.6 for the estimation of ΔE_{FMO} in the plot against ln k gave a poorer correlation. This was explained by assuming that in the transition state the atomic orbital coefficients are enhanced so that the numerator approximates to its maximum value of one. Also, the $E_{\text{HOMO dienophile}} - E_{\text{LUMO diene}}$ term is best omitted. In the transition state the frontier orbitals will tend to become closer in energy and this effect will be more marked for the closest pair of orbitals so it is reasonable to ignore other interactions. In his development of FMO theory, Fukui[8] noted both of these effects and summarised them in his principles of 'growing frontier orbital density along the reaction path' and 'narrowing of interfrontier level separations'.

Other workers[9,10] have also observed correlations between ln k and the HOMO–LUMO energy separation as given by $\text{IP}_{\text{diene}} - \text{EA}_{\text{dienophile}}$. Thus, excellent straight-line correlations were found for ln k versus $(\text{IP}_{\text{diene}} - \text{EA}_{\text{dienophile}})$ for the reaction of cyanoalkenes with cyclopentadiene and with 9,10-dimethylanthracene.[10]

Regioselectivity

Regioselectivity in cycloadditions between unsymmetrical polyenes is determined by the numerator in equations 4.5 to 4.7. The HOMO–LUMO separation is the same for the regioisomeric transition states shown in figs. 4.18 and 4.30 but

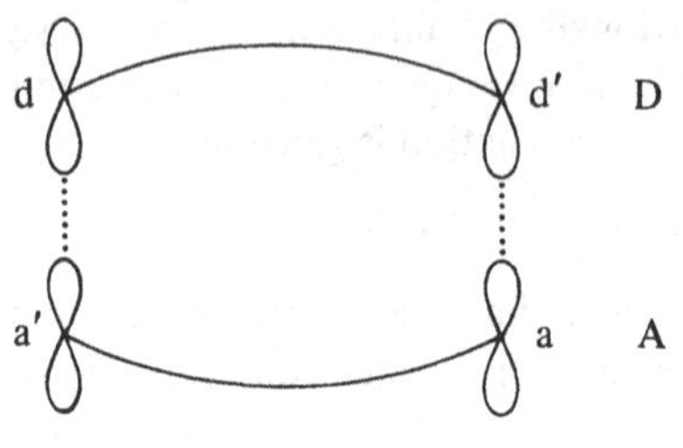

Fig. 4.30

from equation 4.6 the numerators are $(c_{d}^{HO}c_{a}^{LU}\Delta\beta_{ad} + c_{d'}^{HO}c_{a'}^{LU}\Delta\beta_{a'd'})^2$ and $(c_{d}^{HO}c_{a'}^{LU}\Delta\beta_{a'd} + c_{d'}^{HO}c_{a}^{LU}\Delta\beta_{ad'})^2$, respectively.

The preferred regioisomer will be that one for which the numerator is larger. If the $\Delta\beta$ values are fairly constant (this is reasonable for a concerted reaction involving combining atoms of the same element) the favoured orientation will be such that the sum of the squares of the interacting orbital coefficients is larger. Because of the sum of squares relationship the optimum combination of orbital coefficients is large–large and small–small rather than large–small and small–large.

We can now rationalise the regioselectivity observed in the Diels–Alder reaction, posed as a problem on p. 85 and illustrated by the specific examples shown (4.8[11] and 4.9[12]).

OMe, CHO → OMe, CHO (4.8)

CO_2H, CO_2H → CO_2H, CO_2H (4.9)

In order to do this, one has to identify the dominant HOMO–LUMO interaction. Frequently this is clearcut and the complementary HOMO–LUMO pair can be ignored. In other cases where the energy separations are closely balanced (~ 1 to 1.5 eV) then both interactions may have to be taken into account. Having identified the dominant interaction one then has to match the appropriate orbitals so that the two larger coefficients (lobes) are paired.

Thus in the first example, $\Delta E_{\text{diene HOMO-dienophile LUMO}} = 8.5$ eV compared with $\Delta E_{\text{diene LUMO-dienophile HOMO}} = 13.4$ eV; the former interaction is clearly the more important. The orbital coefficients are as in (**12**) and pairing large with large and small with small leads to the greater stabilisation and explains the observed regioselectivity. The tendency of the very much weaker diene LUMO–dienophile HOMO interaction (**13**) to favour the alternative orientation is overridden.

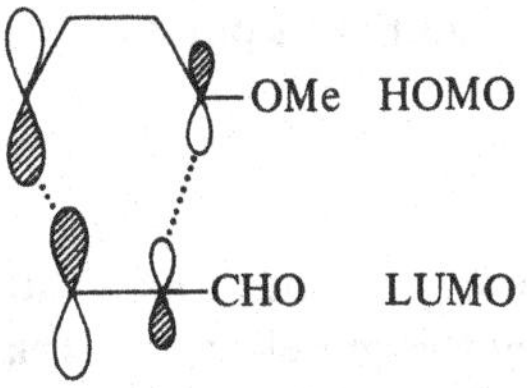

-OMe HOMO
-CHO LUMO
(12)

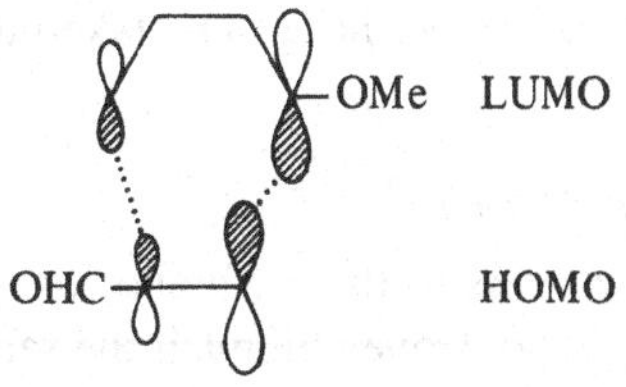

-OMe LUMO
OHC- HOMO
(13)

In the second example $\Delta E_{\text{diene HOMO-dienophile LUMO}} = 9.5$ eV is much closer to $\Delta E_{\text{diene LUMO-dienophile HOMO}} = 10.4$ eV but both interactions **(14)** and **(15)** favour formation of the observed product.

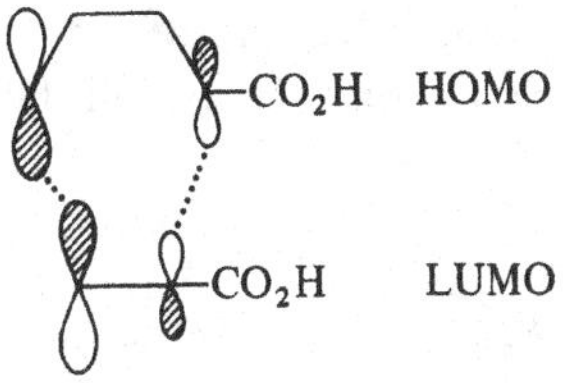

$-CO_2H$ HOMO
$-CO_2H$ LUMO
(14)

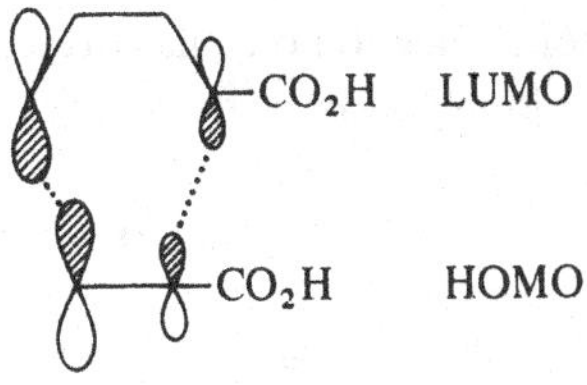

$-CO_2H$ LUMO
$-CO_2H$ HOMO
(15)

It is interesting to note that while the first of these examples might have been rationalised by intuition (a concerted reaction involving unequal bond formation in the transition state such that the process has some of the character of a conjugate addition of a nucleophile to an α,β-unsaturated carbonyl compound) this is not so for the second where combination of two electrophilic centres occurs. Another interesting observation is that when the salts of the acids in the second example combine in a Diels-Alder reaction, equal amounts of the two possible regioisomers **(16)** and **(17)** are produced[13] in spite of the strong electrostatic repulsion operating against the frontier orbital control (4.10).

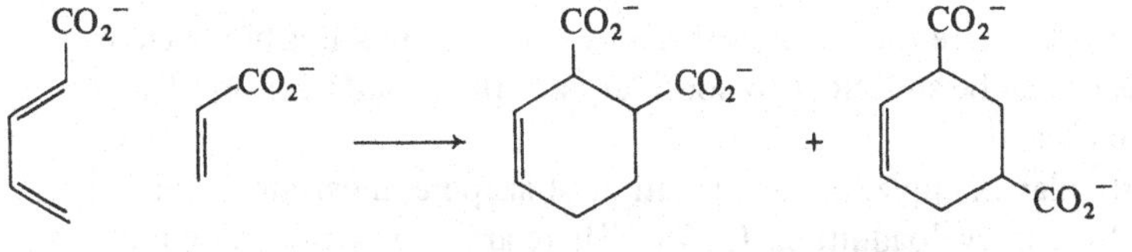

CO_2^- CO_2^- ⟶ CO_2^- CO_2^- + CO_2^- CO_2^- (4.10)

(16) (17)

Regioselectivity in Diels-Alder reactions of heterodienes and dienophiles and in 1,3-dipolar cycloadditions can be treated similarly. However, in these cases the new bonds formed are not all C—C bonds and so the appropriate β

(overlap integral) values have to be taken into account. This point is enlarged upon later.

Periselectivity

Periselectivity refers to the preference of one pericyclic mode over an alternative one. It is controlled in much the same way as regioselectivity. Taking the dimerisation of azepine **(5)**[14] as an example, the HOMO–LUMO separations are the same for the $6\pi + 4\pi$ mode and for the possible Diels–Alder modes and the addition is therefore controlled by the frontier orbital coefficients. Since this is a dimerisation both HOMO–LUMO pairs have to be considered because they must be equally separated in energy. The relative magnitudes for the numerator terms of equation 4.5 are compared in fig. 4.31 for the observed $6\pi + 4\pi$ mode and for the equally allowed $4\pi + 2\pi$ mode shown. The data used are from Hückel molecular orbital calculations,[7] the signs of the coefficients being omitted for simplicity. Clearly the $6\pi + 4\pi$ mode is favoured.

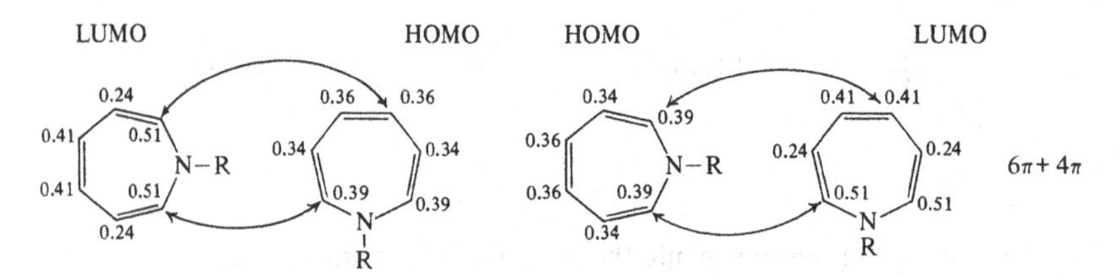

$$\Sigma(c_{\text{HOMO}} \cdot c_{\text{LUMO}})^2 = (0.51 \cdot 0.36 + 0.51 \cdot 0.39)^2 + (0.39 \cdot 0.51 + 0.39 \cdot 0.41)^2 = 0.275$$

N–R N–R $4\pi + 2\pi$

$$\Sigma(c_{\text{HOMO}} \cdot c_{\text{LUMO}})^2 = (0.36 \cdot 0.41 + 0.39 \cdot 0.51)^2 + (0.41 \cdot 0.36 + 0.51 \cdot 0.36)^2 = 0.230$$

Fig. 4.31

In other cases where there is a clear donor–acceptor relationship between the addends it may be sufficient to consider only the HOMO donor–LUMO acceptor coefficients.

This particular example, the dimerisation of azepine, illustrates a further general subtlety in cycloadditions (p. 79). There are two possible orientations for the allowed $6\pi + 4\pi$ mode; the reaction may involve transition state **(18)** leading to *exo*-addition (the interacting π systems oriented away from each other) and transition state **(19)** leading to *endo*-addition (interacting π systems oriented over each other). Only the former is observed[14] and this can be explained by *secondary orbital interactions* which are bonding for the *exo*-

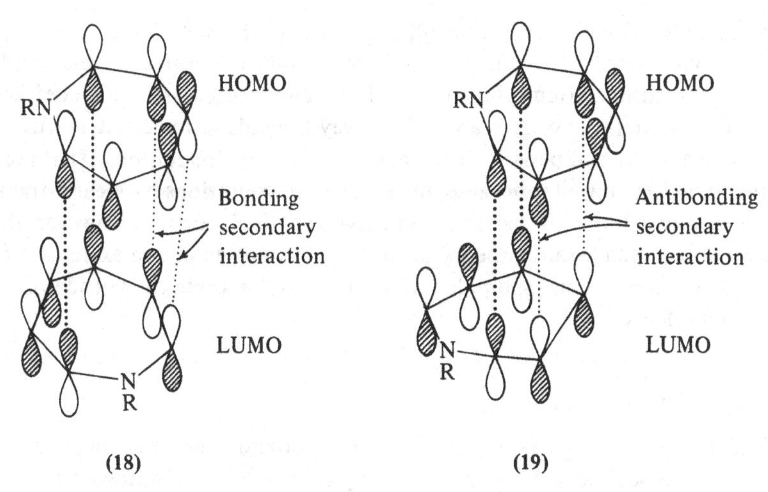

(18) (19)

orientation (**18**) and antibonding for the *endo*-orientation (**19**). Secondary interactions refer to those interactions which are not involved in primary bonding changes and have been used generally to explain such orientational phenomena in cycloadditions and other pericyclic reactions (see later). It has also been suggested that secondary orbital interactions are responsible for regioselectivity in some Diels–Alder reactions.[15] Thus in the addition of butadiene-l-carboxylic acid to acrylic acid (see p. 98), secondary interactions between the diene orbitals on C-2 and C-3 and the carbonyl group are greater for the transition state (**20**) leading to cyclohexene-3,4-dicarboxylic acid than the alternative transition state (**21**) which would give the 3,5-diacid. Calculations indicate that the C-2 coefficient is greater than that at C-3 in the butadiene-l-carboxylic acid HOMO.

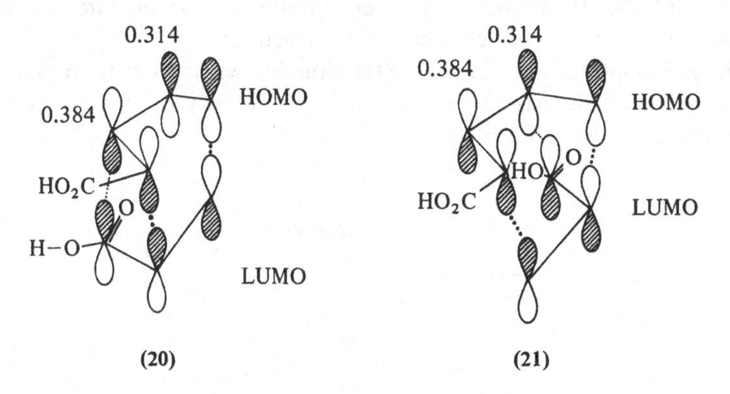

(20) (21)

The question of site selectivity can be rationalised along the same lines. Thus in the example cited earlier, the dimerisation of 2-cyanobutadiene, it can be seen from fig. 4.23 that the largest coefficient is on C-1 in both the HOMO and LUMO. This dominates the $(c_{HOMO} \cdot c_{LUMO})^2$ summation so that 2-cyanobutadiene acts as a dienophile at the cyano-substituted double bond.

In conclusion, given orbital energy levels and molecular orbital coefficients, frontier orbital theory goes a very long way towards rationalising reactivity, regioselectivity and periselectivity phenomena in cycloadditions. That the theory works so well is because, in general, cycloadditions have early transition states in which the reactants have undergone little distortion. However, the gross assumptions made in the theory must not be forgotten and in extreme cases, steric, conformational and polar effects may well operate to override the frontier orbital effects.

4.6. Configuration interaction

The idea that one reactant is an electron donor and another an electron acceptor arises naturally from the FMO picture of cycloadditions. So far we have considered interactions between reactants in their ground state electron configurations but as donor–acceptor tendency increases it becomes reasonable to consider that configurations resulting from electron transfer make a significant contribution. Epiotis[16] has used this idea to develop some very interesting and far-reaching conclusions for cycloadditions and other pericyclic reactions.

His approach is illustrated for 2 + 2 cycloadditions between a donor D and an acceptor A. The transition state for this process is assumed to be a resonance hybrid of ground state AD configurations and charge transfer configurations D^+A^-. Strictly, other charge transfer configurations such as A^+D^- can be included in a full treatment but these are going to be much less important than the configuration arising from formal electron transfer from the donor HOMO to acceptor LUMO. The charge transfer configuration will be more important the greater the donor–acceptor character of the reactants.

An orbital interaction diagram for the situation where there is no clear donor–acceptor relationship, for example between two unsubstituted ethylene molecules, is shown in fig. 4.32.

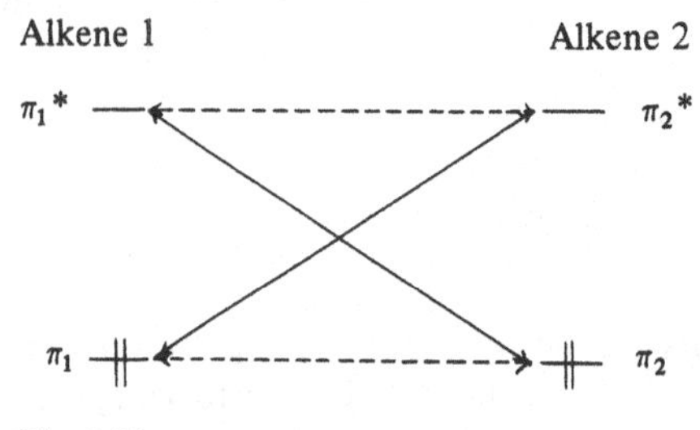

Fig. 4.32

An orbital interaction diagram is not to be confused with a correlation diagram. It indicates which orbitals interact with which in the reaction in question. Thus in the cycloaddition of alkene 1 to alkene 2 the interacting orbitals are HOMO 1-LUMO 2 and LUMO 1-HOMO 2, i.e. π_1-π_2^* and π_2-π_1^* (fig. 4.32), and these are connected by solid arrows. This interaction between filled and vacant orbitals is stabilising; interaction between the filled π_1 and π_2 orbitals and between empty π_1^* and π_2^* orbitals (broken arrows) will not lead to any stabilisation. For a pericyclic interaction between π_1 and π_2^* and π_2 and π_1^* to be stabilising it must, of course, involve a ${}_{\pi}2_s + {}_{\pi}2_a$ interaction for orbital symmetry reasons.

Fig. 4.33 shows a typical situation where alkene 1 is a donor (an alkene with electron-releasing substituents) and reactant 2 is an acceptor (an alkene with electron-withdrawing substituents).

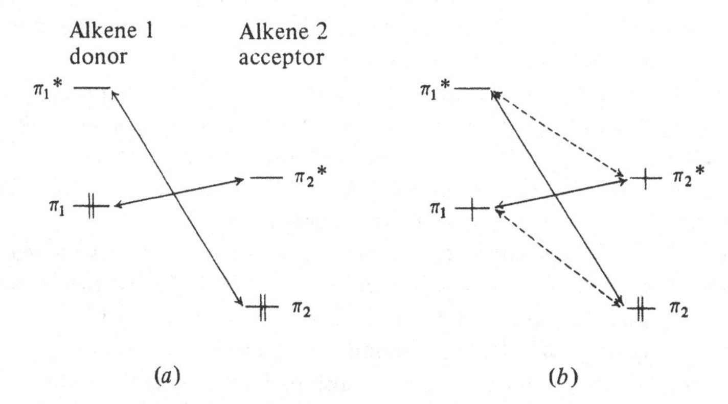

Fig. 4.33. (*a*) Ground configuration; (*b*) charge transfer configuration.

For the ground configuration (fig. 4.33*a*) the major interaction is between π_1 and π_2^*, that between π_2 and π_1^* being less because of the greater orbital energy separation. Both of these interactions are concerned with the ${}_{\pi}2_s + {}_{\pi}2_a$ mode of addition and the substituent effects favour the process by lowering the energy gap between the HOMO and LUMO. In the charge transfer configuration (fig. 4.33*b*) the π_2-π_1^* and π_1-π_2^* interactions involved with the ${}_{\pi}2_s + {}_{\pi}2_a$ mode of combination are again stabilising (interaction of filled with empty, half filled with half filled orbitals) but because π_1 and π_2^* are partially occupied, interactions between π_1 and π_2 and π_1^* and π_2^* are now also stabilising. These interactions are associated with ${}_{\pi}2_s + {}_{\pi}2_s$ union for reasons of orbital symmetry. Moreover, since the geometrical overlap in a ${}_{\pi}2_s + {}_{\pi}2_s$ sense is much more favourable, the overlap integral for this combination will be greater than that for the ${}_{\pi}2_s + {}_{\pi}2_a$ mode. Thus if configuration interaction is included in the PMO equation, for moderately good donor–acceptor pairs the ${}_{\pi}2_s + {}_{\pi}2_s$ concerted mode can become energetically comparable with or even preferable to the ${}_{\pi}2_s + {}_{\pi}2_a$ mode of reaction. Thus a spectrum of mechanisms may be involved in 2 + 2 cycloadditions. At the non-polar extreme the ${}_{\pi}2_s + {}_{\pi}2_a$ process is the preferred pericyclic mode. This derives no appreciable stabilisation from substituent effects and configuration interaction and as it also suffers from poor geometrical overlap such reactions almost certainly proceed via diradical intermediates. At the other polar extreme, configuration interaction combined with the much better geometrical overlap makes the ${}_{\pi}2_s + {}_{\pi}2_s$ process the preferred pericyclic mode and also renders it more favourable than a stepwise pathway though a zwitterionic intermediate. Significantly polar 2 + 2 cycloadditions often show high ${}_{\pi}2_s + {}_{\pi}2_s$ stereoselectivity. Between these two extremes the moderately polar 2 + 2 cycloadditions which are normally non-stereoselective may be cases where concerted ${}_{\pi}2_s + {}_{\pi}2_s$ and ${}_{\pi}2_s + {}_{\pi}2_a$ and/or stepwise modes are in competition. This view of 2 + 2 cycloadditions in which concerted pathways are believed to be important thus differs from the simple picture developed earlier in this chapter. We shall discuss the mechanism of 2 + 2 cycloaddition further in chapter 6.

In an allowed cycloaddition, such as the Diels–Alder $4\pi + 2\pi$ process, configuration interaction again favours both the allowed ${}_{\pi}4_s + {}_{\pi}2_s$ and disallowed ${}_{\pi}4_s + {}_{\pi}2_a$ or ${}_{\pi}4_a + {}_{\pi}2_s$ modes. However, overlap for the first process is geometrically so much better that the effect of configuration interaction is simply to reinforce the predominance of the more favourable process.

Epiotis has also considered other pericyclic processes, electrocyclic reactions and sigmatropic rearrangements and analysed the effect of substituents in terms of their influence on configuration interaction. Conclusions similar to those drawn above for cycloadditions are found. Complementary substituents effects leading to clearcut donor–acceptor relationships favour both allowed and disallowed pathways. Where the former are also least motion pathways (i.e. those involving the least movement of interacting atoms and better orbital overlap, e.g. for 2 + 2 cycloaddition the ${}_{\pi}2_s + {}_{\pi}2_s$ mode is the least motion pathway, the ${}_{\pi}2_s + {}_{\pi}2_a$ the non-least motion pathway) they remain the most favourable processes. However, where the allowed mode is a non-least motion pathway (as in 2 + 2 cycloaddition) then configuration interaction can render the alternative least motion but disallowed pathway energetically more favourable.

In summary the Epiotis view of pericyclic reactions differs from that generally held and developed in chapter 2 and at the beginning of this chapter. Nearly all reactions are considered to be concerted, and lack of stereoselectivity is believed to be the result of competing concerted modes with different stereochemical characteristics rather than of stepwise processes where lack of stereoselectivity is associated with conformational changes within an intermediate. The reactions where intermediates are most likely to be encountered are those where substituents do not close the HOMO–LUMO separation sufficiently for appreciable interaction to occur. Such situations arise for example, with ethylene and ethylene and ethylene and butadiene (the prototypical Diels-Alder reaction where some experimental and sophisticated calculations suggest a diradical process).

References

1. Paquette, L. A., *Principles of modern heterocyclic chemistry,* p. 337, Benjamin, New York, 1968.
2. Huisgen, R., *Angew. Chem., Int. Edn Engl.,* 7, 321 (1968). See also Huisgen, R., Grashey, R. and Sauer, J. in *The chemistry of alkenes,* ed. S. Patai, p. 739, Interscience, London, 1964.
3. Inagaki, S., Fujimoto, H. and Fukui, K., *J. Am. chem. Soc.,* **98**, 4054, 4693 (1976).
4. Woodward, R. B. and Hoffmann, R., *Angew. Chem., Int. Edn Engl.,* **8**, 781 (1969).
5. Hudson, R. F., *Angew. Chem., Int. Edn Engl.,* **12**, 36 (1973); Klopman, J. (ed.), *Chemical reactivity and reaction paths,* Wiley, New York, 1974; Fukui, K., *Acc. chem. Res.,* **4**, 57 (1971); Salem, L., *J. Am. chem. Soc.,* **90**, 543, 553 (1968); Devaquet, N. and Salem, L., *J. Am. chem. Soc.,* **91**, 3793 (1969).
6. Houk, K. N., *Acc. chem. Res.,* **8**, 361 (1975); Houk, K. N., Sims, J., Duke, R. E., Strozier, R. W. and George, J. K., *J. Am. chem. Soc.,* **95**, 7287 (1973).
7. Mok, K. L. and Nye, M. J., *J. chem. Soc. Perkin I,* 1975, 1810.
8. Fukui, K., *Theory of orientation and stereoselection,* Springer-Verlag, Berlin, 1975.
9. Sustmann, R. and Trill, H., *Angew. Chem., Int. Edn Engl.,* **11**, 838 (1972).
10. Houk, K. N., *Acc. chem. Res.,* **8**, 361 (1975).
11. Wichterle, O., *Colln Czech. chem. Commun. Engl. Edn,* **10**, 497 (1938).
12. Alder, K., Schumacher, M. and Wolff, O., *Justus Liebigs Annln Chem.,* **564**, 79 (1949).
13. Alder, K. and Heimbach, K., *Chem. Ber.,* **86**, 1312 (1953).
14. Paul, I. C., Johnson, S. M., Barrett, J. H. and Paquette, L. A., *Chem. Communs,* 1969, 6; Paquette, L. A., Barrett, J. H. and Kuhla, D. E., *J. Am. chem. Soc.,* **91**, 3616 (1969).
15. Alston, P. V. and Ottenbrite, R. M., *J. org. Chem.,* **40**, 1111 (1975).
16. Epiotis, N. D., *Angew. Chem., Int. Edn Engl.,* **13**, 751 (1975); Epiotis, N. D., *J. Am. chem. Soc.,* **94**, 1924, 1935, 1941, 1946 (1972); Epiotis, N. D., *J. Am. chem. Soc.,* **95**, 1191, 1200, 1206, 1214 (1973).

5 Cycloadditions and eliminations involving six electrons

Thermal cycloadditions and eliminations involving six electrons are by far the commonest. Significantly they are in general concerted. This is undoubtedly associated with the allowed nature of concerted $_{\pi}4_s + _{\pi}2_s$ cycloadditions and the fact that such a process is also geometrically favourable.

5.1. The Diels–Alder reaction

Diene Dienophile

Fig. 5.1. The Diels–Alder reaction.

The reaction of conjugated dienes with monoenes (fig. 5.1) to give six-membered ring compounds is the best known of cycloadditions and bears the names of the two workers who first recognised the scope of the reaction and began investigation into its mechanism (about 1928). As we shall see, most of the evidence is consistent with a concerted reaction involving simultaneous formation of the two new σ bonds.

Scope[1]

Some idea of the scope is given by the examples shown in equations 5.1–5.6. Examples of Diels–Alder reactions are known for almost all types of π bond acting as dienophiles (table 5.1). The diene system is commonly all-carbon but again considerable variation in the basic diene structure is possible (table 5.2).

One or both of the diene double bonds may be part of an aromatic system. The reactions can be carried out in the gas phase or in solution, the rate and temperature required depending greatly on the structures of the components. The rate of reaction can be increased by the presence of catalysts and by application of high pressures. High stereoselectivity in the $_{\pi}4_s + _{\pi}2_s$ sense is observed for both diene and dienophile. The considerable variation possible in the structure of both diene and dienophile makes the Diels–Alder reaction a very versatile synthetic route to both carbocyclic and heterocyclic compounds.

Table 5.1. Types of π bonds acting as dienophiles in Diels–Alder reactions

>C=C<	O=O	(singlet O_2)
–C≡C–	–N=O	
>C=N–	–N=S	(R–N=S=O)
–C≡N	S=O	(e.g. SO_3)
>C=O	N=Se	(R–N=Se=N–R)
>C=S	Se=O	
–N=N–	>Si=C<	

Table 5.2. Types of systems acting as dienes in Diels–Alder reactions

–C=C–C=C–	O=C–C=C–	S=C–N⁺=C– [a]
–N=C–C=C–	O=C–C=O	–C=C–C=N⁺– [a]
–C=N–C=C–	–N=C–C=O	–C=C–N⁺=O [a]
–C=N–N=C–	O=N–N=C–	–C=C–N⁺=C– [a]
–C=C–N=N–	O=C–N=C–	
–N=C–N=C–	>P=C–C=O	
O=N–C=C–	(cyclopropyl)–C=C–	

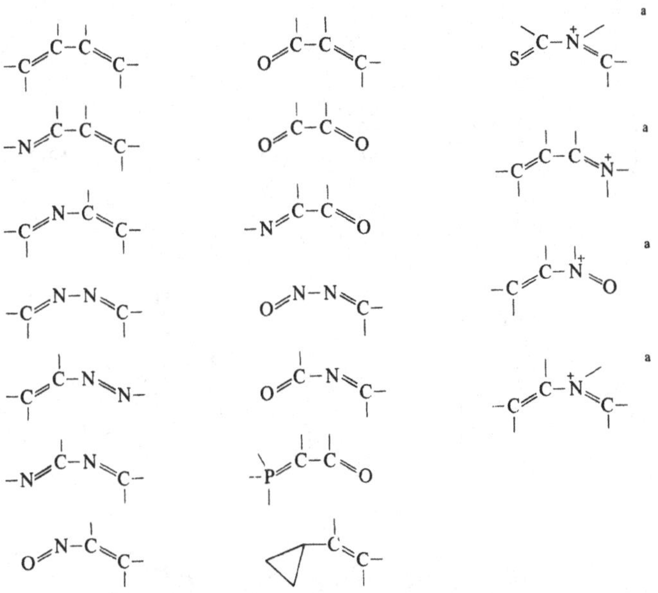

[a]4 + 2 Cycloadditions of such charged dienoid components are referred to as polar 1,4-cycloadditions (ref. 27*b*).

Me; CO_2Me; CO_2Me —150 °C, Toluene→ H, Me, CO_2Me, CO_2Me 85 % (5.1)

NC, CN, NC, CN —Below room temperature→ CN, CN, CN, CN (5.2)

CO_2Et, N, N, EtO_2C —10 °C, Ether→ $N-CO_2Et$, N, CO_2Et 100 % (5.3)

O; O —80 °C, Benzene→ CHO, H, O 45 % (5.4)

O, NC, CN —20 °C→ O, CN, CN 92 % (5.5)

MeO_2C, O, O; CO_2Et, CO_2Et → [MeO_2C, O, O, CO_2Et, CO_2Et]

$-CO_2$→ MeO_2C, CO_2Et, CO_2Et (5.6)

The fact that the retro-Diels–Alder reaction is also general (§5.2) considerably extends the synthetic utility of the Diels–Alder reaction since an adduct may undergo a retro-reaction to give components different from those from which it was formed. (See equations 5.6 and 5.11 and §5.2 for several examples.)

An increasing number of intramolecular Diels–Alder reactions is appearing[2] (e.g. equation 5.7 and 5.8). These are also highly stereoselective but frequently

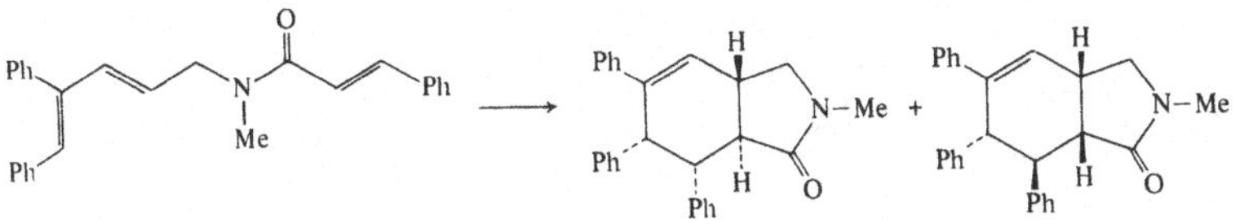

(5.7)

proceed under considerably milder conditions than corresponding intermolecular additions; the entropy of activation is less for the intramolecular reaction in which the two components are constrained within the same molecule. This type of reaction has found considerable use in the synthesis of complex natural products as, for example, in the conversion of the aryl propargyl ester (**1**) into the lignan (**2**).[3] This example also illustrates the diene activity of a styrene unit, the initial adduct undergoing spontaneous aromatisation. It is probable that facile intramolecular Diels-Alder reactions occur in biosynthetic pathways.

The diene can only react in a *cisoid* configuration; molecules in which the diene system is fused *transoid* do not react, and the rate of reaction with open-chain dienes depends on the equilibrium proportions of *cisoid* conformer present. Thus, substituents in the diene can affect the rate not only by their electronic character but by their influence on conformer proportion. *cis*-1-Substituted butadienes (**3**) are less reactive than their *trans*-isomers (**4**) since a bulky R group disfavours the *cisoid* conformation (5.9). Large 2-substituents in the diene favour the *cisoid* conformation and the diene is correspondingly more reactive.

cis Fused cyclic dienes are usually the most reactive, particularly if the ring size is such that the ends of the diene are a convenient distance apart for bonding to the dienophile. Thus cyclopentadiene is somewhat more reactive than cyclohexadiene, both being much more reactive than larger ring dienes, for example cyclo-octadiene. Cyclic dienoid systems such as furan also take part in the Diels–Alder reaction but as expected the dienoid activity falls in the series furan > pyrrole > thiophen as the diene character is replaced by aromatic character. Pyrroles frequently undergo alternative side reactions and Diels–Alder additions to thiophen have so far only been observed with arynes, which are exceptionally reactive dienophiles.

A similar situation occurs with aromatic hydrocarbons. Benzene acts as a diene only with arynes and highly reactive acetylenes, but naphthalene and particularly anthracene, where there is more 'bond fixation', function more

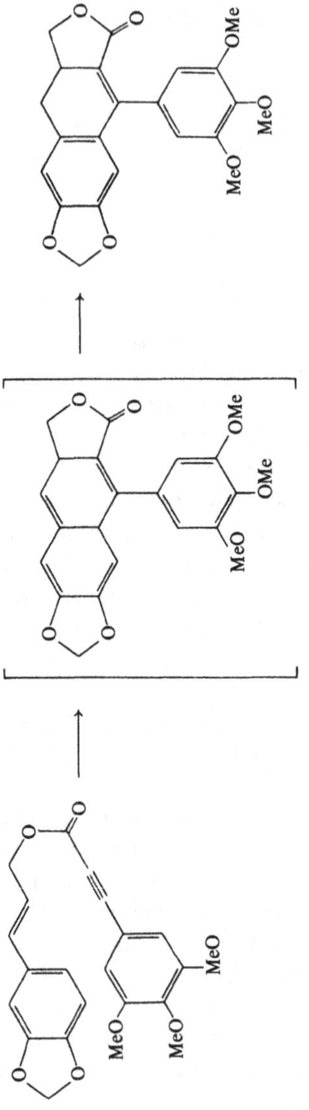

(1)

(2)

(5.8)

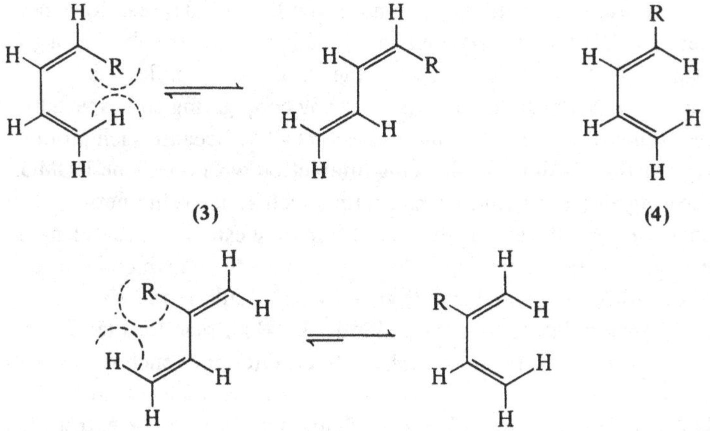

(3) (4)

(5.9)

commonly as dienes. Addition to anthracene normally occurs across the 9,10-positions so as to leave two benzenoid aromatic rings (5.10). Aromatic heterocyclic systems such as tetrazines also function as dienes (5.11). In this case spontaneous loss of nitrogen by a retro-Diels–Alder reaction generates a new heterocycle.[4]

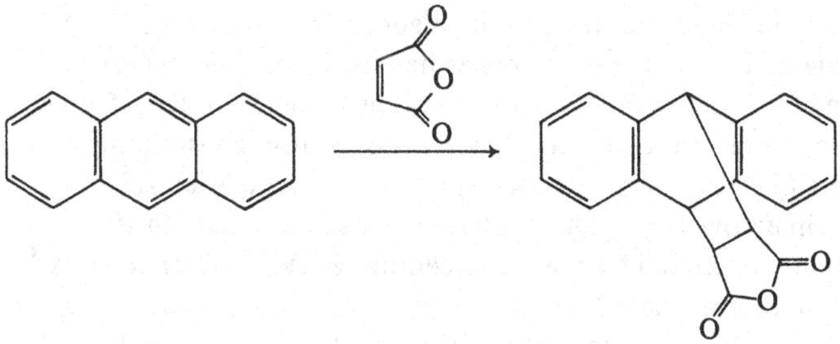

(5.10)

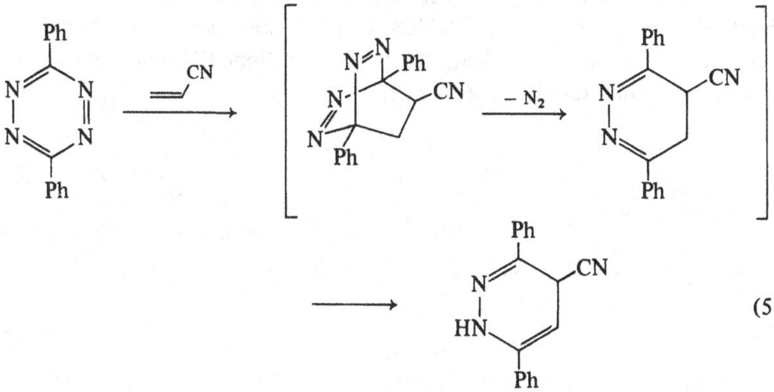

$-N_2$

(5.11)

Considerable variation in reactivity is found in the Diels–Alder reaction, the prototypical reaction between butadiene and ethylene being particularly sluggish. The effect of substituents on the reactivity of dienes and dienophiles has been discussed in chapter 4. Normally dienophiles bearing conjugating and especially conjugating electron-withdrawing groups are more reactive because such groups lower the energy of the LUMO, so enhancing interaction with the diene HOMO. Thus alkenes bearing electron-withdrawing groups, such as acrolein, methyl acrylate, *p*-benzoquinone, styrene, maleate and fumarate esters and acetylenedicarboxylic esters, are more reactive than ethylene and acetylene which require strongly forcing conditions. Tetracyanoethylene is a particularly reactive dienophile. The common dienes encountered in Diels–Alder reactions tend to be activated by electron-releasing groups (alkyl, NR_2, OR) since such substituents raise the diene HOMO energy and therefore further close the diene HOMO–dienophile LUMO energy separation. The complementary nature of electron-rich diene with electron-deficient dienophile leading to greater reactivity is very general and such Diels–Alder reactions are often referred to as normal Diels–Alder reactions. On the other hand, for some electron-deficient dienes, for example hexachlorocyclopentadiene, the electronic requirements of the dienophile are reversed and electron-donating substituents favour the reaction. Several examples of such Diels–Alder reactions with so-called inverse electron demand are now known.

Any factor which increases the strain in the dienophile so raising its HOMO energy and lowering its LUMO energy increases its reactivity. Thus cyclopropenes are more reactive than corresponding olefins – there is relief of angle strain in the cyclopropene as the reaction proceeds. Similarly, constraining a triple bond in a ring makes it more reactive. Thus cyclo-octyne is an isolable but reactive dienophile; lower cycloalkynes have not been isolated but their existence has been demonstrated by their interception as Diels–Alder adducts.[5] Similarly the reactive intermediate benzyne is very readily intercepted by dienes such as furan, cyclopentadiene, or tetraphenylcyclopentadienone.[5] Cyclic azo compounds with electron-withdrawing substituents are highly reactive dienophiles due to their low energy LUMOs. Unstable species such as (5) and (6) can be readily trapped and 4-phenyltriazoline-3, 5-dione (7) is an even more reactive dienophile than tetracyanoethylene.

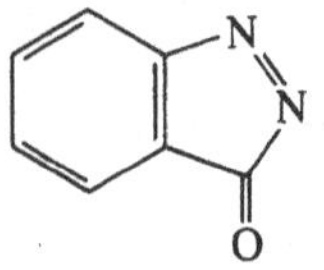

(5) 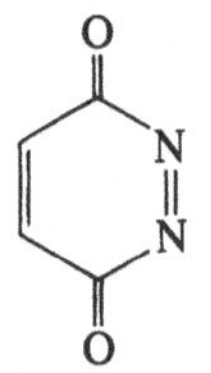 (6) 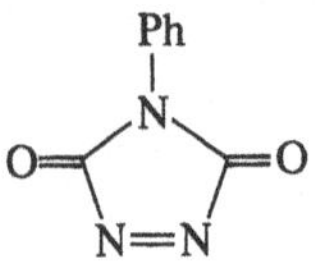 (7) 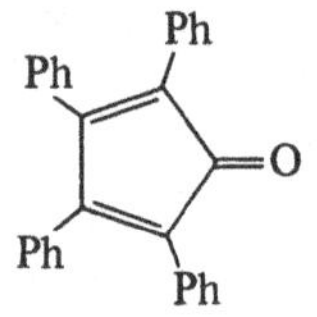 (8)

Oxygen has a triplet ($^3\Sigma g$) ground state in which the degenerate, orthogonal, antibonding π orbitals (π^*_{O-O}) are singly occupied (fig. 5.2). Two low-lying

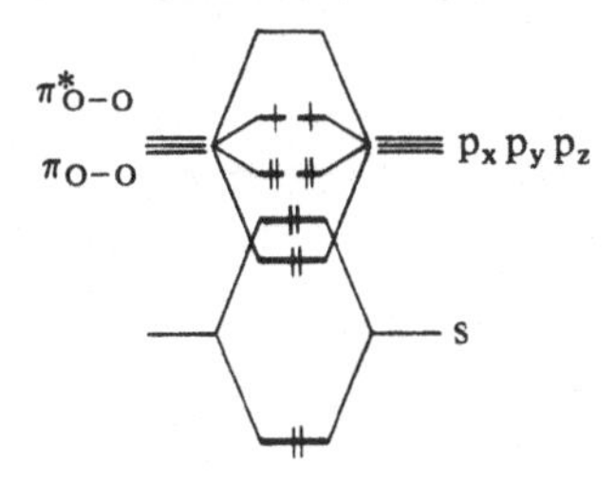

Fig. 5.2. Molecular orbitals of $^3\Sigma g$ oxygen.

singlet states are also possible: $^1\Delta g$ in which the two electrons are spin paired in one of these π^*_{O-O} orbitals and $^1\Sigma g$ in which the spins are paired but the electrons are in the separate orbitals. The lower $^1\Delta g$ state is readily accessible in solution by dye-sensitised irradiation of solutions of oxygen and from certain chemical reactions in which oxygen is produced necessarily initially in a singlet spin paired state from spin paired precursors.[6] This singlet oxygen is then a reactive dienophile behaving like a very electron-deficient alkene. The LUMO which will be involved in the most important frontier orbital interaction is antisymmetric (empty π^*_{O-O}) and therefore compatible in phase with the diene HOMO. The less important $^1O_2$ HOMO (filled π^*_{O-O}) is orthogonal to the $^1O_2$ and diene LUMOs and is also antisymmetric and therefore incompatible in phase with the diene LUMO (9). The significant occupied orbital will be the NHOMO (next to highest occupied MO) (π_{O-O}) which is in the same plane as the LUMO.[7]

Ordinary triplet oxygen is unreactive in cycloadditions since the molecule would have to correlate with an excited state of the product.

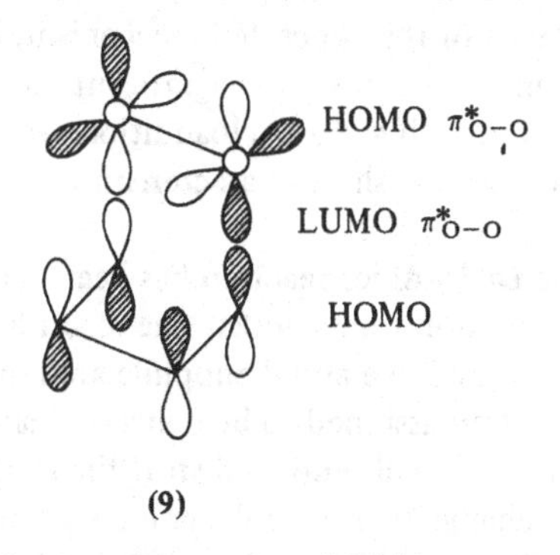

(9)

Cyclopentadienones are highly reactive dienes; indeed they are so reactive towards dimerisation (one molecule acting as diene, the other as dienophile) that only highly substituted derivatives, such as the tetraphenyl derivative (8), are stable as monomers. This reactivity can be explained by a small energy difference between the HOMO and LUMO of cyclopentadienones. A similar situation arises with cyclobutadienes which also show a great tendency to dimerise (see p. 118).

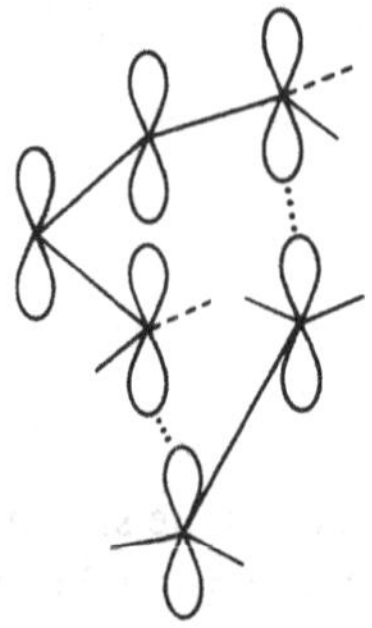

(10)

Mechanism [1a, c, 8, 9]

The most widely accepted and satisfactory picture of the Diels–Alder reaction is that of a concerted reaction in which both new σ bonds are formed to significant if not necessarily equal extents in the transition state which resembles **(10)**. A symmetrical $_{\pi}4_{s} + _{\pi}2_{s}$ transition state **(10)** is in line with orbital symmetry considerations but it does not seem unreasonable that where substituents cause distortion of the diene and dienophile orbitals without changing their overall symmetry properties, bonding at one terminus may be more important than at the other. Indeed this picture arises naturally out of the perturbation FMO approach (chapter 4 and p. 120) and provides a satisfactory explanation for the regioselectivity observed in Diels–Alder reactions–originally one of the major problems associated with acceptance of the concerted mechanism. In extreme cases the formation of the second bond may be insignificant compared to the first in the transition state and such 4 + 2 cycloadditions are then truly stepwise. Such cases are, however, rare and show characteristics atypical of the normal Diels–Alder reaction.

Most of the mechanistic work on the Diels–Alder reaction has been carried out with dienes and dienophiles with all-carbon frameworks. The reaction also proceeds quite generally when atoms in both diene and dienophile are replaced by heteroatoms and these additions are often assumed to be concerted also. This is reasonable since although heteroatoms will tend to distort the simple polyene orbitals they will not generally change their overall symmetry. However, heteroatoms will be likely to stabilise polar intermediates so that hetero-Diels–Alder reactions are more likely to be non-concerted or at least to proceed through unsymmetrical transition states. The following mechanistic discussion therefore strictly applies to all-carbon Diels-Alder reactions between reactants which are not so heavily and unsymmetrically substituted as to destroy the basic character of the process.

The similarity of the rates of the reaction in the gas phase and in solvents of widely differing polarity is hardly consistent with a stepwise reaction through a zwitterionic intermediate. The observed regioselectivity of addition for substituted dienes and dienophiles (fig. 5.3) does not depend on the nature of R and X and is therefore in many cases inconsistent with such a mechanism. There is an encouraging correlation between reactivity and frontier orbital energy separation using the concerted reaction model. Also, the orientation of addition can be satisfactorily explained by FMO theory on the basis of a concerted mechanism (see p. 94).

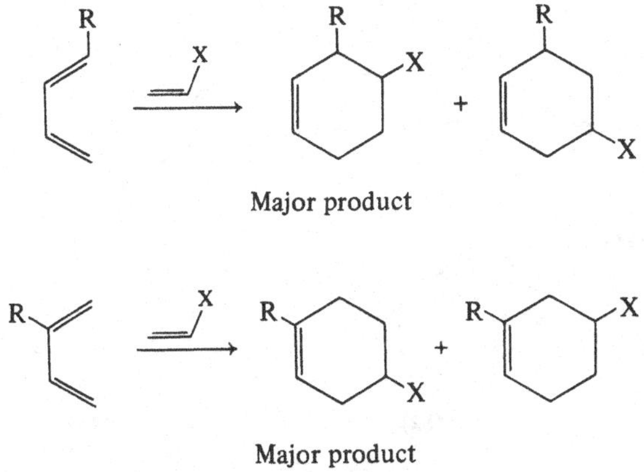

Major product

Major product

Fig. 5.3

A stepwise reaction through a diradical intermediate has been most commonly considered as the alternative to the concerted mechanism. Such a mechanism would explain the regioselectivity observed with substituted dienes and dienophiles (fig. 5.3) which is consistent with combination of the diene and dienophile through the most stable diradical. It would also fit in with the lack of solvent effect. However, it does pose some serious problems. Formation of a diradical intermediate should not be the prerogative of a *cisoid* diene molecule; *transoid* diene molecules should equally well form such intermediates. However, there is considerable evidence that these intermediates from *transoid* dienes would lead to four-membered ring products (2 + 2 addition). Since most dienes and monoenes only react through the *cisoid* diene molecules (even when the diene is open chain and exists mainly in a *transoid* conformation) to give exclusively six-membered ring Diels-Alder adducts, this is good evidence for a concerted mechanism (see p. 179 for a full discussion of this point; see also p. 122).

Any diradical mechanism, therefore, has to be very special and clearly different from that involved in 2 + 2 cycloaddition. The diradicals **(11)** must be formed from *cisoid* dienes in just the right conformation for very rapid ring closure to the six-membered ring prior to any competing bond rotations (fig. 5.4). Any other diradicals which must be formed with comparable ease but which do not have this ideal conformation, e.g. **(12)** or **(13)**, must therefore dissociate to diene and monoene extremely rapidly so that they do not have a chance to collapse to the four-membered ring. Whilst the case for this mechanism has been extensively argued it is not generally accepted and seems less satisfactory than the concerted mechanism.

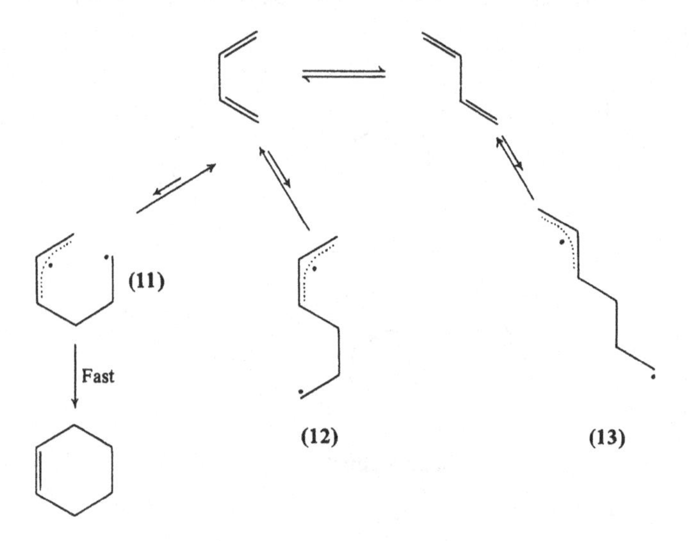

Fig. 5.4. Diradical mechanism for Diels–Alder reaction. Diradicals **(12)** or **(13)** must revert to diene and monoene faster than internal bond rotations and collapse to four-membered ring.

Diels–Alder reactions show high stereospecificity with respect to both diene and dienophile. The specificity is that of ${}_{\pi}4_{s} + {}_{\pi}2_{s}$ addition, the relative orientation of substituents on both diene and dienophile being retained in the adduct (5.12).

R X Y X Y R ⟶ R Y X Y X R (5.12)

This is further strong evidence for the concerted reaction since non-stereospecific addition should result from bond rotations competing with ring closure in a diradical or zwitterionic intermediate. The addition of hexachloro-cyclopentadiene to α-methylstyrene is a case where a diradical intermediate (**14**) should be particularly favoured since one radical centre would be tertiary and benzylic and the other allylic and stabilised by an α-chlorine. However, even this addition is completely stereoselective.[10†]

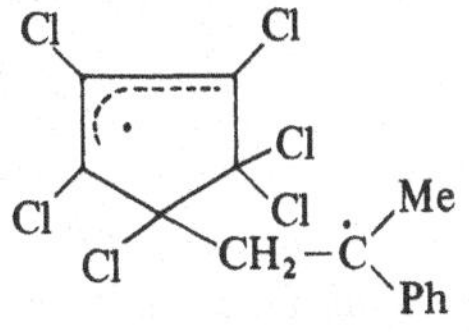

(14)

Small inverse secondary H/D isotope effects, $k_D/k_H > 1$, for both diene and dienophile are in agreement with simultaneous but small changes in hybridisation at all four terminal atoms in the transition state. As expected, with highly unsymmetrical dienophiles the isotope effects indicate that although bond formation begins simultaneously it has proceeded to different extents in the transition state.[11] The retro-Diels-Alder reaction has frequently been studied to give information of its microscopic reverse - the forward Diels-Alder reaction.[12] Again secondary H/D isotope effects for the retro-reaction of the 2-methylfuran-maleic anhydride adduct indicate synchronous cleavage of both bonds.[13]

Diels-Alder reactions uniformly show large negative entropies of activation in line with the rigid transition state of the concerted mechanism. Several determinations of activation volume have also been interpreted in favour of a concerted mechanism.[14] The large negative values for activation volume observed indicate that the transition state is smaller than the reactants, again in line with the restricted transition state of a concerted reaction. One important consequence of the large negative activation volume is that Diels-Alder reactions are favoured by increasing pressure. Extremely high pressures are required in practice (8-20 kbar, 1 kbar = 1000 atm) but under these conditions Diels-Alder additions between normally unreactive substrates can be effected at room temperature as illustrated for the formation of (**15**) from furan and acrylonitrile (5.13).[15] The technique is particularly useful for producing thermally labile adducts where higher reaction temperatures are precluded.

† In the particular modification of the diradical mechanism discussed above, ring closure or reversion to reactants is faster than internal bond rotation and so stereoselective addition would be accommodated.

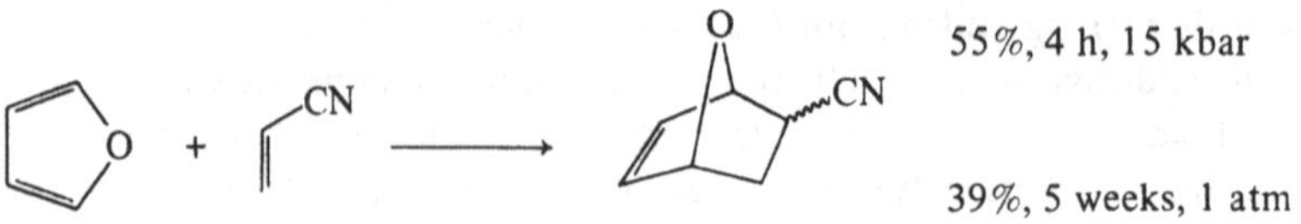

55%, 4 h, 15 kbar

39%, 5 weeks, 1 atm

(5.13)

(15)

Thus the experimental evidence is consistent with the mechanism being that of the concerted ${}_{\pi}4_s + {}_{\pi}2_s$ cycloaddition. Moreover, FMO theory provides a satisfactory explanation for the effect of substituents on reactivity and orientation in the Diels–Alder reaction.

Reactivity

The electronic effect of substituents and the importance of configurational and conformational factors on the ease of the Diels–Alder reaction have already been discussed in detail (pp. 94, 109, 112). Apart from their electronic effects substituents can exert steric effects which can be severe in some cases. For example, cyclobutadiene (**16**) is a highly reactive, transient molecule which undergoes Diels–Alder dimerisation even at very low temperatures to give (**17**) and (**18**) (5.14). The ease of this dimerisation can be related to the

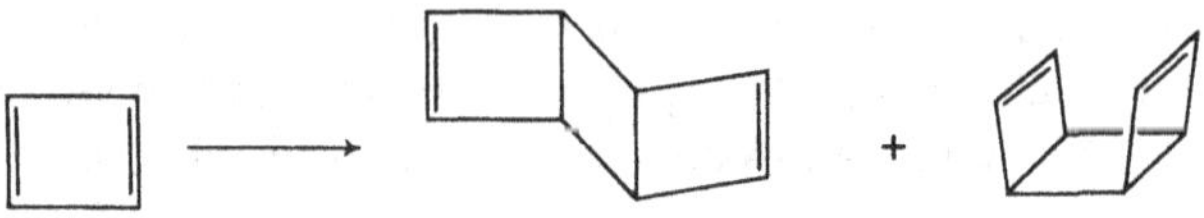

(5.14)

(16) **(17)** **(18)**

closeness in energy of the cyclobutadiene HOMO and LUMO. (Simple Hückel theory actually predicts that the HOMO and LUMO are degenerate so that the molecule may have a triplet diradical ground state. Slight Jahn Teller distortion will, however, remove this degeneracy so that it is very likely that cyclobutadiene is a singlet molecule with closely spaced HOMO and LUMO.) In contrast, tri-t-butylcyclobutadiene (**19**) in which the bulky t-butyl groups hinder dimerisation can be isolated at room temperature.

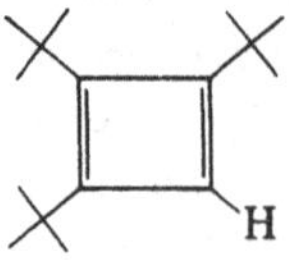

(19)

Stereoselectivity and endo-*effect*

The high suprafacial, suprafacial stereoselectivity observed with the Diels–Alder reaction is a logical consequence of the concerted mechanism.

A further important stereochemical consideration in the Diels–Alder reaction is illustrated by the addition of maleic anhydride to cyclopentadiene. Two orientations are possible, that leading to *exo*-addition (**20**) and that to *endo*-addition (**21**). In this and other similar Diels–Alder reactions, *endo*-adducts normally predominate, often almost to the exclusion of *exo*-adducts. Initially formed *endo*-adducts may, however, isomerise to the less sterically hindered, more stable *exo*-adducts by a retro-Diels–Alder reaction followed by recombination (see p. 131).

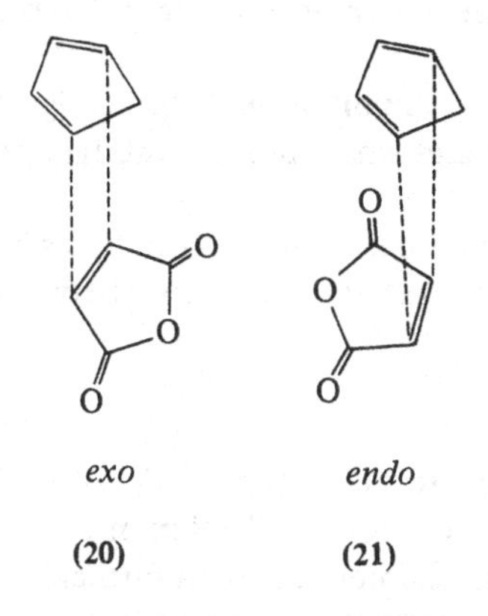

exo (**20**) *endo* (**21**)

Woodward and Hoffmann have put forward an explanation for this widely observed preference for *endo*-addition.[16] Both *endo*- and *exo*-addition is symmetry-allowed but for dienophiles with additional π bonds or other available orbitals, secondary orbital interactions operate so as to favour the *endo* transition state. Thus for cyclopentadiene dimerisation, the *endo* transition state is stabilised by secondary interactions between appropriate highest occupied and lowest vacant orbitals (fig. 5.5).

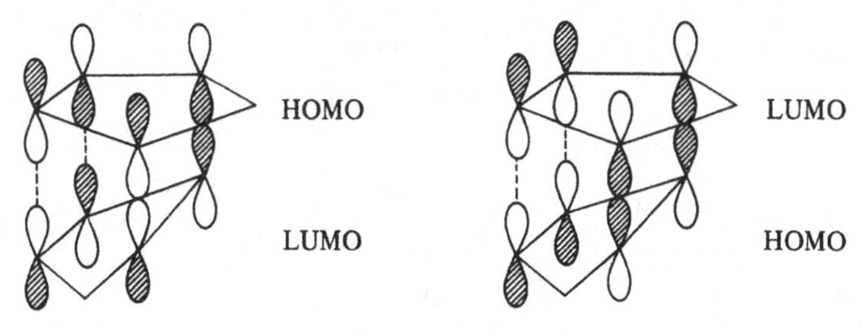

Fig. 5.5

An alternative explanation by Herndon and Hall is that *endo*-addition results from more efficient geometrical overlap for *endo*-addition than for *exo*-addition.[17] This is based on calculations of overlap integrals, for example in cyclopentadiene dimerisation, assuming transition states in which the two molecular planes are parallel for *exo*-addition but inclined at 60° for *endo*-addition ($\Delta\beta$ *endo* > $\Delta\beta$ *exo* in equations 4.2-4.4, chapter 4). This theory better explains the predominant *endo*-addition of such dienophiles as cyclopropene and cyclopentene (which do not possess the orbitals necessary for secondary interactions). Experiments have been quoted as favouring one or other of these explanations and a clear distinction is not yet possible. Usually, however, the *endo* preference is greatest in those cases where Woodward and Hoffmann secondary interactions are most readily conceivable. Such secondary interactions can also account for the preferred orientations observed in other cycloadditions[18] (see p. 100).

Secondary orbital interactions may also be important in determining regioselectivity in Diels-Alder reactions in cases where primary orbital overlap effects are finely balanced (p. 101).

Other explanations for the *endo*-effect have been put forward. It has been attributed to dipole-induced dipole or charge transfer interactions between the diene and dienophile.[19]

Regioselectivity

Regioselectivity was originally one of the greatest unsolved problems concerning the Diels-Alder reaction. The very promising application of FMO theory to this problem has been discussed in chapter 4. Several further examples of Diels-Alder reactions which illustrate the regiochemical phenomena summarised in fig. 5.3 are given in equations 5.15 to 5.19. All of these can be accommodated as outlined in chapter 4; the dominant interactions are indicated and the sites of largest orbital coefficients are asterisked.

Example 5.19 is particularly interesting because the major product is predicted by FMO theory but is not consistent with combination via the more stabilised diradical intermediate.[20]

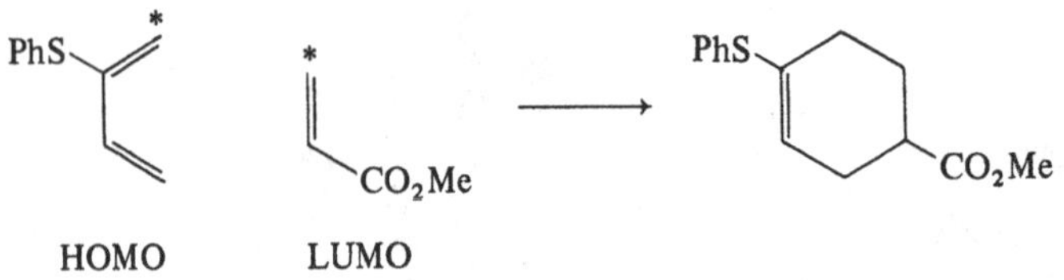

PhS * HOMO + * CO_2Me LUMO → PhS CO_2Me (5.15)

OAc * HOMO + O=N(*)–Ar LUMO → OAc O N Ar (5.16)

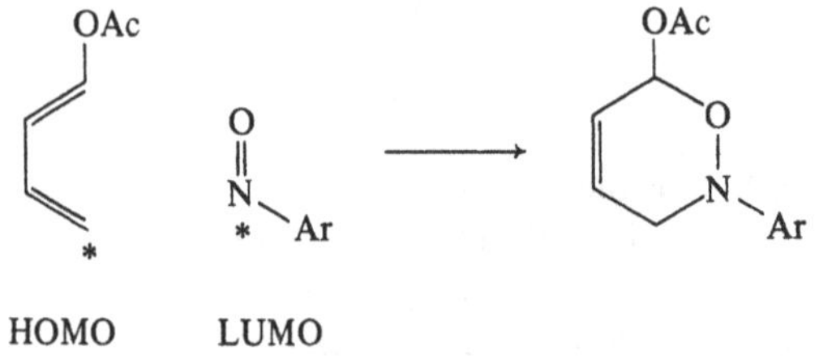

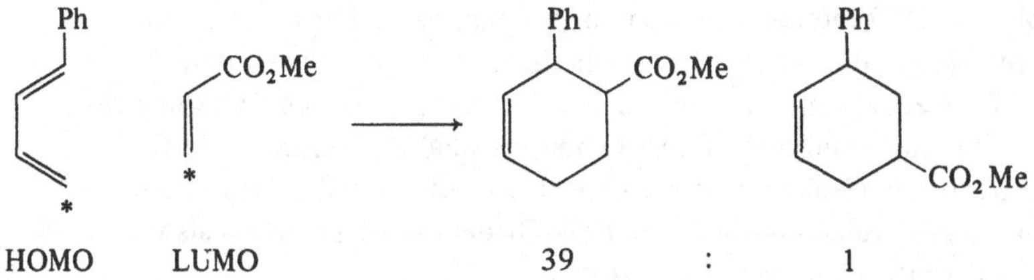

Ph, CO_2Me → Ph/CO_2Me adducts (5.17)

HOMO LUMO 39 : 1

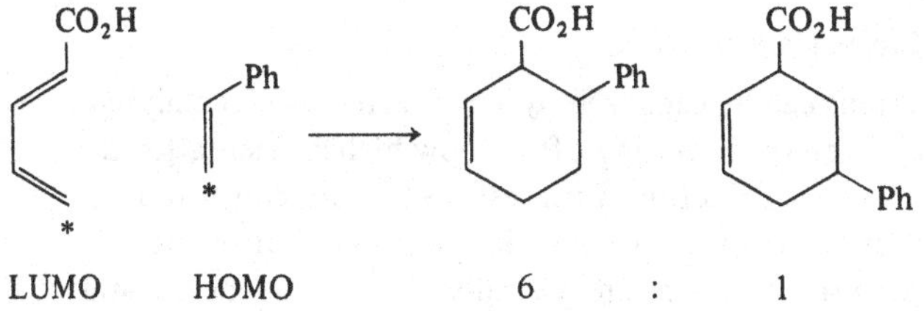

CO_2H, Ph → CO_2H/Ph adducts (5.18)

LUMO HOMO 6 : 1

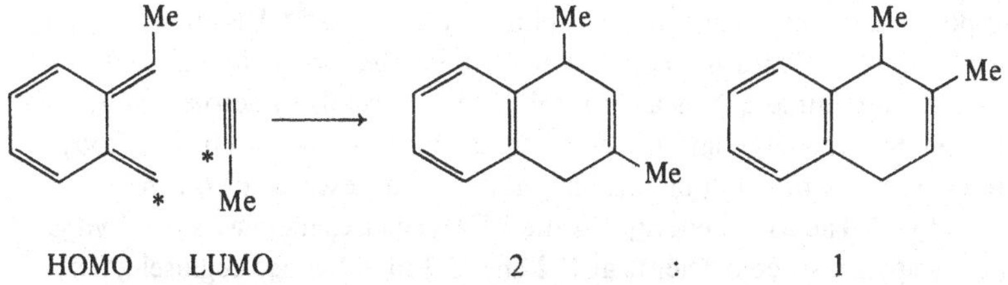

Me, Me → Me/Me adducts (5.19)

HOMO LUMO 2 : 1

The regioselectivity of dimerisation of acrolein (5.4) has been a problem. It involves bonding between two formally electrophilic carbons, and initial applications of FMO theory using coefficients derived from extended Hückel theory failed. However, calculations better suited to systems containing heteroatoms[21] give a set of coefficients (fig. 5.6) which account for the observed orientation. Both diene HOMO–dienophile LUMO and diene LUMO

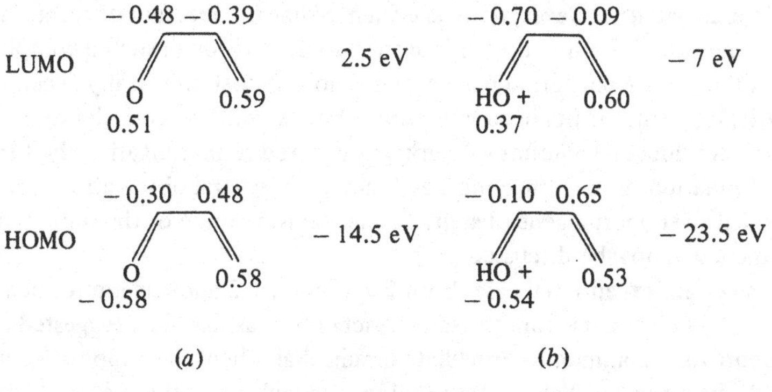

LUMO: (*a*) −0.48, −0.39; O 0.51, 0.59; 2.5 eV. (*b*) −0.70, −0.09; HO+ 0.37, 0.60; −7 eV

HOMO: (*a*) −0.30, 0.48; O −0.58, 0.58; −14.5 eV. (*b*) −0.10, 0.65; HO+ −0.54, 0.53; −23.5 eV

(*a*) (*b*)

Fig. 5.6. Frontier orbital energies and coefficients for (*a*) acrolein and (*b*) protonated acrolein.

-dienophile HOMO interactions are equally important. The latter clearly favours the observed product whereas the former shows no preference because of the equal terminal coefficients. However, since the β (resonance integral) value for formation of a C—O bond is smaller than that of a C—C bond at the internuclear separations expected in the transition state, the maximum $\Sigma(c\beta)^2$ value for the diene HOMO–dienophile LUMO is also consistent with the observed orientation.

Catalysis in Diels–Alder reactions

Many Diels–Alder reactions can be catalysed by Lewis acids. Such catalysed additions are also highly stereospecific and often show higher orientational selectivity than the uncatalysed reaction, with respect to both *endo*- and *exo*-distribution and the orientation of substituents in the adducts. It seems likely, therefore, that such catalysed additions are basically the same as the uncatalysed; the catalytic action is probably due to complex formation between the Lewis acid and polar substituents on the dienophile components.[8a] The origin of the catalysis and the reason for increased selectivity can be seen from calculations performed on acrolein as a typical dienophile. The orbitals of acrolein (fig. 5.6*a*) are lowered in energy and the coefficients are modified (as in fig. 5.6*b*) by coordination of a proton (an extreme model for a Lewis acid) on the carbonyl oxygen. The lower energy for the LUMO causes increased reactivity, the greater disparity of coefficients at C-2 and C-3 in enhanced regioselectivity, and the large C-1 coefficient in greater *endo*-selectivity (greater secondary interaction in *endo* transition state) for the complexed acrolein.[21,22]

Other mechanistic observations

Charge transfer complexes can be formed between dienes and dienophiles. Colours are often generated on mixing, the colour fading as the adduct is formed. The role of such complexes in Diels–Alder reactions has not yet been fully elucidated; however, there is evidence that the reaction of tetracyano-ethylene with 9,10-dimethylanthracene proceeds through such a complex.[23]

Although the concerted mechanism is now almost universally accepted, particularly since it fits in so well with orbital symmetry considerations, in the past a considerable amount of work was devoted to investigating the timing of the formation of the new bonds, and inevitably several observations were claimed to support a general stepwise mechanism. Some of the more significant of these will now be discussed.

In certain exceptional cases both 2 + 4 and 2 + 2 adducts are formed in comparable amounts from the same reactants. This has been suggested as support for a common intermediate diradical and hence as support for a radical mechanism for the Diels–Alder reaction, particularly in those cases where the relative proportions of 2 + 2 and 2 + 4 additions did not vary with temperature or solvent.[24] However, it now seems likely that these cases involve competing stepwise radical 2 + 2 addition and concerted 2 + 4 addition, where the two reactants respond similarly to such changes (see p. 179).

Woodward and Katz observed an intramolecular rearrangement of the Diels-Alder adduct **(22)** at 140 °C which could be explained by cleavage of bond *a* followed by recombination of the resulting allylic fragments as shown.[25]

(22)

(5.20)

Since, at slightly higher temperatures, both bonds *a* and *b* cleave in a retro-Diels-Alder reaction it was assumed that in this retro-Diels-Alder reaction bond *a* cleaved before bond *b*. Thus the microscopic reverse - the forward Diels-Alder reaction - was assumed to be a two-stage process. To explain the stereospecificity of the Diels-Alder reactions secondary interactions were assumed to be responsible for preventing rotations about C—C single bonds. This type of argument for a two-stage mechanism has been severely criticised since the rearrangement observed at 140 °C is merely a Cope rearrangement (see §7.4) which need have no connection with the retro-Diels-Alder reaction.[9]

Various sophisticated calculations aimed at estimating the lowest energy pathway have been carried out for the Diels-Alder reaction. However, the results are conflicting and the conclusions drawn range from a fully concerted mechanism with a symmetrical transition state to one in which one bond is formed completely before formation of the other begins so that the transition state resembles a diradical.

There is thus no evidence to support a *general* stepwise mechanism for the Diels-Alder reaction, although adducts of the Diels-Alder type may be formed by stepwise routes when intermediates are especially stabilised (p. 124).

The concerted ${}_{\pi}4_a + {}_{\pi}2_a$ mode of addition is also formally allowed for the Diels-Alder reaction, but it is geometrically much less favourable than the s,s mode. The a,a mode of addition has tentatively been suggested in a few cases; for example, the thermal rearrangement of octamethylcyclo-octatetraene **(23)** to the semibullvalene derivative **(24)**.[26]

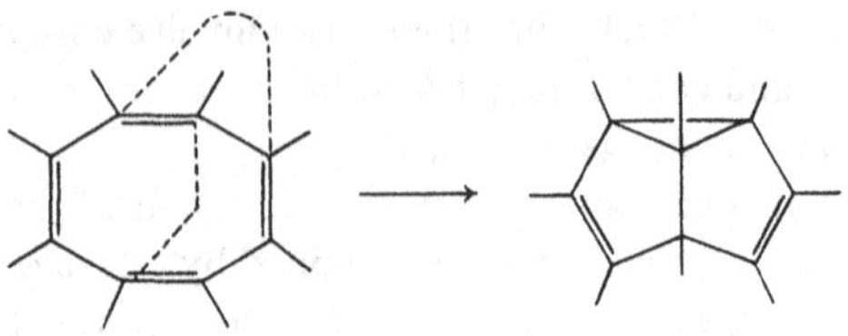

(5.21)

(23) (24)

Stepwise 4 + 2 cycloadditions[27]

Although the majority of 4 + 2 cycloadditions are concerted there are a few cases where zwitterionic intermediates are so stabilised that a stepwise reaction becomes a viable alternative. This applies particularly for heterodienes and heterodienophiles. Formal Diels-Alder reactions which proceed through zwitterionic intermediates are illustrated by equations 5.22 and 5.23.

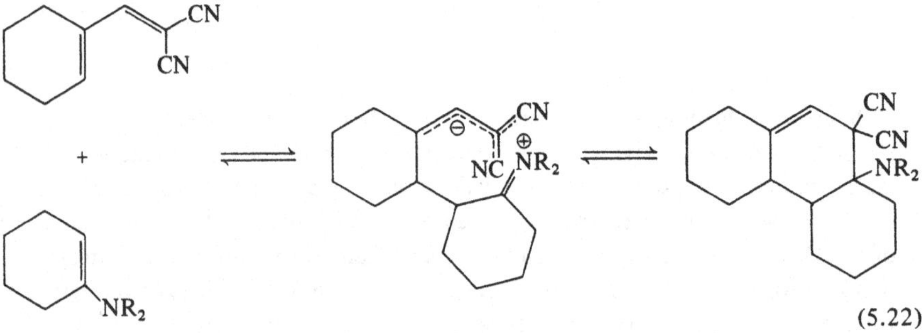

(5.22)

(25)

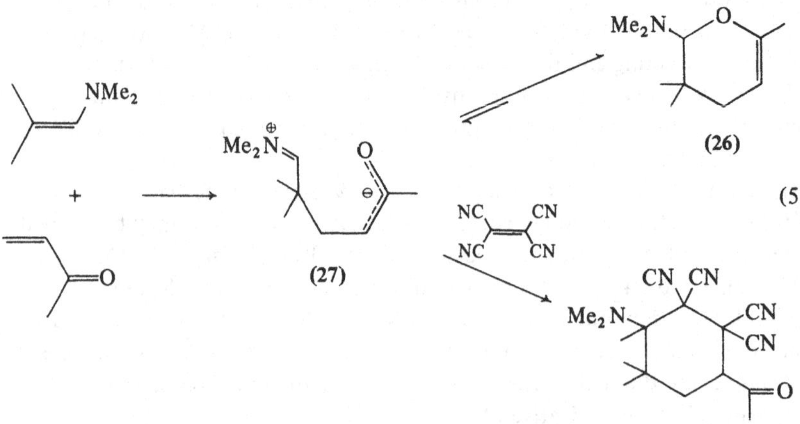

(26)

(5.23)

(27)

In the first example, the zwitterionic intermediate (**25**) can actually be isolated at –40 °C.[27a] In the second example, the adduct (**26**) formed from dimethylaminoisobutene and methyl vinyl ketone is in equilibrium with the zwitterion (**27**) which can be trapped by tetracyanoethylene.[28]

Thus for 4 + 2 cycloadditions we can consider a whole spectrum of mechanisms ranging from the fully concerted process, through the asymmetric concerted, to the truly stepwise reaction. In some cases formation of the second bond has become so unimportant in the transition state that it is more realistic to consider the process as stepwise. The vast majority, however, comes into the category of symmetrical or unsymmetrical concerted reactions.

There are no clear examples of stepwise 4 + 2 cycloadditions involving a diradical intermediate (see p. 179 for a full discussion). Diradical intermediates close exclusively or largely to four-membered ring products which are stable under the reaction conditions. Since the zwitterionic reactions, such as those shown above, are readily reversible it may be that the more stable six-ring adducts are isolated because thermodynamic control is operating.

5.2. The retro-Diels–Alder reaction[9]

The retro-Diels–Alder reaction (fig. 5.7) or retro-diene reaction has been known for almost as long as the Diels–Alder reaction itself. Since most Diels–Alder adducts will dissociate if heated sufficiently strongly the scope is comparable with that of the forward reaction.

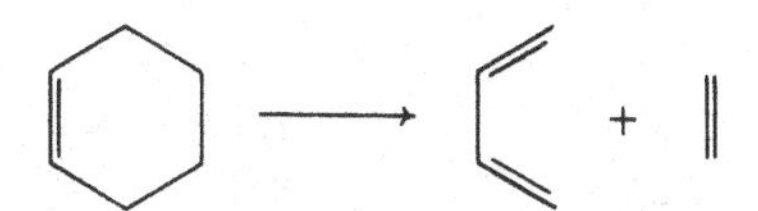

Fig. 5.7. The retro-Diels–Alder reaction.

Retro-Diels–Alder reactions occur more easily when one or both of the fragments are particularly stable. If the diene and dienophile react readily in the forward reaction the retro-reaction is often less favourable. For example, retro-Diels–Alder reactions which involve elimination of nitrogen usually occur very readily, but nitrogen has never been observed to act as a dienophile.

The retro-Diels–Alder reaction has considerable synthetic utility. The prototypical reaction, the conversion of cyclohexene into butadiene and ethylene, is used as a laboratory preparation of butadiene.[29] Another important example is the thermal cracking of dicyclopentadiene to give cyclopentadiene (5.24).[30]

Frequently in synthesis the retro-Diels–Alder reaction is used in conjunction with the Diels–Alder reaction. Benzoquinone epoxide cannot be obtained by direct epoxidation of benzoquinone since this leads to the anhydride (**28**).

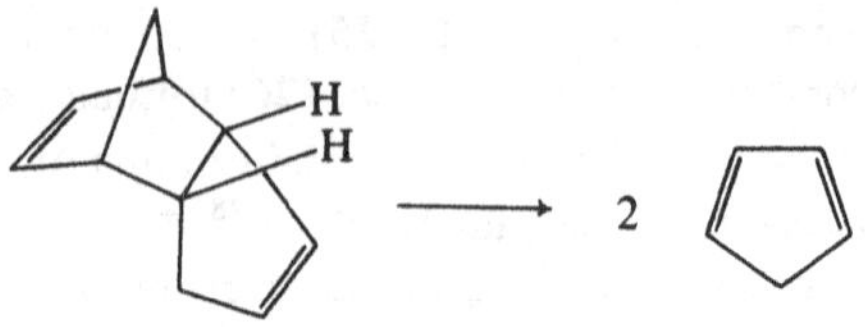

(5.24)

However, epoxidation of the Diels–Alder adduct of cyclopentadiene and benzoquinone followed by retro-Diels–Alder reaction of the product provides a route to the monoepoxide (5.25).[31] A useful preparation of diimide for reductions involves the Diels–Alder addition of azodicarboxylic ester to anthracene. Hydrolysis of the adduct causes decarboxylation to **(29)** which on heating to ~ 80 °C smoothly decomposes to give diimide. Reductions can therefore be carried out conveniently in refluxing ethanol (5.26).[32] Vogel's synthesis of benzocyclopropene also makes use of Diels–Alder and retro-Diels–Alder reactions. The cyclodecapentaene **(30)** undergoes Diels–Alder addition via its valence tautomer **(31)** (5.27).[33]

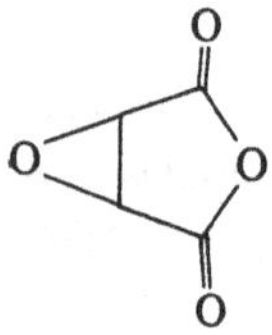

(28)

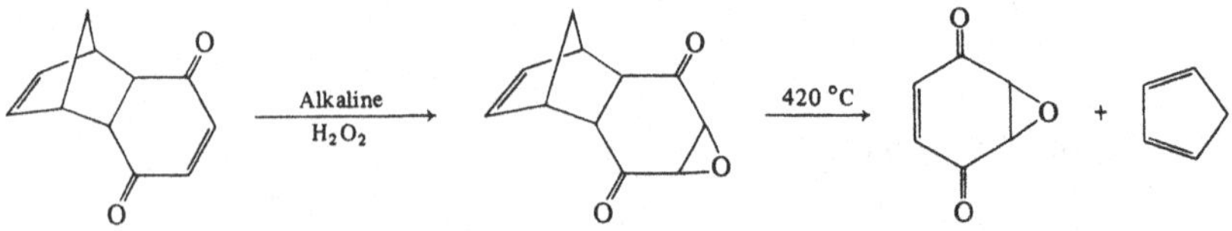

Alkaline H_2O_2

420 °C

(5.25)

cis-Dichlorocyclobutene may be formed from cyclo-octatetraene by the sequence illustrated in equation 5.28.[34]

The *trans*-dichlorocyclobutene formed undergoes conrotatory ring opening to 1,4-dichlorobutadiene. However, steric hindrance to conrotatory opening of the *cis*-dichloro-isomer renders it stable.

As can be seen such retro-Diels–Alder reactants are designed so as to ensure that the fragment formed besides the desired product is particularly stable and often aromatic.

The retro-Diels–Alder reaction may be used in conjunction with other cycloadditions as shown by equations 5.29 and 5.30.[9]

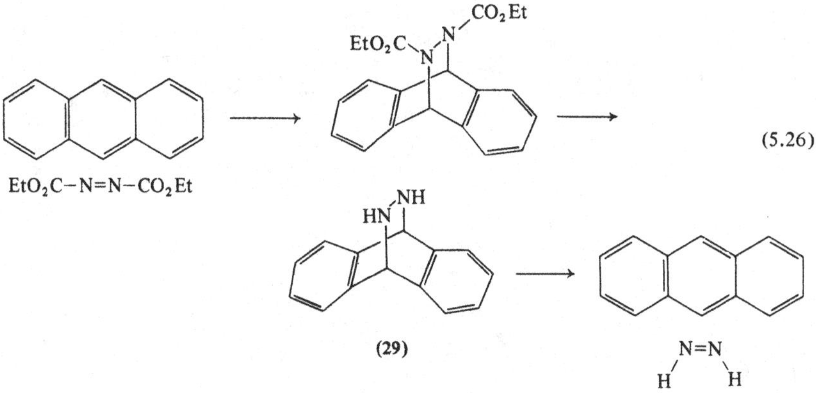

$EtO_2C-N=N-CO_2Et$

CO_2Et, EtO_2C, N, N

(5.26)

HN, NH

(29)

N=N, H, H

(30) **(31)**

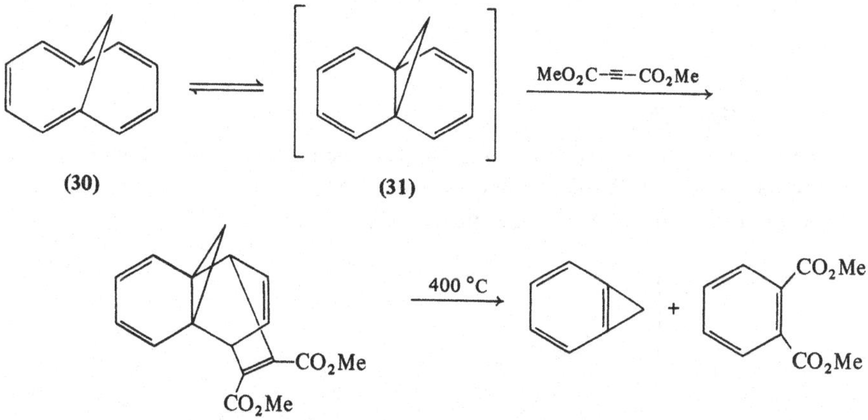

$MeO_2C-\equiv-CO_2Me$

400 °C

CO_2Me CO_2Me + CO_2Me CO_2Me

(5.27)

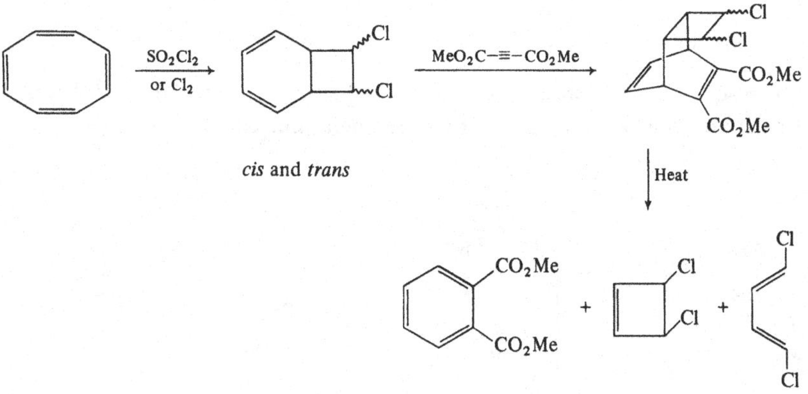

SO_2Cl_2 or Cl_2

Cl, Cl

cis and *trans*

$MeO_2C-\equiv-CO_2Me$

Cl, Cl, CO_2Me, CO_2Me

Heat

CO_2Me, CO_2Me + Cl, Cl + Cl, Cl

(5.28)

Ph–C̊=N–N̊Ph, 135 °C; 98% (5.29)

H–C̊=N–O̊, ~150 °C; 90% (5.30)

Mechanism

The retro-Diels–Alder reaction is the microscopic reverse of the forward reaction; and the bulk of experimental evidence is in support of the expected concerted allowed ${}_{\pi}2_s + {}_{\sigma}2_s + {}_{\sigma}2_s$ process (fig. 5.8).

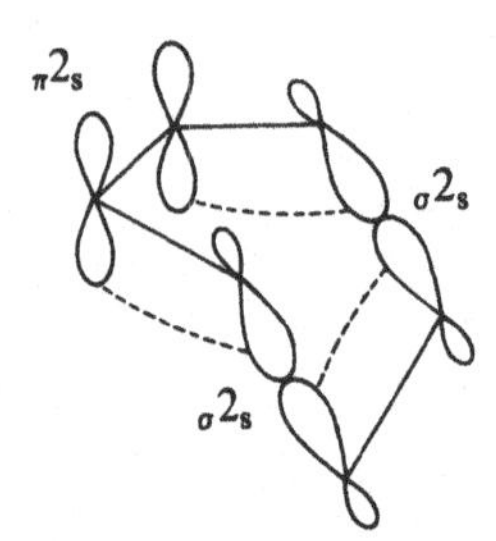

Fig. 5.8

The reaction is stereospecific as shown in equations 5.31 and 5.32.[35] These particular reactions go smoothly at room temperature. The *cis*-isomer **(32)**

Oxidation (5.31)

(32)

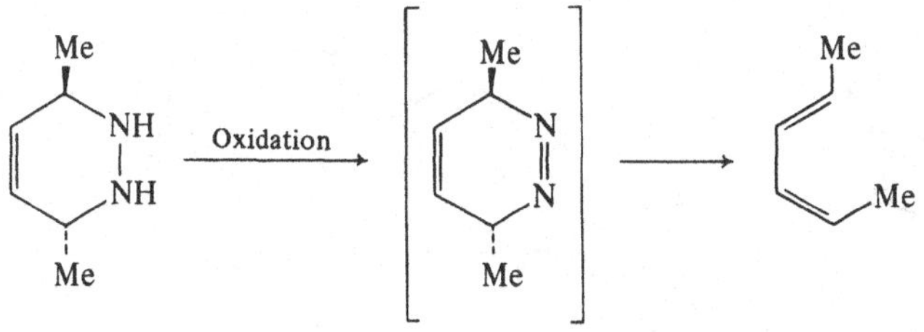

(5.32)

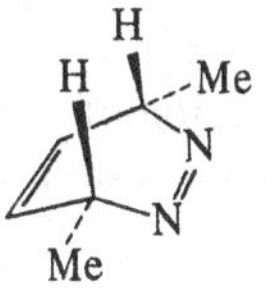

(33) 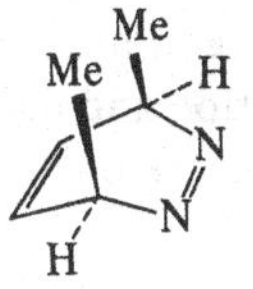 (34)

only gives the *trans, trans* diene since the conformation necessary for a $_{\pi}2_s + _{\sigma}2_s + _{\sigma}2_s$ process, leading to *cis, cis* diene, would be too sterically hindered. The reaction takes place through the preferred conformation **(33)** rather than **(34)**.

The activation entropy for the retro-Diels–Alder reaction is small ($\Delta S^{\ddagger} = +2.1$ to -14.6 J mol^{-1} K^{-1} in solution, +35.1 to -23 J mol^{-1} K^{-1} in the gas phase). A stepwise process would involve a large release in ordering and should have a more positive entropy of activation. The small change of ordering is therefore consistent with a concerted process with a cyclic transition state close in character to the adduct.[9] Steric effects of substituents also support this. The forward reaction is greatly retarded by bulky substituents but the retro-reaction is only slightly accelerated as expected if the transition state lies close to the adduct on the reaction coordinate.

Isotope effects show that both bonds are cleaved simultaneously although not necessarily at the same rate. Secondary H/D isotope effects for the decomposition of the 2-methylfuran–maleic anhydride adduct **(35)** indicate equal cleavage of bonds *a* and *b* in the transition state. However, primary ^{13}C and ^{18}O isotope effects in the rather exceptional retro-Diels–Alder reaction illustrated in equation 5.33 are consistent with weakening of only bond *a* in the transition state.[36]

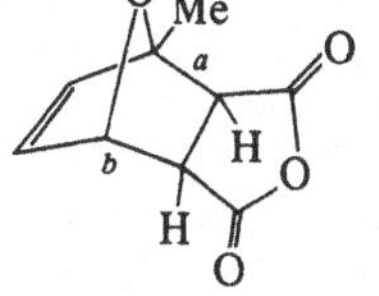

(35)

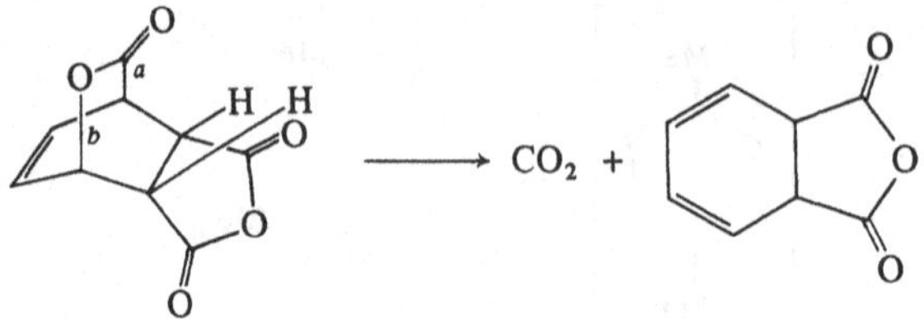

$\longrightarrow$ CO_2 + (5.33)

Just as *endo*-adducts are often formed more easily than the *exo*-isomers in the forward reaction so *endo*-adducts undergo retro-Diels–Alder reactions more easily. The transition state for formation and decomposition of *exo*-adducts is higher than that of *endo*-adducts (see p. 118 for a discussion of the origin of the effect).

Retro- $_{\pi}4_{a} + _{\pi}2_{a}$ addition is also thermally allowed but the unfavourability of this, compared to the normal retro- $_{\pi}4_{s} + _{\pi}2_{s}$ mode, is underlined by the behaviour of the two isomers **(36)** and **(37)**.[16] The isomer **(36)** for which retro- $_{\pi}4_{s} + _{\pi}2_{s}$ addition is possible, readily gives butadiene (5.34). However, elimination of butadiene from **(37)** would necessarily either involve a concerted retro- $_{\pi}4_{a} + _{\pi}2_{a}$ addition (since the methyl group and hydrogen in the resulting monoene fragment are constrained to be *cis*) or a stepwise fragmentation. Significantly this isomer does not give butadiene even at 400 °C.

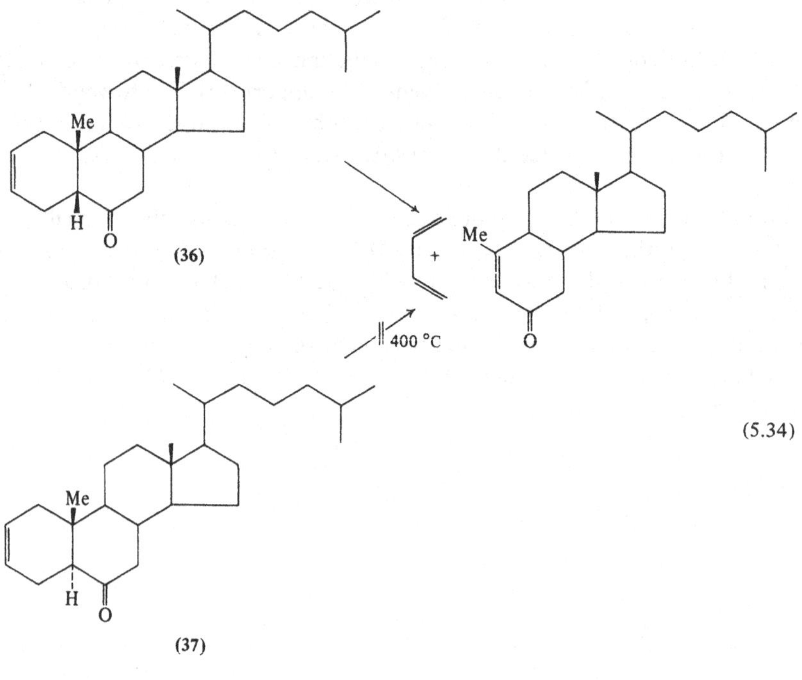

(36)

400 °C

(5.34)

(37)

Fragmentation of the cyclic azo compound (**38**), which is a retro-Diels-Alder reaction, is much more facile than that of the saturated analogue (**39**) which does not have the necessary π bond. Compound (**38**) decomposes as fast as it is formed at -10 °C (5.35)[37] whereas (**39**) only loses nitrogen on heating to 200 °C (5.36).[38] In the retro-Diels-Alder reaction as in many other cases, a cyclopropane ring can play the same role as a π bond (see §7.2). The cyclopropane derivative (**40**) is intermediate in stability and loses nitrogen smoothly at 25 °C (5.37), an estimated 10^{17} times faster than (**39**).[38, 39] Compound (**41**) also loses nitrogen more than 10^{11} times faster than (**42**).[40]

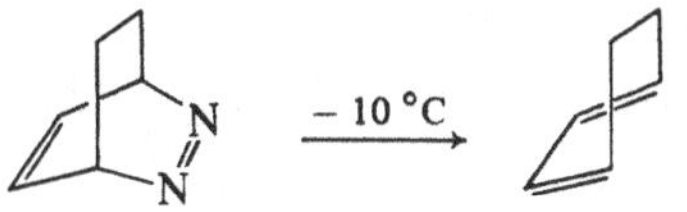

(**38**) $\xrightarrow{-10\ °C}$ (5.35)

(**39**) $\xrightarrow{200\ °C}$ (5.36)

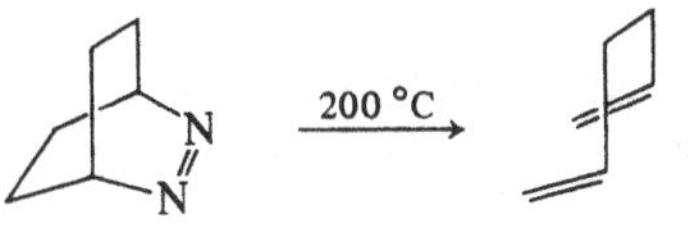

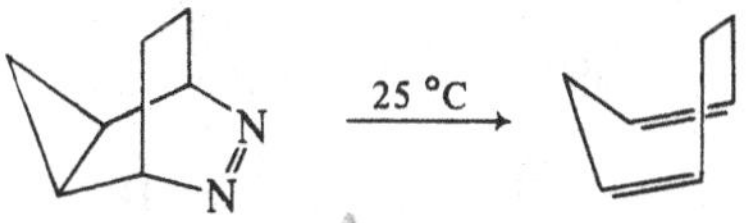

(**40**) $\xrightarrow{25\ °C}$ (5.37)

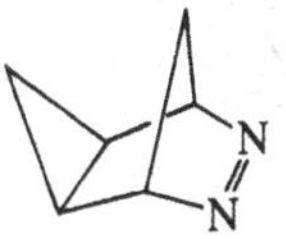

(**41**) (**42**)

Such processes (5.37) are retro-homo-Diels-Alder reactions (for an example of the forward reaction see p. 162). They are also highly stereospecific (5.38-5.40). Nitrogen leaves *anti* to the methylene group of the cyclopropane even though in (**43**) this involves severe methyl-methyl repulsion as the reaction proceeds through transition state (**44**). The necessary orbital overlap is, however, more easily achieved in transition state (**44**) than in the alternative (**45**) in which the leaving nitrogen and cyclopropane ring are *syn*.[35] Loss of nitrous oxide (N_2O) from cyclic azoxy compounds occurs much less readily than from the corresponding cyclic azo compounds.[41]

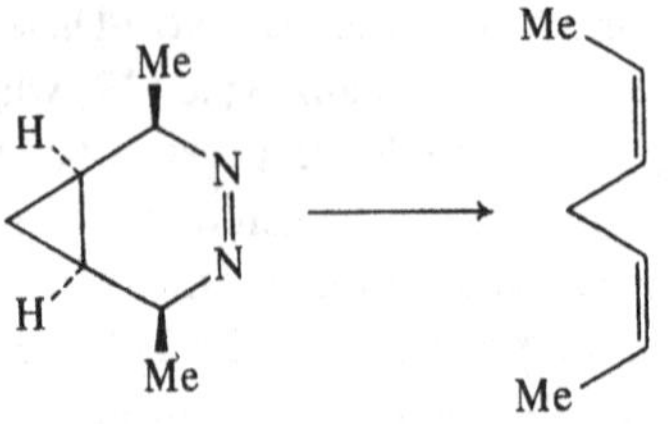

(5.38)

(43)

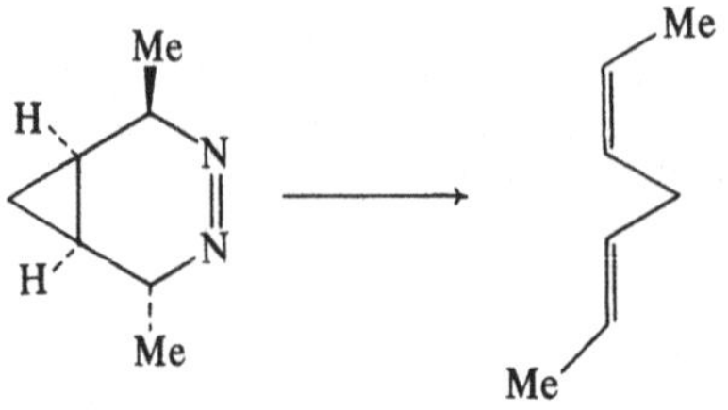

(5.39)

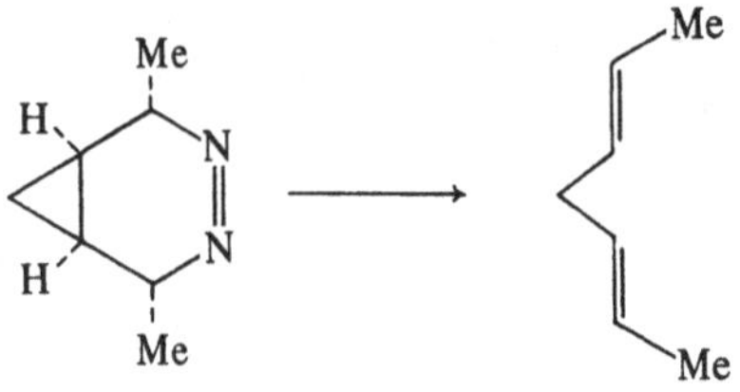

(5.40)

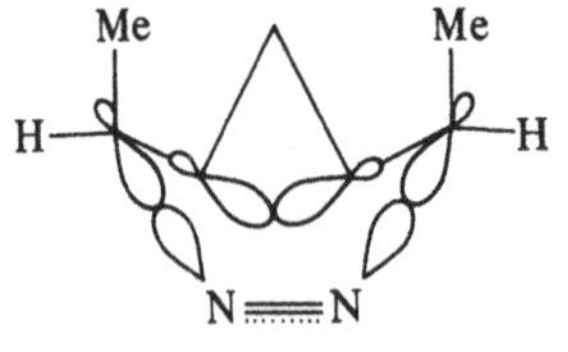

(44) 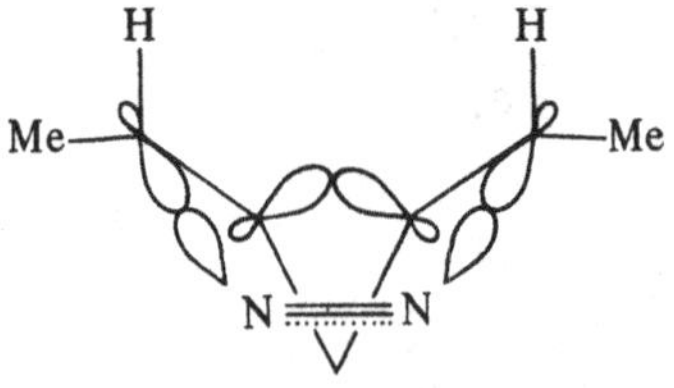 **(45)**

The mechanisms of the *endo*- to *exo*-rearrangement of *endo*-Diels–Alder adducts has received considerable attention. There has in the past been considerable controversy as to whether the rearrangement is intramolecular. However, it now seems clear that it involves a retro-Diels–Alder reaction followed by recombination.[9]

5.3. 1,3-Dipolar cycloadditions (fig. 5.9)[42]

In the Diels–Alder reaction, the $_{\pi}4_{s}$ component is a diene in which the four π electrons are distributed over four carbon atoms. An allyl anion has four π

electrons distributed over three carbon atoms but the symmetry of the HOMO (fig. 5.10) is the same as that of a diene so that concerted $_{\pi}4_s + _{\pi}2_s$ addition to a monoene should be allowed.

⊕ a b c ⊖ + d=e ⟶ a b c d—e

Fig. 5.9

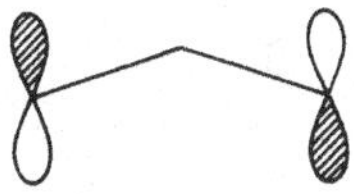

Fig. 5.10

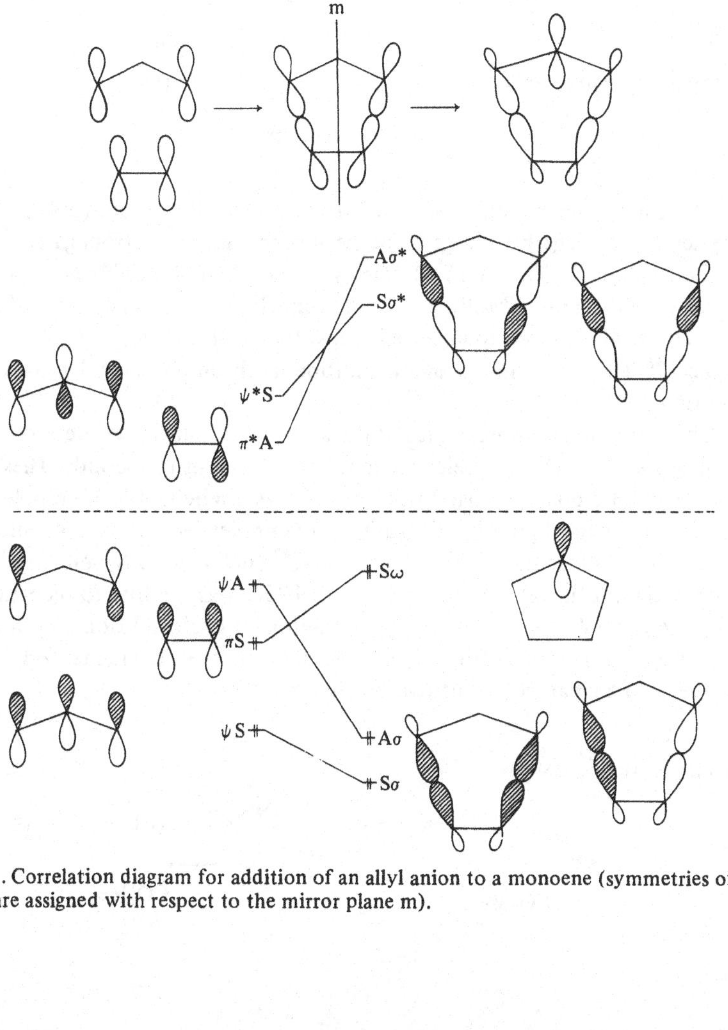

m

Aσ*
Sσ*
ψ*S
π*A
ψA
πS
Sω
ψS
Aσ
Sσ

Fig. 5.11. Correlation diagram for addition of an allyl anion to a monoene (symmetries of orbitals are assigned with respect to the mirror plane m).

The correlation diagram for addition of an allyl anion to a monoene is very similar to that for the Diels–Alder reaction, the difference being that a lone pair orbital, rather than a π bond, appears in the product (fig. 5.11).

The simple addition of an allyl anion to a monoene has only rarely been observed[43] probably because the allylic delocalised negative charge becomes localised on a single carbon atom of the cyclopentanyl anion produced. However, the reaction has been utilised for stereospecific pentannelation (5.41).[44]

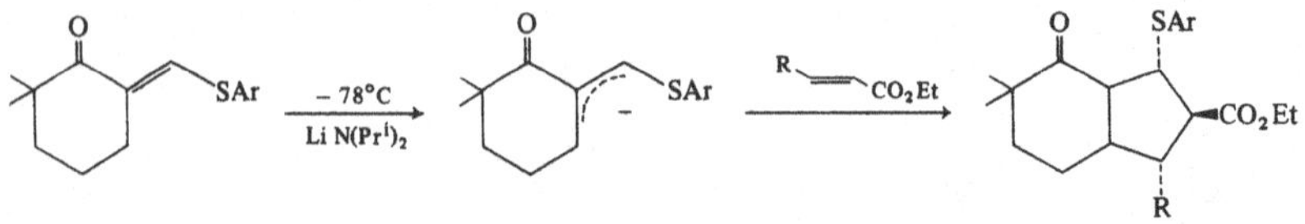

(5.41)

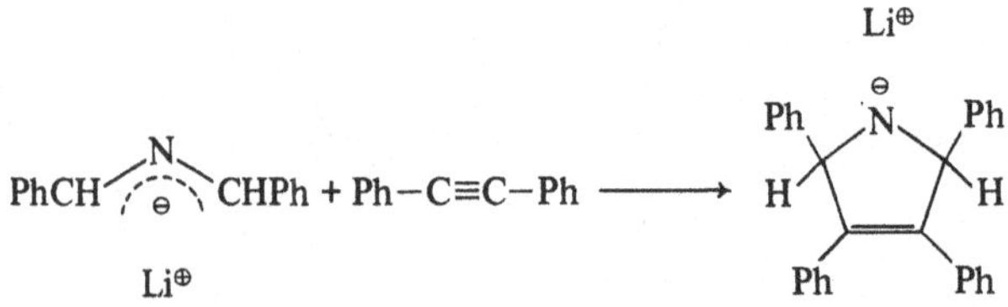

(5.42)

There are more examples of the addition of aza-allyl anions (5.42) where the developing negative charge is localised onto a more electronegative N atom in the product.[45] [Note in (5.41) the cyclopentyl anion produced is particularly well stabilised by the carbonyl group.] The stereochemistry of addition of such aza-allyl anions to alkenes is consistent with a concerted ${}_{\pi}4_s + {}_{\pi}2_s$ process.[46] This type of alkyl anion addition has been termed 1,3-anionic addition.

There are numerous examples of the addition of neutral 4π electron, three atom systems which are isoelectronic with allyl anions to π bonds. These are known as 1,3-dipolar cycloadditions, the 4π system being the 1,3-dipole and the 2π system the dipolarophile. Although examples (equations 5.43 and 5.44) were known before the turn of the century,[47] such reactions were largely neglected until the late 1950s. Huisgen, in 1958, was the first to recognise fully the general concept and scope of 1,3-dipolar cycloaddition and since that time, largely due to his efforts, it has become a most valuable method for the synthesis of a great variety of five-ring heterocycles.

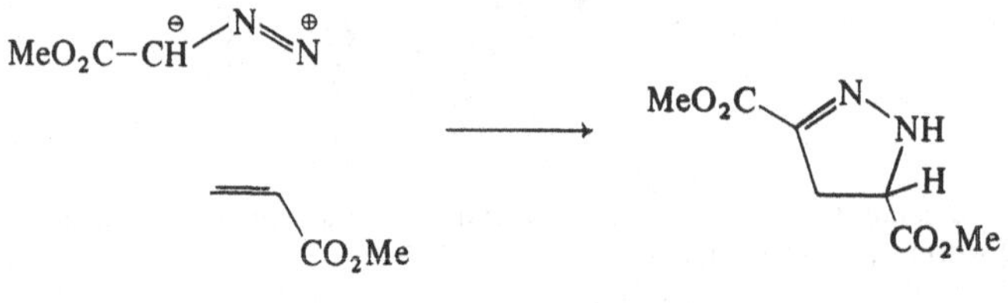

(5.43)

$$\mathrm{Ph{-}\overset{\ominus}{N}{-}N{=}\overset{\oplus}{N}} \quad + \quad \mathrm{RO_2C{-}C{\equiv}C{-}CO_2R} \longrightarrow \text{1-phenyl-1,2,3-triazole-4,5-}(\mathrm{CO_2R})_2 \qquad (5.44)$$

A 1,3-dipole is basically a system of three atoms amongst which are distributed four π electrons as in an allyl anion system (fig. 5.12). The three atoms can be a wide variety of combinations of C, O and N. The dipolarophile can be virtually any double or triple bond.

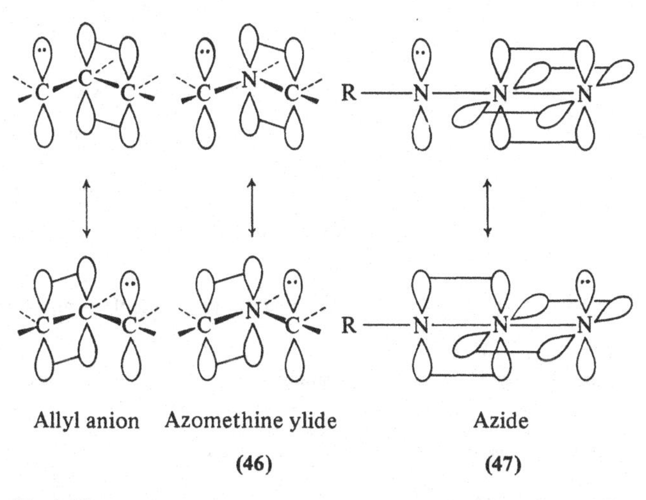

Fig. 5.12

The term 1,3-dipole arose because in valence bond theory such compounds can only be described in terms of dipolar resonance contributors, as shown for diazomethane (5.45).

$$\underset{\textbf{(48a)}}{:\overset{\ominus}{C}H_2{-}\overset{\oplus}{N}{=}N:} \leftrightarrow CH_2{=}\overset{\oplus}{N}{=}\overset{\ominus}{\underset{\cdot\cdot}{N}}: \leftrightarrow \underset{\textbf{(48b)}}{:\overset{\ominus}{C}H_2{-}\overset{\oplus}{N}{\equiv}N:} \leftrightarrow \underset{\textbf{(48c)}}{\overset{\oplus}{C}H_2{-}\ddot{N}{=}\overset{\ominus}{\underset{\cdot\cdot}{N}}:} \leftrightarrow CH_2{=}\ddot{N}{-}\underset{\cdot\cdot}{\ddot{N}} \qquad (5.45)$$

It is important not to misinterpret this picture of 1,3-dipoles. They must not be confused with zwitterions in which the positive and negative charges are localised at specific sites. In 1,3-dipoles, the formal charges are interchangeable so the fact that dipolar forms can be written does not imply a high dipole moment; indeed dipolar forms such as **(48b)** and **(48c)** largely cancel so that such molecules generally have low dipole moments. Diphenyldiazomethane, for

Table 5.3. 1,3-Dipoles with octet stabilisation

Name	Structure	Stability and generation
1,3-Dipoles with an orthogonal double bond		
Nitrile ylides[48,49]	$-\overset{+}{C}=N-\overset{-}{C}\langle \leftrightarrow -C\equiv\overset{+}{N}-\overset{-}{C}\langle$	*In situ* from 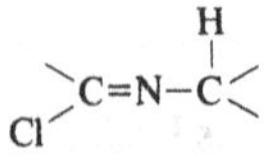 $(Cl)\rangle C=N-C(H)\langle$ and NEt_3 [48] or by photolysis of azirines[49]
Nitrile imines[50]	$-\overset{+}{C}=N-\overset{-}{N}- \leftrightarrow -C\equiv\overset{+}{N}-\overset{-}{N}-$	*In situ* from 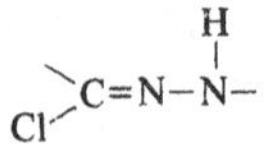 $(Cl)\rangle C=N-N(H)-$ and NEt_3 or by photolysis of 2*H*-tetrazoles
Nitrile oxides[51]	$-\overset{+}{C}=N-\overset{-}{O} \leftrightarrow -C\equiv\overset{+}{N}-\overset{-}{O}$	*In situ* from 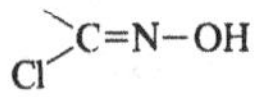 $(Cl)\rangle C=N-OH$ and NEt_3 (some are isolable)
Nitrile sulphides[52]	$-\overset{+}{C}=N-\overset{-}{S} \leftrightarrow -C\equiv\overset{+}{N}-\overset{-}{S}$	*In situ* from thermolysis of oxathiazolones 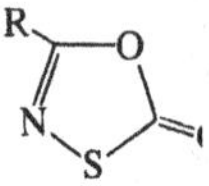 (R-substituted 1,3,4-oxathiazol-2-one ring: R, O, N, S)
Diazo compounds[53]	$\overset{+}{N}=N-\overset{-}{C}\langle \leftrightarrow N\equiv\overset{+}{N}-\overset{-}{C}\langle$	Variable stability, isolable in many cases
Azides[53,54]	$\overset{+}{N}=N-\overset{-}{N}- \leftrightarrow N\equiv\overset{+}{N}-\overset{-}{N}-$	Isolable
Nitrous oxide[55]	$\overset{+}{N}=N-\overset{-}{O} \leftrightarrow N\equiv\overset{+}{N}-\overset{-}{O}$	Stable, isolable
1,3-Dipoles without an orthogonal double bond		
Azomethine ylides[50,53]	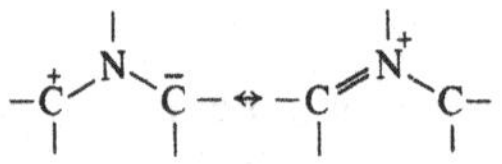 $-\overset{+}{C}-N(-)-\overset{-}{C}- \leftrightarrow -C=\overset{+}{N}(-)-\overset{-}{C}-$	*In situ* from 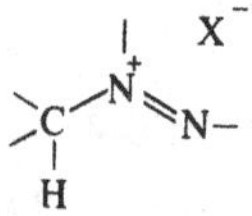 $\rangle C(H)-\overset{+}{N}(-)=N-\ X^-$ and NEt_3 or by electrocyclic ring opening of aziridines
Azomethine imines[50,53]	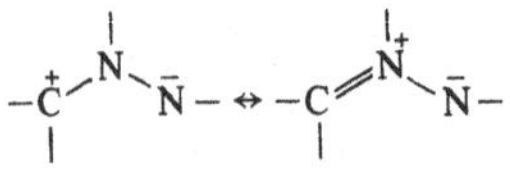 $-\overset{+}{C}-N(-)-\overset{-}{N}- \leftrightarrow -C=\overset{+}{N}(-)-\overset{-}{N}-$	*In situ* from 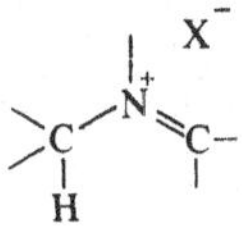 $\rangle C(H)-\overset{+}{N}(-)=C-\ X^-$ and NEt_3 or from aldehydes and *N*-acyl-*N*-alkylhydrazines[56]
Nitrones[57]	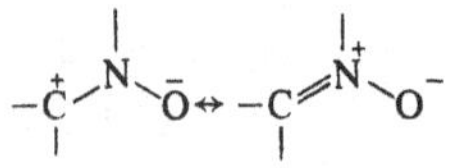 $-\overset{+}{C}-N(-)-\overset{-}{O} \leftrightarrow -C=\overset{+}{N}(-)-O^-$	Isolable
Azimines[58]	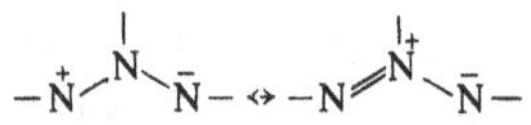 $-\overset{+}{N}-N(-)-\overset{-}{N}- \leftrightarrow -N=\overset{+}{N}(-)-\overset{-}{N}-$	Variable stability; several are isolable

Table 5.3 continued

Name	Structure	Stability and generation
1,3-Dipoles without an orthogonal double bond		
Azoxy compounds[59]	$-\overset{+}{N}-N(-)-\bar{O} \leftrightarrow -N=\overset{+}{N}(-)-O^-$	Stable, isolable
Carbonyl ylides[53,60]	$-\overset{+}{C}-O-\bar{C}- \leftrightarrow -C=\overset{+}{O}-\bar{C}-$	*In situ* by photolysis or thermolysis of oxiranes
Carbonyl imines	$-\overset{+}{C}-O-\bar{N}- \leftrightarrow -C=\overset{+}{O}-\bar{N}-$	Unknown
Carbonyl oxides[53,61]	$-\overset{+}{C}-O-\bar{O} \leftrightarrow -C=\overset{+}{O}-\bar{O}$	*In situ* from cleavage of ozonides or by reaction of carbenes with oxygen
Nitroso imines	$-\overset{+}{N}-O-\bar{N}- \leftrightarrow -N=\overset{+}{O}-\bar{N}-$	Unknown
Nitroso oxides	$-\overset{+}{N}-O-\bar{O} \leftrightarrow -N=\overset{+}{O}-\bar{N}-$	Unknown
Ozone[61]	$\overset{+}{O}-O-\bar{O} \leftrightarrow O=\overset{+}{O}-\bar{O}$	Stable
Thiocarbonyl ylides[53,62]	$-\overset{+}{C}-S-\bar{C}- \leftrightarrow -C=\overset{+}{S}-\bar{C}-$	*In situ* by thermolysis of thiadiazolines (N=N, S ring)
Thiocarbonyl imines[53,63]	$-\overset{+}{C}-S-\bar{N}- \leftrightarrow -C=\overset{+}{S}-\bar{N}-$	A few stable examples known
Sulphur di-imides[64]	$-N=S=N- \leftrightarrow -N=\overset{+}{S}-\bar{N}-$	Isolable
Sulphinylamines[64]	$-N=S=O \quad -N=\overset{+}{S}-\bar{O}$	Isolable

example, has a dipole moment of 1.42 D whereas the form corresponding to **(48b)** has a calculated dipole moment of 6 D. To a first approximation it is impossible to assign a nucleophilic or electrophilic end to the dipole. However, for unsymmetrical 1,3-dipoles normally one end tends to be nucleophilic and the other electrophilic; the extreme resonance hybrids do not carry the same weighting in the overall description of the species. Thus the hybrid **(48b)** is the most important contributor to the structure of diazomethane and this accounts for the well-known nucleophilic and basic character of the diazomethane carbon. This 'electronic asymmetry' of 1,3-dipoles also arises naturally from a molecular orbital description (see p. 144).

A selection of 1,3-dipoles is presented in table 5.3 which also gives some indication of their stability and mode of formation and leading literature references. 1,3-Dipoles fall into two categories, those without and those with a π bond orthogonal to the 4π allyl system. The former are bent, for example azomethine ylides (fig. 5.12, **46**) and the latter linear, for example azides (fig. 5.12, **47**). All of the dipoles in table 5.3 are said to have internal octet stabilisation, since the central atom has a lone pair of electrons which is utilised to increase the number of possible resonance contributors (see diazomethane above). The stability of the dipoles varies considerably, for example phenyl azide or ozone can be isolated, but nitrile imines or nitrile ylides can only be generated *in situ.* Nitrile oxides are of intermediate stability, some being isolable and others not, depending on the substituent on carbon. Broadly, the stability trend of 1,3-dipoles follows the HOMO energy which decreases as more electronegative elements are incorporated into the system. The stability of dipoles is discussed in more detail on p. 144.

The 1,3-dipole may be part of a heterocyclic system.[66] Sydnones (**49**) and oxazolones (**50**) are masked azomethine imines and azomethine ylides, respectively. They give initial adducts which lose carbon dioxide after rate-determining 1,3-dipolar addition (see p. 153 for examples).

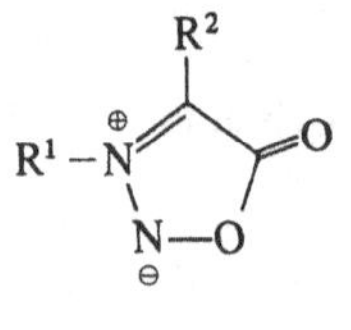

(49) 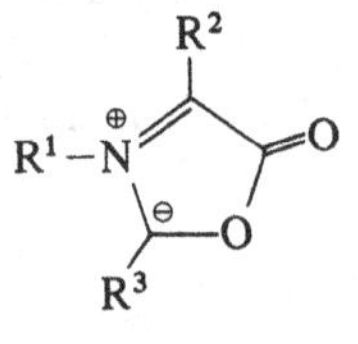 **(50)**

Azines (**51**) are interesting molecules in that they function not as dienes but as 1,3-dipoles to give 'criss-cross' bis-adducts in cycloadditions with typical dienophiles (5.46).[67]

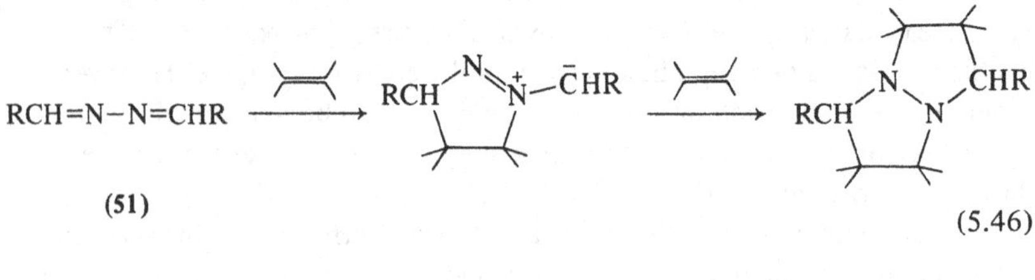

RCH=N–N=CHR

(51)

(5.46)

Probably because of lone pair–lone pair repulsion, azines appear to behave as though the diene π bonds are orthogonal so that the system has two orthogonal azomethine imine moieties (**52**).

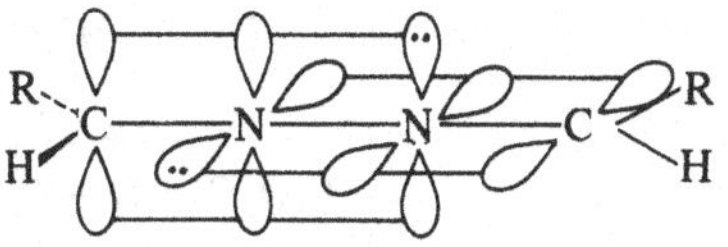

(52)

If the central atom is carbon, octet stabilisation is impossible (see table 5.4). These types of 1,3-dipole are highly reactive species with very short lifetimes, often showing both the reactions of the carbene or nitrene and of the 1,3-dipole. Not all of the possible 1,3-dipolar systems have yet been generated.

Table 5.4. 1,3-Dipoles without octet stabilisation[a]

Vinyl carbenes	$-\overset{+}{C}=C(-)-\overset{-}{C}<$	↔	$-\ddot{C}-C(-)=C<$
Iminocarbenes	$-\overset{+}{C}=C(-)-\overset{-}{N}-$	↔	$-\ddot{C}-C(-)=N-$
Ketocarbenes	$-\overset{+}{C}=C(-)-\overset{-}{O}$	↔	$-\ddot{C}-C(-)=O$
Vinyl nitrenes	$\overset{+}{N}=C(-)-\overset{-}{C}<$	↔	$\ddot{N}-C(-)=C<$
Iminonitrenes	$\overset{+}{N}=C(-)-\overset{-}{N}-$	↔	$\ddot{N}-C(-)=N-$
Ketonitrenes	$\overset{+}{N}=C(-)-\overset{-}{O}$	↔	$\ddot{N}-C(-)=O$

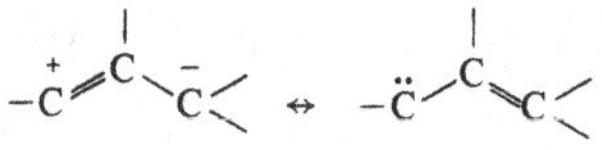

[a]All are highly reactive intermediates.[65]

The dipolarophile can be almost any double or triple bond; $>C=C<$,

$-C\equiv C-$, $>C=N-$, $-C\equiv N$, $-N=N-$, $>C=O$, $>C=S$, $-N=O$,

$-N=S=$ (for example in $Ph-N=S=O$). The π bond may be isolated, conjugated or part of a cumulene system; it can even be the enol form of a ketone or similar compound (5.47).

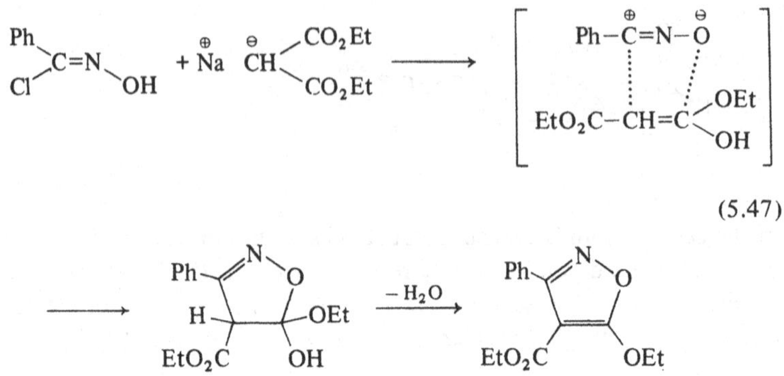

(5.47)

It is the great structural variety of both 1,3-dipole and dipolarophile that makes 1,3-dipolar cycloaddition so valuable and versatile in heterocyclic synthesis. Some additional examples are given in equations 5.48–5.52.

Many intramolecular 1,3-dipolar cycloadditions are also known.[2,68]

The question of reactivity and regioselectivity in 1,3-dipolar cycloaddition can be approached using FMO theory in much the same way as in the Diels-Alder reaction. However, because of the much greater structural variety of 1,3-dipoles, simple generalisations are not possible. We shall therefore return to this problem on pp. 144 and 150.

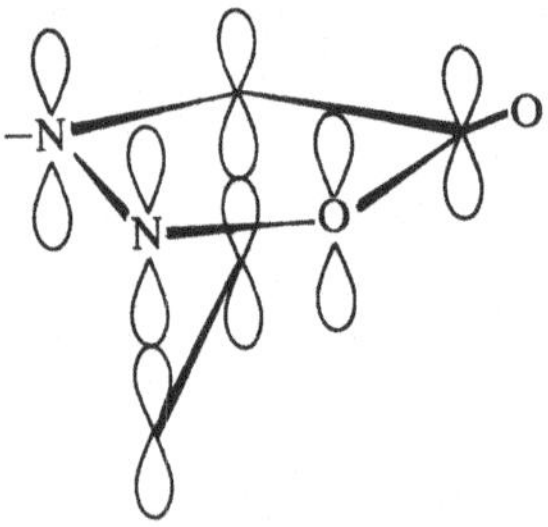

(53)

Mechanism[42,69]

Huisgen and others have systematically studied the mechanism of 1,3-dipolar cycloaddition. In most cases the evidence, like that for the Diels–Alder reaction, points to a concerted reaction in line with orbital symmetry considerations. Thus the rates of 1,3-dipolar additions are largely insensitive to solvent polarity, the reactions have large negative $\Delta S^{\ddagger}$ and only moderate $\Delta H^{\ddagger}$ values and also show high stereoselectivity in addition to pairs of *cis* and *trans* substituted dipolarophiles. Isotope effects, in the cases studied, indicate a one-step mechanism.[70] The fact that sydnones and oxazolones behave like other 1,3-dipoles is good evidence that the dipole and dipolarophile approach each other in parallel planes ($_{\pi}4_s + _{\pi}2_s$) since approach of dipolarophiles to these planar molecules can only be from above or below as shown in **(53)**.

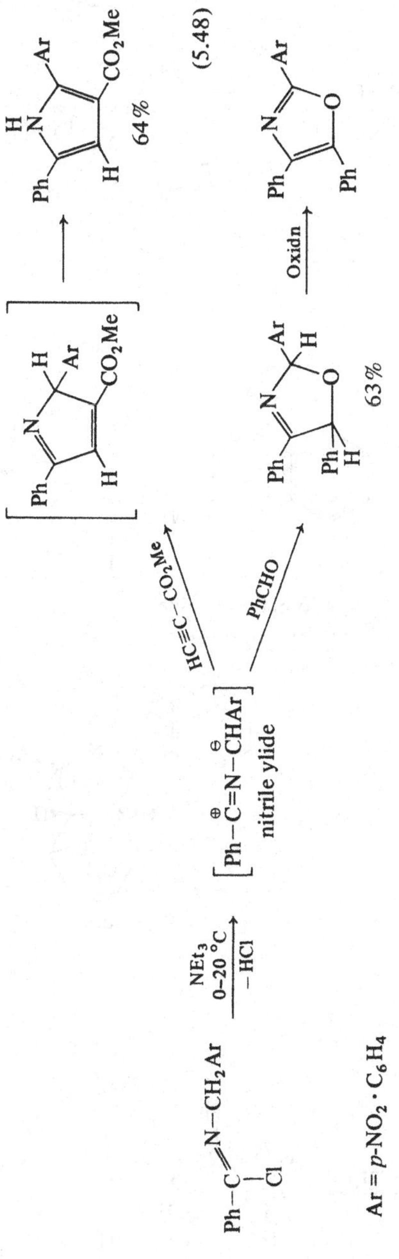

$$\mathrm{Ph{-}C(Cl){=}N{-}CH_2Ar} \xrightarrow[-\mathrm{HCl}]{\mathrm{NEt_3},\ 0\text{–}20\,^\circ\mathrm{C}} \left[\mathrm{Ph{-}\overset{\oplus}{C}{=}N{-}\overset{\ominus}{C}HAr}\right]$$

nitrile ylide

$\xrightarrow{\mathrm{HC{\equiv}C{-}CO_2Me}}$ [2-Ar-2-H-3-CO_2Me-4-H-5-Ph-2H-pyrrole] $\longrightarrow$ 5-Ph-1H-pyrrole, 2-Ar, 3-CO_2Me, 4-H (64%)

$\xrightarrow{\mathrm{PhCHO}}$ 4-Ph-5-Ph-5-H-2-Ar-2-H-oxazoline (63%) $\xrightarrow{\mathrm{Oxidn}}$ 2-Ar-4,5-diphenyloxazole

(5.48)

$Ar = p\text{-}NO_2 \cdot C_6H_4$

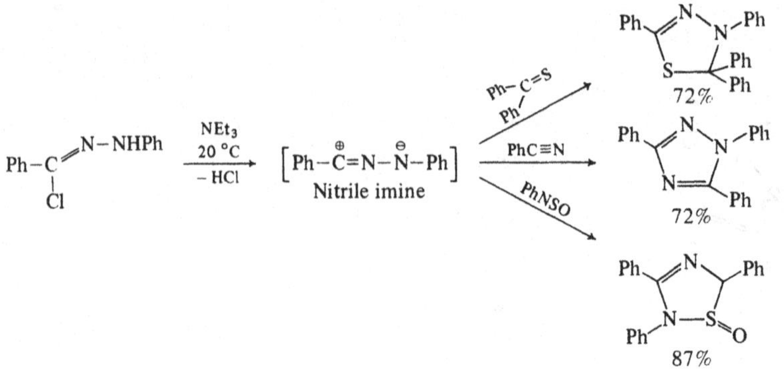

Ph−C(Cl)=N−NHPh →(NEt$_3$, 20 °C, − HCl) [Ph−C$^{\oplus}$=N−N$^{\ominus}$−Ph] Nitrile imine

Ph$_2$C=S → Ph, N, N, Ph, S, Ph, Ph 72%

PhC≡N → Ph, N, N, Ph, N, Ph 72%

PhNSO → Ph, N, Ph, Ph, N, S, O 87%

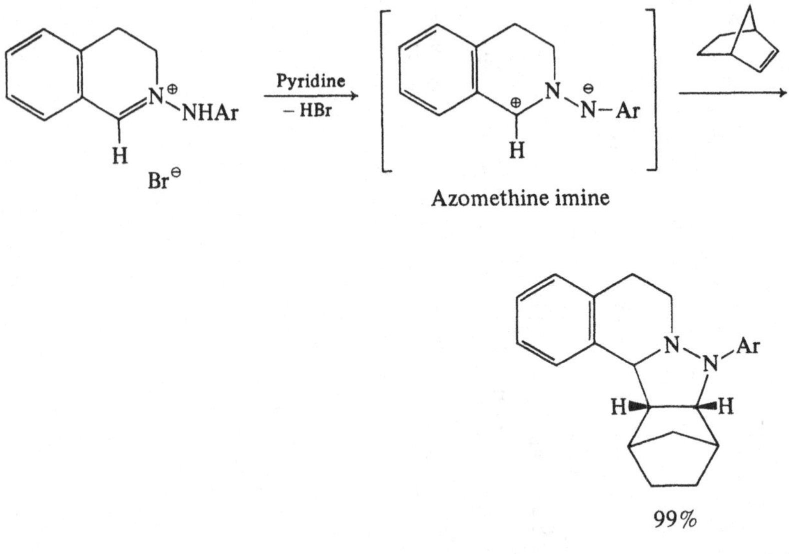

(5.49)

N$^{\oplus}$ NHAr, H, Br$^{\ominus}$ →(Pyridine, − HBr) [$\oplus$, N, N$^{\ominus}$−Ar, H] Azomethine imine →

N, N, Ar, H, H

99%

(5.50)

H, Ph, C=N$^{\oplus}$, Ph, O$^{\ominus}$ →(CH$_2$=CH−CH$_2$OH) Ph, N, O, Ph, H, H, H, H, CH$_2$OH

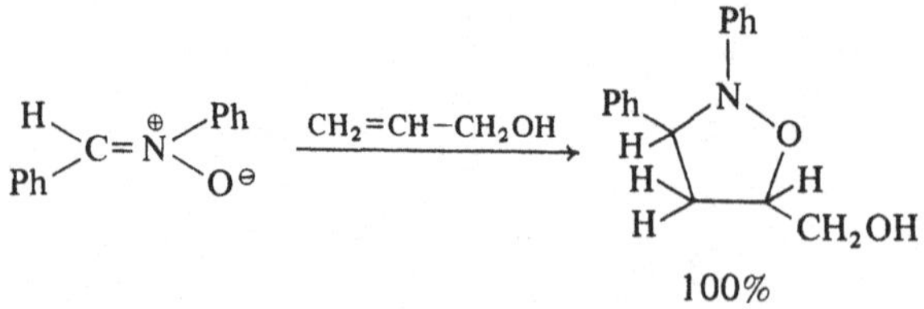

100%

(5.51)

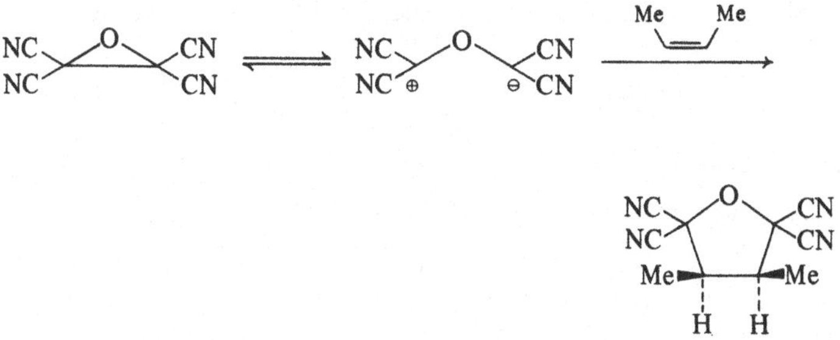

(5.52)

In order to achieve this type of transition state the linear dipoles must first bend; this involves breaking of the orthogonal π bond but leaves the allyl anion π system undisturbed and since such a change is probably not very advanced in the transition state it is not too energy demanding. There is no such problem with the bent dipoles which do not have the orthogonal π bond.

Until recently no really satisfactory explanation for reactivity and regioselectivity in 1,3-dipolar additions was available on the basis of the concerted mechanism. Early explanations considered that although the formation of the new σ bonds was concerted it was not necessarily symmetrical. This led therefore to some degree of charge build-up in the transition state and the effect of substituents in the dipolarophile on regioselectivity were rationalised in terms of their ability to stabilise such partial charges. In some cases this approach did not give the right answer and steric effects had to be invoked. This rather unsatisfactory state of affairs led to the proposal of an alternative mechanism for 1,3-dipolar additions.[69b] This involved formation of an intermediate singlet diradical which rapidly closes stereospecifically if generated in the right conformation for ring closure but reverts to reactants if not (cf. the diradical mechanism for the Diels-Alder reaction). Regioselectivity is then explained in terms of the reaction proceeding preferentially through the most stabilised diradical. This theory has been criticised on many grounds and is not generally considered as a viable alternative to the concerted mechanism.[69a] Moreover, application of FMO theory now appears to give a satisfactory explanation for reactivity and regioselectivity on the basis of the concerted $_{\pi}4_{s} + _{\pi}2_{s}$ mechanism, making the latter the currently most satisfactory picture of the process.

Although most 1,3-dipolar additions are probably concerted, it is quite possible that some are stepwise since the highly unsymmetrical structure of most dipoles and many dipolarophiles is such that zwitterionic intermediates would be highly stabilised. However, no clear examples of stepwise 1,3-dipolar cycloadditions have yet been reported.[71]

The isolation of open-chain hydrazones (**54**) as well as pyrazoles (**55**) in the addition of nitrile imines to aryl acetylenes could be evidence for a stepwise reaction with a common intermediate zwitterion or diradical, although the two types of product are more likely to be formed independently by separate mechanisms.[72] A similar situation exists in the addition of nitrile oxides to aryl acetylenes.[73]

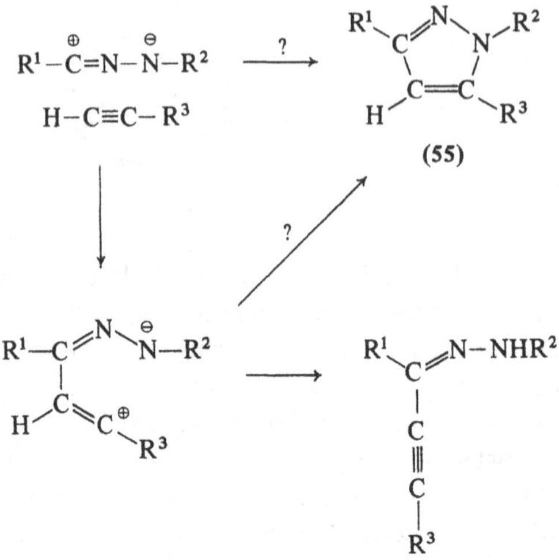

(55)

(54)

Reactivity[74]

In order to discuss reactivity and regioselectivity in 1,3-dipolar additions it is necessary to know the dipole HOMO and LUMO energy levels and coefficients. These have been estimated by Houk[74] for a selection of the more common 1,3-dipoles and are presented in tables 5.5 and 5.6. The data which are derived from a combination of experiment and calculation are given for the unsubstituted dipoles and for representative substituted analogues. In general, substituents affect the orbital energies and coefficients in much the same way as they do for dienes. However, only in cases where the terminal coefficients are almost equal for the unsubstituted dipole do substituents cause a reversal of the relative magnitudes of these coefficients from that in the unsubstituted dipole. In such cases, the site of the larger coefficient can depend on the substituent; this occurs, for example, with nitrile oxides and diazo compounds.

The orbital coefficients are expressed as $(c\beta)^2/15$ values. This requires some explanation; β the resonance integral, appears because it depends on the type of bond being formed and is different for C—C, C—N and C—O bonds. In dipolar additions bond formation does not always involve C—C bond formation, as it does in the normal Diels-Alder reaction, and so β cannot be cancelled in equations 4.2–4.4. It similarly has to be taken into account in hetero-Diels-Alder reactions. The $(c\beta)^2/15$ values are calculated assuming overlap of the dipole with a C—C dipolarophile and a reasonable developing bond length (β

Table 5.5. Frontier orbital energies of 1,3-dipoles[74]

Dipole	HOMO energy (eV)	LUMO energy (eV)
$HC\equiv\overset{+}{N}-\bar{C}H_2$	−7.7	0.9
$PhC\equiv\overset{+}{N}-\bar{C}H_2$	−6.4	0.6
$HC\equiv\overset{+}{N}-\bar{N}H$	−9.2	0.1
$PhC\equiv\overset{+}{N}-\bar{N}Ph$	−7.5	−0.5
$HC\equiv\overset{+}{N}-\bar{O}$	−11.0	−0.5
$PhC\equiv\overset{+}{N}-\bar{O}$	−10.0	−1.0
$N\equiv\overset{+}{N}-\bar{C}H_2$	−9.0	1.8
$N\equiv\overset{+}{N}-\bar{N}H$	−11.5	0.1
$N\equiv\overset{+}{N}-\bar{N}Ph$	−9.5	−0.2
$N\equiv\overset{+}{N}-\bar{O}$	−12.9	−1.1
$CH_2=\overset{+}{N}-\bar{C}H_2$	−6.9	1.4
$ECH=\overset{+}{N}(Ar)-\bar{C}HE$[a]	−7.7	−0.6
$CH_2=\overset{+}{N}-\bar{N}H$	−8.6	−0.3
$PhCH=\overset{+}{N}-\bar{N}Ph$	−5.6	−1.4
$CH_2=\overset{+}{N}-\bar{N}COR$	−9.0	−0.4
$CH_2=\overset{+}{N}H-\bar{O}$	−9.7	−0.5
$PhCH=\overset{+}{N}H-\bar{O}$	−8.0	−0.4
$CH_2=\overset{+}{O}-\bar{C}H_2$	−7.1	0.4
$(CN)_2C=\overset{+}{O}-\bar{C}(CN)_2$	−9.0	−1.1
$O=\overset{+}{O}-\bar{O}$	−13.5	−2.2

[a]E = CO_2R.

Table 5.6. Orbital coefficients $[(c\beta)^2/15]$ for terminal atoms[a] of 1,3-dipoles[74]

Dipole	HOMO		LUMO	
$HC\equiv\overset{+}{N}-\bar{C}H_2$	1.07	1.50	0.69	0.64
$HC\equiv\overset{+}{N}-\bar{N}H$	0.90	1.45	0.92	0.36
$HC\equiv\overset{+}{N}-\bar{O}$	0.81	1.24	1.18	0.17
$N\equiv\overset{+}{N}-\bar{C}H_2$	0.85	1.57	0.56	0.66
$N\equiv\overset{+}{N}-\bar{N}H$	0.72	1.55	0.76	0.37
$N\equiv\overset{+}{N}-\bar{O}$	0.67	1.33	0.96	0.19
$CH_2=\overset{+}{N}-\bar{C}H_2$	1.28	1.28	0.73	0.73
$CH_2=\overset{+}{N}-\bar{N}H$	1.15	1.24	0.87	0.49
$CH_2=\overset{+}{N}H-\bar{O}$	1.11	1.06	0.98	0.32
$CH_2=\overset{+}{O}-\bar{C}H_2$	1.29	1.29	0.82	0.82

[a]Values are given in the same order as the atoms appear in the left-hand column; e.g. for the HOMO of $HC\equiv\overset{+}{N}-\bar{O}$ the values are 0.81 on carbon and 1.24 on oxygen.

also depends on the nuclear separation); the denominator 15 is introduced to bring the values near to unity. The LUMO orbitals quoted in table 5.5 are not always strictly the lowest of the unoccupied orbitals. They are, however, the lowest of the allyl anion-type orbitals which are the important ones involved in cycloaddition. Thus, in the 1,3-dipoles having an orthogonal double bond, such as nitrile ylides, the lowest unoccupied orbital actually lies in the plane of molecule (**56**) and is a combination orbital involving the orthogonal π bond and CH_2 σ-type orbitals. The important vacant orbital for cycloaddition is the

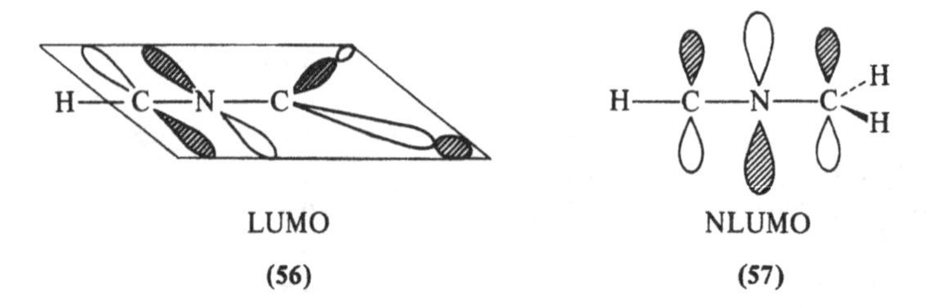

LUMO (**56**)　　NLUMO (**57**)

π allyl type (**57**) which is strictly the NLUMO (next to lowest unoccupied orbital). A similar situation applies for nitrile imines, azides and diazo compounds. Where the terminal atom is O as in formonitrile oxide $H{-}C{\equiv}\overset{+}{N}{-}\bar{O}$ the orthogonal LUMO and NLUMO become degenerate. The degeneracy is removed in benzonitrile oxide, $Ph{-}C{\equiv}N^{+}{-}O^{-}$, because only one of the degenerate orthogonal pair (the π allyl one) can be conjugated with the benzene ring. For the occupied orbitals of dipoles with an orthogonal double bond the HOMO (**58**) is the π allyl type but an orthogonal combination NHOMO (**59**) of slightly lower energy occurs. Again these two orthogonal orbitals are degenerate for $H{-}C{\equiv}N^{+}{-}O^{-}$ and $Me{-}C{\equiv}N^{+}{-}O^{-}$ but not $Ph{-}C{\equiv}N^{+}{-}O^{-}$.

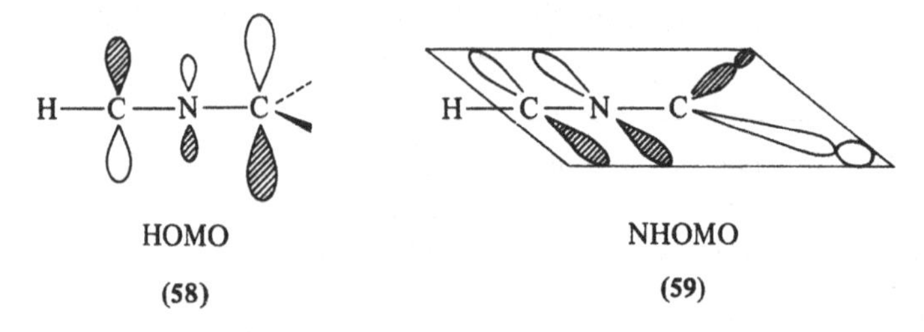

HOMO (**58**)　　NHOMO (**59**)

From table 5.5 it can be seen that the general trend is for the HOMO energies to decrease as more electronegative elements are incorporated in the dipolar system. The LUMOs also decrease in energy but less markedly and therefore the HOMO energy can be taken as a rough guide to the stability and reactivity of the dipoles. For example, nitrile ylides can only be produced as transient intermediates, nitrile imines are less reactive transient intermediates and nitrile oxides are often isolable or dimerise slowly. For the diazonium betaine series, diazoalkanes are isolable but react with most alkenes at 0 °C, azides are somewhat less reactive and require heating, while nitrous oxide only reacts with alkenes above 200 °C and at pressures of several hundreds of atmospheres. For the azomethinium betaine series, azomethine ylides are transient intermediates, azomethine imines react readily with dipolarophiles or dimerise reversibly whereas nitrones can be isolated.

As for the Diels–Alder reaction, the rate of 1,3-dipolar additions is related to the HOMO-LUMO energy separation of the reactants. The dipole HOMO, the dipole LUMO or both, may be involved in the dominant interaction, depending on the relative energies of the dipole and dipolarophile orbitals (figs. 5.13–5.15).

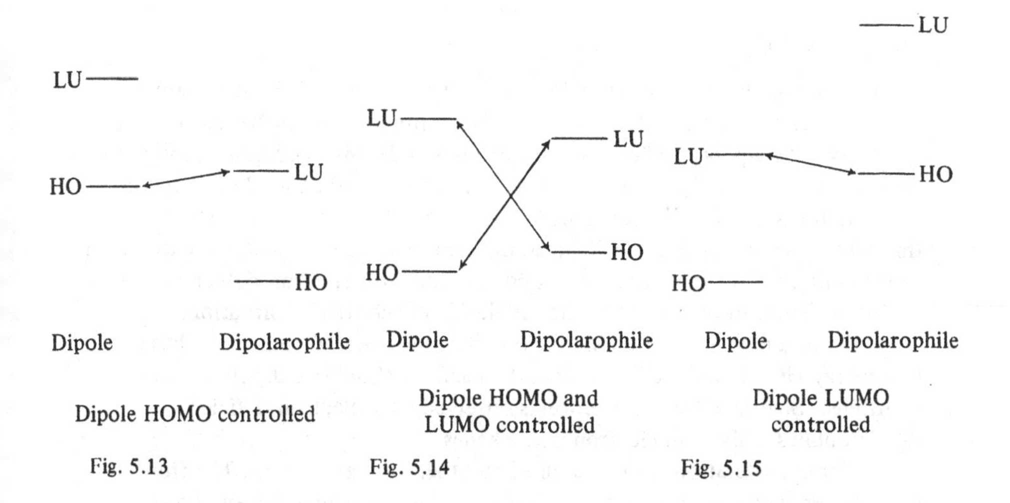

Fig. 5.13 Fig. 5.14 Fig. 5.15

Substituents which raise dipole HOMO energies (electron-releasing R and x, conjugating c) and which lower dipolarophile LUMO energies (c, and electron-withdrawing z) will accelerate dipole HOMO controlled reactions, fig. 5.13, but slow down dipole LUMO controlled reactions, fig. 5.15. On the other hand, substituents which lower dipole LUMO (c, z) and raise dipolarophile HOMO energies (x, R, c) will accelerate dipole LUMO controlled reactions and slow down dipole HOMO controlled reactions. Substituents which increase either frontier orbital interaction will favour reactions of the type shown in fig. 5.14. Usually it is found that electron-deficient dipolarophiles lead to dipole HOMO controlled reactions and electron-rich dipolarophiles to dipole LUMO controlled reactions. Thus the rate of addition of phenyl azide with alkenes is faster for both electron-rich and electron-deficient dipolarophiles than for ethylene (fig. 5.16).[75]

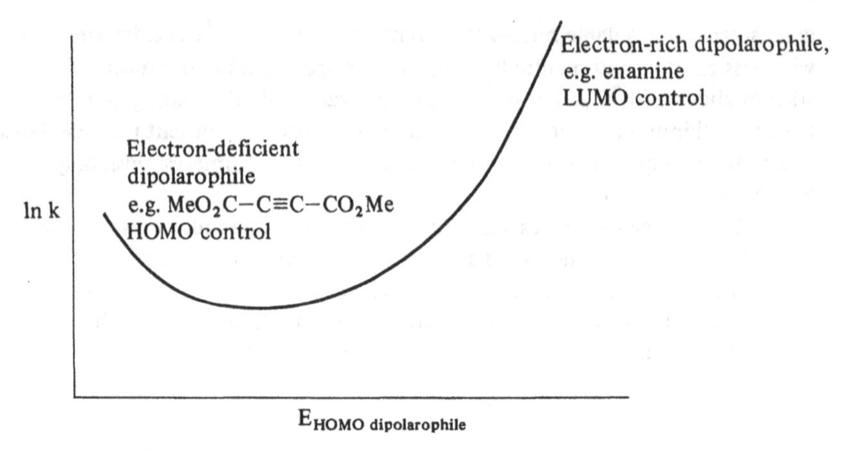

Fig. 5.16

Diazoalkanes have higher HOMO energies than azides and therefore are more generally dipole HOMO controlled. Thus simple diazoalkanes show high reactivity with dipolarophiles having electron-withdrawing and conjugating substituents but only low reactivity with alkylolefins and enol ethers. The order of reactivity found for diazoalkanes $MeCHN_2 > CH_2N_2 \gg EtO_2CCHN_2$ follows this pattern. However, for the diazo compounds with strong electron-withdrawing substituents, such as diazoketones, rapid reaction with electron-rich enamines is observed indicating reversal to the dipole LUMO controlled situation.

At the other extreme of the diazonium betaine series, nitrous oxide has a low energy HOMO and LUMO so that its reactions should be dipole LUMO controlled. Significantly, no reaction is observed with electron-deficient dipolarophiles, only with electron-rich alkenes.

Similar considerations can be applied to other dipolar systems. The frontier orbital energy separations are summarised in table 5.7 which therefore gives some idea of which types of dipolarophiles react most rapidly with the different dipolar systems. With the limited data available the table is far from perfect and while it reflects reactivity trends for each dipolar system with a series of dipolarophiles quite well it is less good for comparisons between different dipolar systems. For example, the lack of reactivity of nitrous oxide is not apparent from the low value of ΔE (dipole HOMO–dipolarophile LUMO).

The above simple treatment of reactivity has neglected the coulombic contribution to the overall energy of interaction (equations 4.2–4.4). Coulombic effects must contribute appreciably towards reactivity in dipolar cycloadditions because all 1,3-dipoles have an excess of negative charge on the terminal atoms and so there is a 'coulombic bias' favouring reactions with electron-deficient dipolarophiles. Thus it is found that frontier orbital energy considerations tend to underestimate reactivity for dipole HOMO controlled reactions with electron-deficient dipolarophiles.

Table 5.7. Frontier orbital energy separations for 1,3-dipoles and dipolarophiles

ΔE	(dipole LUMO–dipolarophile HOMO) (eV)				(dipole HOMO–dipolarophile LUMO) (eV)			
Dipole	$=\backslash_x$	$=\backslash_c$	$=$	$=\backslash_z$	$=\backslash_x$	$=\backslash_c$	$=$	$=\backslash_z$
PhN_3	7.8	8.8	10.3	10.7	12.5	10.5	11.0	9.5
CH_2N_2	9.8	10.8	12.3	12.7	12.0	10.0	10.5	9.0
N_2O	6.9	7.9	9.4	9.8	15.9	13.9	14.4	12.9
$RO_2C\underset{+}{C}H\overset{Ar}{N}\underset{-}{C}HCO_2R$	7.4	8.4	9.9	10.3	10.7	8.7	9.9	7.7
$H_2C{=}\underset{+}{\overset{H}{N}}{-}\underset{-}{N}H$	7.7	8.7	10.2	10.6	11.6	9.6	10.1	8.6
$Ph{-}\underset{H}{C}{=}\underset{+}{\overset{R}{N}}{-}\underset{-}{N}Ar$	6.6	7.6	9.1	9.5	8.6	6.6	7.1	5.6
$CH_2{=}\underset{+}{\overset{R}{N}}{-}\underset{-}{N}COR$	7.6	8.6	10.1	10.5	12.0	10.0	10.5	9.0
$CH_2{=}\underset{+}{\overset{H}{N}}{-}\underset{-}{O}$	7.5	8.5	10.0	10.4	12.7	10.7	11.2	9.7
$CH_2{=}\underset{+}{\overset{R}{N}}{-}\underset{-}{O}$	8.3	9.3	10.8	11.2	11.7	9.7	10.2	8.7
$Ph{-}CH{=}\underset{+}{\overset{Me}{N}}{-}\underset{-}{O}$	7.6	8.6	10.1	10.5	11.0	9.0	9.5	8.0
$HC{\equiv}\overset{+}{N}{-}\bar{O}$	7.5	8.5	10.0	10.4	14.0	12.0	12.5	11.0
$Ph{-}C{\equiv}\overset{+}{N}{-}\bar{O}$	7.0	8.0	9.5	9.9	13.0	11.0	11.5	10.0
$H{-}C{\equiv}\overset{+}{N}{-}\bar{N}H$	8.1	9.1	10.6	11.0	12.2	10.2	10.7	9.2
$Ph{-}C{\equiv}\overset{+}{N}{-}\bar{N}Ph$	7.5	8.5	10.0	10.4	10.5	8.5	9.0	7.5
$HC{\equiv}\overset{+}{N}{-}\bar{C}H_2$	8.9	9.9	11.4	11.8	10.7	8.7	9.2	7.7
$Ph{-}C{\equiv}\overset{+}{N}{-}\bar{C}H_2$	8.6	9.6	11.1	11.5	9.4	7.4	7.9	6.4
$CH_2{=}\overset{+}{O}{-}\bar{C}H_2$	8.4	9.4	10.9	11.3	10.1	8.1	8.6	7.1
$Ar\underset{CN}{C}{=}\overset{+}{O}{-}\underset{CN}{\bar{C}}Ar$	7.4	8.4	9.9	10.3	9.5	7.5	8.0	6.5
$(NC)_2C{=}\overset{+}{O}{-}\bar{C}(CN)_2$	6.9	7.9	9.4	9.8	12.0	10.0	10.5	9.0
$O{=}\overset{+}{O}{-}\bar{O}$	5.8	6.8	8.3	8.7	16.5	14.5	15.0	13.5

x = electron releasing OR, NR_2, etc.
c = conjugating Ph, vinyl, etc.
z = electron withdrawing CO_2R, CN, etc.

A further point to be considered is the stability of reactants and products. From the point of view of HOMO and LUMO energies, benzene, for example, should be an acceptable dipolarophile. However, the loss of aromatic stabilisation would make cycloaddition to benzene highly endothermic and such reactions are not observed.

Regioselectivity[74]

In order to rationalise regioselectivity the dominant frontier orbital interaction must first be identified (table 5.7). The orbital coefficients [strictly $(c\beta)^2/15$ values [are then matched so as to maximise the numerator in equations 4.2–4.4 (i.e. combination of large coefficient with large coefficient is preferred). This is illustrated for several examples.

Diazoalkanes. Diazoalkane additions are generally dipole HOMO controlled. The HOMO of diazomethane is as shown (5.53) so that the observed mode of addition to alkenes with conjugating electron-withdrawing groups to give 3-substituted Δ^1-pyrazolines is as expected. The much slower addition to alkyl-alkenes and enol ethers to give 4-substituted pyrazolines is also as anticipated (5.54) since although both dipole HOMO–dipolarophile LUMO and dipole LUMO–dipolarophile HOMO interactions are now comparable the coefficients in the diazomethane LUMO are almost equal so that the HOMO coefficients, which are clearly different, control regioselectivity.

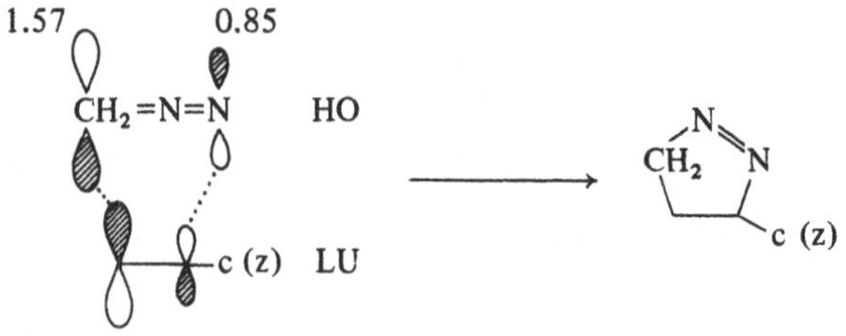

1.57 0.85
$CH_2{=}N{=}N$ HO
c (z) LU
$\longrightarrow$ (5.53)

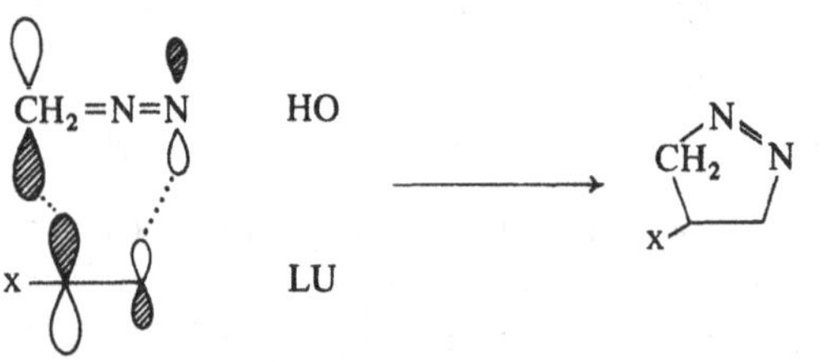

$CH_2{=}N{=}N$ HO
X LU
$\longrightarrow$ (5.54)

With diazoketones and enamines the observed orientation is as shown (5.55). In this case the dipole LUMO energy is lowered by the electron-withdrawing group and the reaction is dipole LUMO controlled. Moreover the electron-withdrawing group will diminish the LUMO coefficient at its point of attachment to the dipole so that the terminal coefficients (almost equal in diazomethane) are now sufficiently different for the regioselectivity prediction to be clearcut.

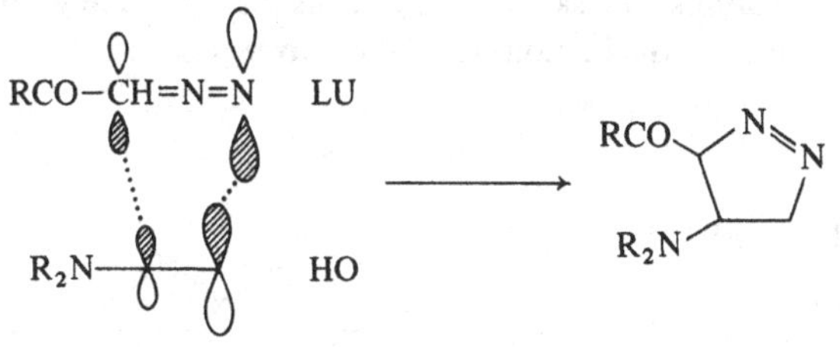

(5.55)

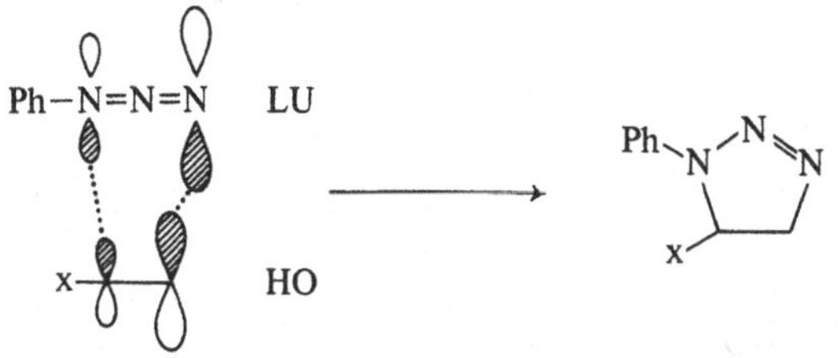

(5.56)

Azides. For azides (e.g. PhN_3), reactions with electron-rich dipolarophiles are dipole LUMO controlled and the preferred orientation (5.56) is readily explained. With electron-deficient dipolarophiles the situation is more finely balanced but in general the reactions appear to be just dipole HOMO controlled since the products are normally the 4-substituted triazolines (5.57). With conjugated alkenes, the phenyl azide LUMO and dipolarophile HOMO interaction becomes dominant although the complementary interaction cannot be completely neglected. In line with this, preferential formation of 5-substituted triazoline is observed (5.58) but the regioselectivity is not high.

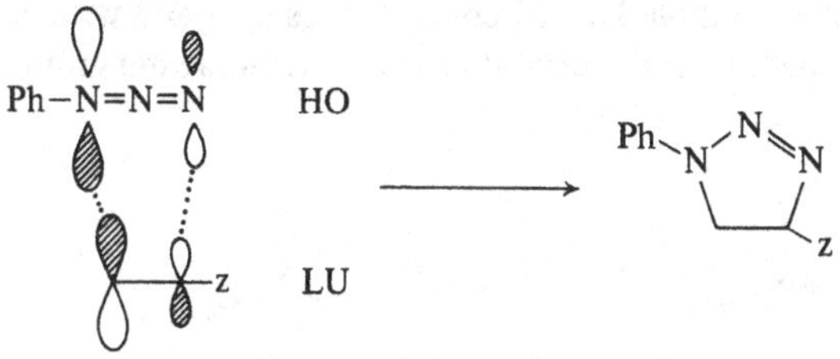

(5.57)

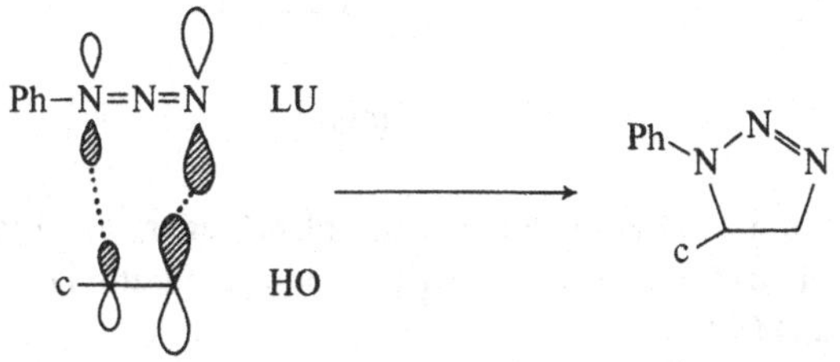

(5.58)

The balance between dipole HOMO and LUMO control is very fine for addition of an aryl azide to an acetylene with a conjugating substituent because the acetylene HOMO is ~ 0.5 eV lower than that of the corresponding alkene whereas the LUMO is of almost the same energy. Thus phenyl azide with phenylacetylene gives almost equal amounts of the two regioisomers (5.59).

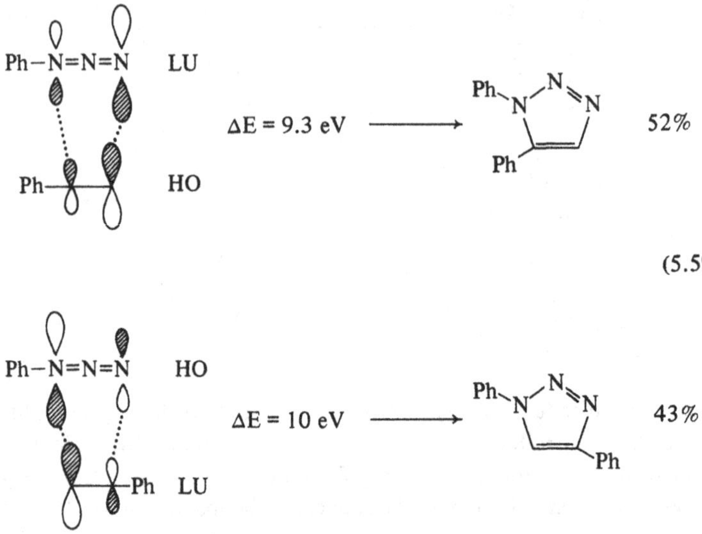

Ph–N=N=N LU

Ph– HO

ΔE = 9.3 eV ⟶ 52%

Ph–N=N=N HO

–Ph LU

ΔE = 10 eV ⟶ 43%

(5.59)

Azomethine imines. The FMO energies of a typical diaryl azomethine imine (**60**) are such that both dipole HOMO and LUMO control is important with conjugated dipolarophiles and not unexpectedly approximately equal amounts of the two possible regioisomers are formed with, for example, styrene. With electron-deficient alkenes, dipole HOMO control takes over and with acrylonitrile, the adduct (**61**) is formed. Introduction of electron-withdrawing groups at either

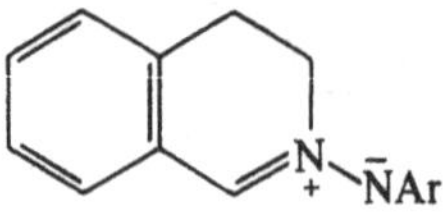

(**60**) 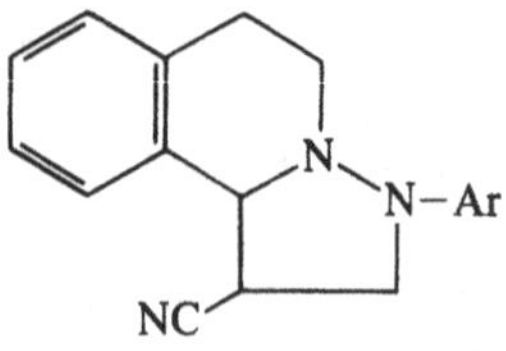 (**61**)

end of the azomethine imine system will lower the LUMO energy and favour dipole LUMO control. Thus the dipole (**62**) with a conjugated dipolarophile such as styrene gives (**63**) (5.60).

LU $CH_2-N(R)-N-COR$ (62) → (63) (5.60)

HO Ph

Sydnones can be considered as azomethine imines with an acyl group on carbon and again dipole LUMO control operates. The larger LUMO $(c\beta)^2/15$ term is on carbon although this will be reduced by the electron-withdrwaing aryl group making it more nearly equal to that on nitrogen and causing reduced regioselectivity. The fact that sydnones react with all classes of dipolarophile to give predominantly but not always exclusively the regioisomers shown (5.61) is therefore explained.

C, X, Z — $-CO_2$ — H tautomerism — (5.61)

Nitrones. In the case of nitrones, dipole LUMO control is to be expected in all additions except those with electron-deficient dipolarophiles. The nitrone coefficients are quite different in the LUMO (larger on carbon) and almost equal in the HOMO and so the preferred regioisomers are as expected (5.62).

$RCH=\overset{+}{N}(R')-\bar{O}$ → C, X, Z (5.62)

Even in the addition to electron-deficient dipolarophiles the less important dipole LUMO–dipolarophile HOMO interaction controls regioselectivity because the HOMO coefficients are equally balanced. With very electron-deficient dipolarophiles, dipole HOMO control does predominate and the reaction becomes non-regioselective. In the C-phenyl substituted nitrones the conjugating phenyl tends to decrease the HOMO coefficient on carbon making that on oxygen the larger one. Thus with very electron-deficient dipolarophiles, dipole HOMO control now leads to complete reversal of regiochemistry and 4-substituted isoxazolidines are formed.

Nitrile oxides. Nitrile oxide additions are clearly dipole LUMO controlled for electron-rich and conjugated dipolarophiles so that only 5-substituted isoxazolines are formed (5.63). For electron-deficient dipolarophiles, dipole HOMO and LUMO control is finely balanced and since the former tends to produce the 4-substituted isoxazoline the regioselectivity falls.

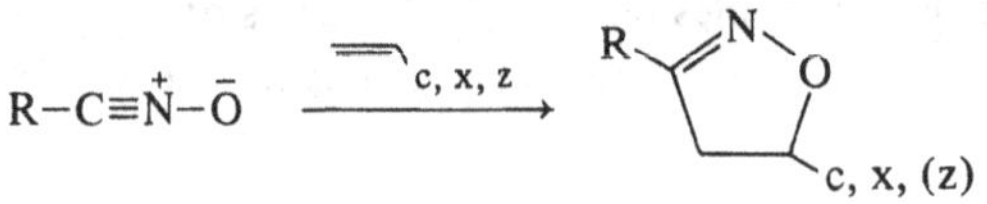

$$R{-}C{\equiv}\overset{+}{N}{-}\overset{-}{O} \xrightarrow{\text{c, x, z}} \text{isoxazoline (R at C-3; c, x, (z) at C-5)} \qquad (5.63)$$

Nitrile imines. Diphenylnitrilimine shows high regioselectivity with electron-rich dipolarophiles where dipole LUMO control predominates, leading to 5-substituted Δ^2-pyrazolines (5.64). With conjugated dipolarophiles both HOMO and LUMO control is important but the greater disparity in the dipole LUMO coefficients tends to favour formation of the 5-substituted pyrazoline. This applies also with weakly electron-deficient dipolarophiles but reversal of regiochemistry should and does occur with strongly electron-deficient dipolarophiles.

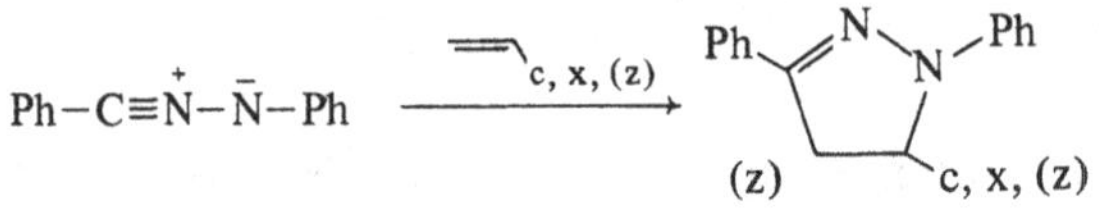

$$Ph{-}C{\equiv}\overset{+}{N}{-}\overset{-}{N}{-}Ph \xrightarrow{\text{c, x, (z)}} \text{pyrazoline (Ph at C-3, Ph on N-1; (z) at C-4; c, x, (z) at C-5)} \qquad (5.64)$$

Nitrile ylides. Nitrile ylide cycloadditions are generally dipole HOMO controlled and in fact, addition to electron-rich dipolarophiles has not yet been observed. Using the calculated coefficients in table 5.5, the predicted regioisomers in nitrile ylide additions are the 3-substituted Δ^1-pyrrolines **(64)** but only the 4-substituted isomers **(65)** are actually observed. More sophisticated calculations, however, give a different picture with the relative magnitudes of

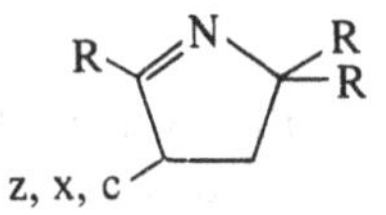

R, R, R, N, z, x, c

(64)

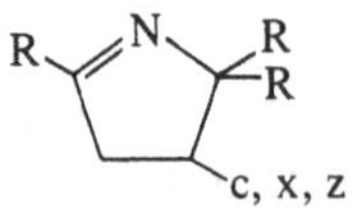

R, R, R, N, c, x, z

(65)

114–116°
H, C, N, C, H, H

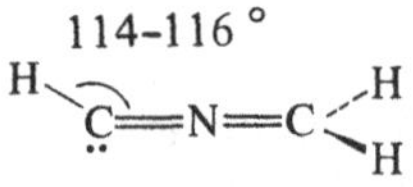

(66)

the $(c\beta)^2/15$ values reversed, thus removing the anomaly.[76] Electrophilic additions to nitrile ylides also confirm that the sp-hydridised carbon is the more nucleophilic (higher $(c\beta)^2/15$ term in the HOMO).[77] Interestingly these calculations also indicate that nitrile ylides and nitrile imines are not strictly linear as in our simple model but that the substituent on the sp-hybridised carbon is bent out of line as shown **(66)**.[76]

Additions to heterodipolarophiles

The regioselectivity of addition to heterodipolarophiles can also be adequately accounted for by frontier orbital theory. The larger HOMO coefficient in the dipolarophile will be on the more electronegative element and the larger LUMO coefficient on the less electronegative, and their energies will be similar to those of electron-deficient alkenes.

Secondary orbital interactions

The *endo*-effect and its rationalisation in terms of secondary interactions involving diene orbital lobes on C-2 and C-3 has been discussed for the Diels-Alder reaction on p. 119. In principle a similar situation arises in some dipolar cycloadditions. For a 'bent' dipole the central lobe of the dipole LUMO can overlap with the orbitals of a conjugating substituent in the dipolarophile in a bonding fashion, giving rise to a favourable secondary interaction for the *endo* transition state **(67)**.[78]

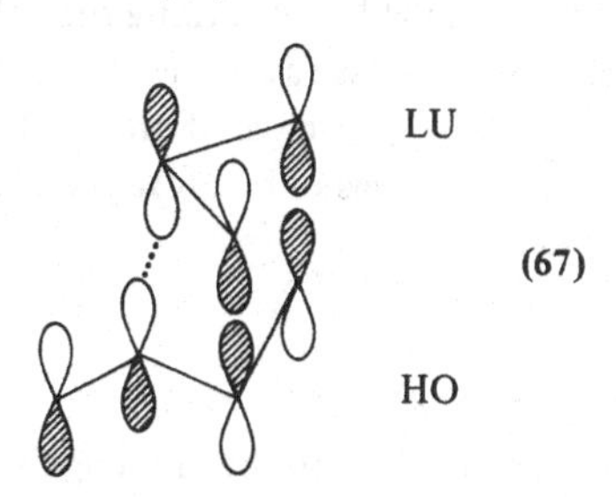

(67)

Clearly such interactions will only be important in dipole LUMO controlled additions since in the dipole HOMO there is a node or near node at the central atom and no *endo*-effect is to be expected.[79]

(5.65)

With some linear dipoles similar orientational selectivity has been observed, for example in the nitrile ylide addition (5.65) the aryl and ester groups appear preferentially *syn* in the adduct.[80] It has been suggested that, in addition to the normal interaction between the dipolarophile π and dipole π allyl orbitals, there is a secondary interaction between the dipolarophile HOMO and the orthogonal π^* orbital of the dipole. This secondary interaction leads to a twisted transition state (**68**) rather than the expected 'parallel planes' approach. In such a transition state, steric interactions will be minimised if the nitrile ylide hydrogen approaches hydrogen rather than the larger substituent on the dipolarophile.[74]

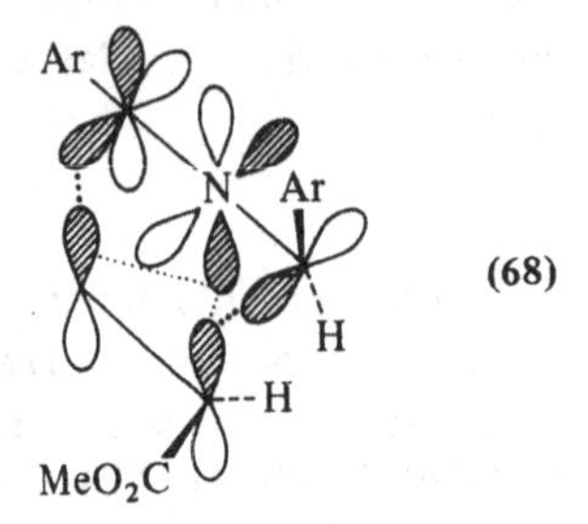

(68)

FMO theory is clearly far from perfect in rationalising reactivity and regioselectivity in dipolar cycloadditions. This is not surprising in view of the rather sweeping approximations and the limited data available concerning orbital energy levels and coefficients and our inability to assess easily the steric and coulombic effects also involved. Nevertheless the approach is of great value since it allows us to discuss in a coherent way what was only a few years ago a puzzling array of apparently unconnected facts.

5.4. Retro-1,3-dipolar cycloadditions

Retro-1,3-dipolar cycloaddition (fig. 5.17) has not hitherto been recognised as a general reaction class. It is far less well known than the retro-Diels–Alder reaction. Some systems that could in principle undergo this reaction are aromatic and therefore stable; in others, at least one of the fragments (the 1,3-dipole) is not very stable compared with, for example, the fragments from a retro-Diels–Alder reaction. Even so one might anticipate a future growth in the scope of the retro-reaction comparable with that which has taken place for the forward reaction over recent years.

a–b–c–e–d (ring) ⟶ $\overset{\oplus}{a}$–b–$\overset{\ominus}{c}$ + d=e

Fig. 5.17

The prototypical retro-1,3-anionic cycloaddition (1,3-anionic cycloelimination) has been observed in the rapid loss of nitrogen from the pyrazolin-4-yl anion **(70)** to give the allyl anion **(71)** (5.66). This loss of nitrogen from the anion occurs more than 10^{12} times faster than thermal elimination of nitrogen from the neutral pyrazoline **(69)**.[81]

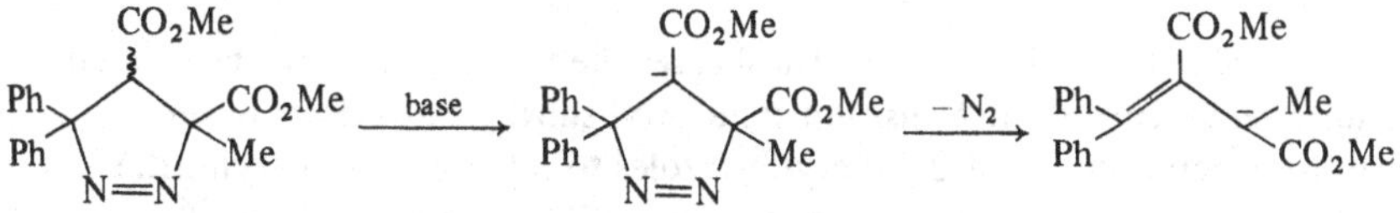

(69) —base→ **(70)** —$-N_2$→ **(71)**

(5.66)

Other examples[82, 83, 84] of this type of process include the base-induced fragmentation of dioxolanes[82] (72, X = O) and dithiolanes[83] (72, X = S) (5.67). The former process has been used as the basis of an olefin synthesis.[85]

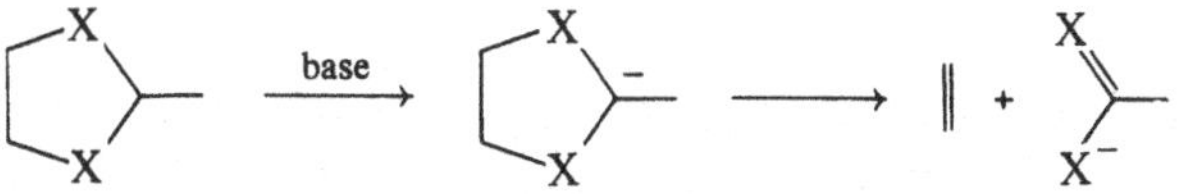

(72) —base→ (5.67)

Several examples of neutral retro-1,3-dipolar cycloadditions also exist. Thus some 1,3-dipolar systems have been generated by thermal decomposition of heterocyclic systems (equations 5.68 and 5.69). The initial adducts of 1,3-dipoles to sydnones spontaneously lose carbon dioxide in what is formally a retro-1,3-dipolar cycloaddition.[86] The new cyclic 1,3-dipole isomerises or undergoes further reaction to give a stable product (5.61) (see p. 153).

Little mechanistic study has been carried out for retro-1,3-dipolar cycloaddition. However, the concerted process is allowed and would seem likely to operate except in cases where intermediates would be highly stabilised. As in

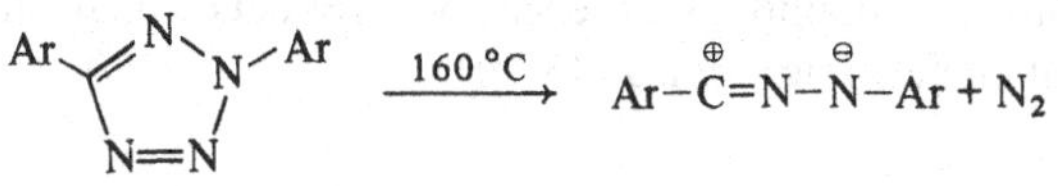

—160 °C→ $Ar-\overset{\oplus}{C}=N-\overset{\ominus}{N}-Ar + N_2$ (5.68)

$\longrightarrow$ $\longleftrightarrow$ $+ CO_2$ (5.69)

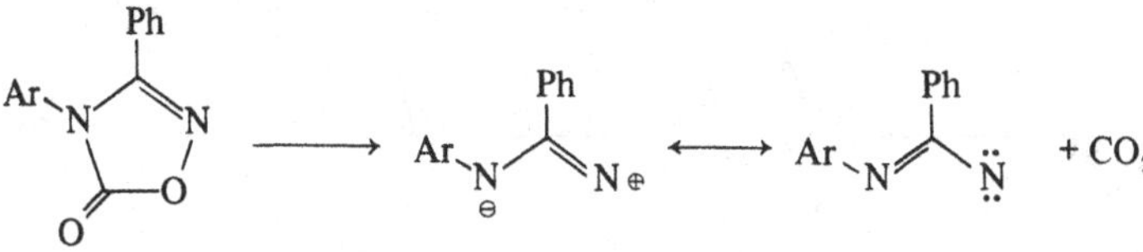

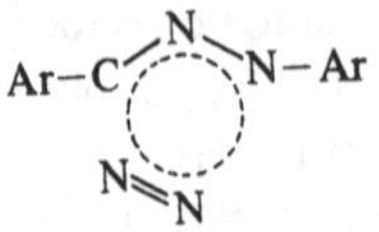

Ar–C N N–Ar N N

(73)

the forward reaction, in unsymmetrical cases the two σ bonds may be broken to unequal extents in the transition state. Aryl substituent effects in the thermal decomposition of 2,5-diaryltetrazoles to nitrogen and diarylnitrile imines (5.68) confirm that the transition state **(73)** is unsymmetrical.[87]

5.5. Cheletropic reactions involving six π electrons[1d,9,16]

Six-electron cycloadditions of the type in fig. 5.18, where both new σ bonds are formed to the same atom of one component, are quite limited in scope. The corresponding retro-cycloadditions (extrusion of X) occur more widely.[88] This is because X is normally a small inorganic molecule of high thermodynamic stability.

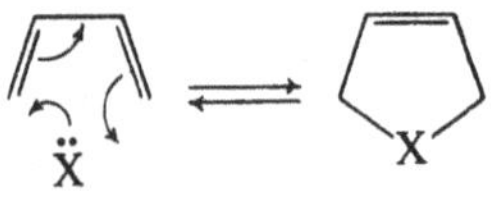

X ⇌ X

Fig. 5.18

The forward reaction is only well known for sulphur dioxide but even in this case the reverse process takes place readily. This has been put to use in the separation and purification of dienes. The addition of sulphur monoxide has also been reported.[89] Selenium dioxide adds to dienes, but by a Diels–Alder reaction in which the Se=O bond acts as dienophile.[90] *N*-Sulphinylamines (R–N=S=O)[91] and sulphines (R_2C=S=O)[92] also add to dienes by the Diels–Alder rather than by the cheletropic mode.

Trivalent phosphorus compounds and dienes give products which can be explained by an initial cheletropic reaction (5.70).

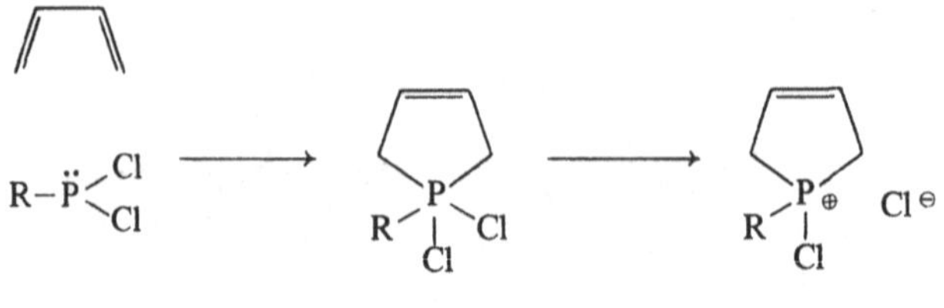

R–P̈Cl Cl ⟶ P R Cl Cl ⟶ P⊕ R Cl Cl⊖ (5.70)

The elimination (extrusion) reaction is well known for X = SO_2,N_2 and CO but is surprisingly inefficient for N_2O (compare the very ready loss of nitrous oxide from *N*-nitrosoaziridines, chapter 6).

The theory of cheletropic reactions has been discussed in chapter 4. These six π electron additions and eliminations are apparently concerted and thus come into the category of linear cheletropic process ($_{\pi}4_s + {}_{\omega}2_s$ cycloaddition and $_{\pi}2_s + {}_{\sigma}2_s + {}_{\sigma}2_s$ cycloelimination). They are therefore mechanistically closely related to the Diels–Alder and retro-Diels–Alder reactions, the π orbital of the dienophile being replaced by a lone pair orbital on X.

The thermal addition and extrusion of sulphur dioxide are both stereospecific with disrotatory motions of the terminal methylene groups of the diene (5.71)[93] Photolysis of these sulphones, though not completely stereospecific, is mainly conrotatory as predicted for the photochemical reaction.[94]

Me, Me, S, O_2, Heat, $h\nu$, SO_2 + Me, Me

(5.71)

It is interesting to note that the thermal loss of sulphur dioxide from cyclic β,γ-unsaturated sulphinic esters (**74**) by a retro-Diels–Alder reaction occurs even more readily than the cheletropic extrusion from the isomeric sulphones, and indeed such sulphinic esters rearrange to sulphones by retro-Diels–Alder reaction followed by cheletropic addition (5.72).[95] A case has been found where, for a very reactive diene, the kinetically controlled adduct with sulphur dioxide is the Diels–Alder adduct, the thermodynamically controlled adduct being the sulphone.[96]

S=O, O, 0 °C, SO_2 +, O_2S

(5.72)

(**74**)

Extrusion of nitrogen from diazenes (equations 5.73 and 5.74) is spontaneous and also stereospecific and disrotatory.[97] Diazenes can be considered as *N*-nitrenes, and as such can be generated by oxidation of the corresponding *N*-amino compounds or by the action of base on the tosyl derivatives. In the example shown the *N*-nitrenes were generated directly from the NH compounds

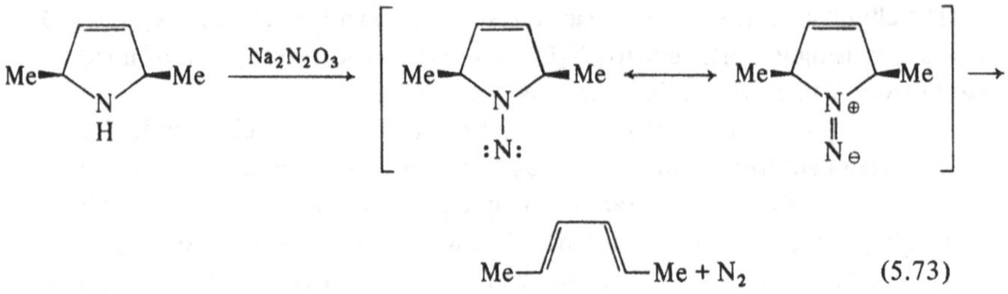

Me Me N H $\xrightarrow{Na_2N_2O_3}$ [Me Me N :N: $\longleftrightarrow$ Me Me $N^{\oplus}$ $N^{\ominus}$] $\longrightarrow$

Me Me + N_2 (5.73)

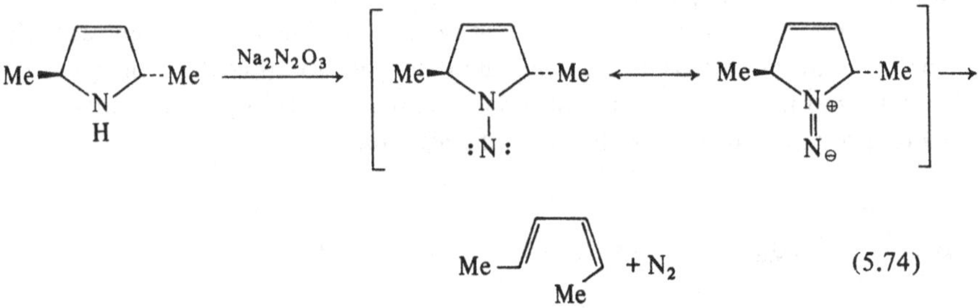

Me ---Me N H $\xrightarrow{Na_2N_2O_3}$ [Me ---Me N :N: $\longleftrightarrow$ Me ---Me $N^{\oplus}$ $N^{\ominus}$] $\longrightarrow$

Me Me + N_2 (5.74)

by treatment with Angeli's salt ($Na_2N_2O_3$). Difluoramine (HNF_2) effects the same conversion.

Loss of carbon monoxide from cyclopent-3-enones proceeds much more readily than loss of carbon monoxide from cyclopropanones and cyclopenta-

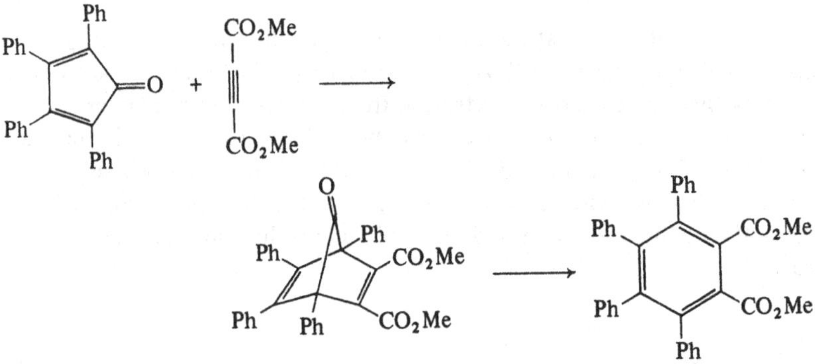

Ph Ph Ph Ph =O + CO_2Me CO_2Me $\longrightarrow$

O Ph Ph CO_2Me Ph Ph CO_2Me $\longrightarrow$ Ph Ph CO_2Me Ph CO_2Me Ph

(75) (5.75)

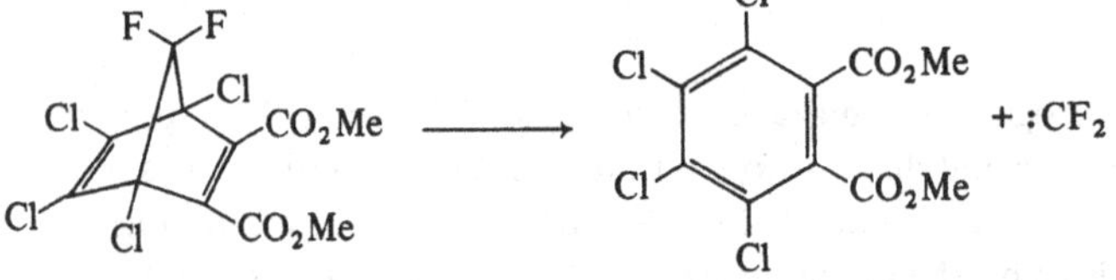

F F Cl Cl CO_2Me Cl Cl CO_2Me $\longrightarrow$ Cl Cl CO_2Me Cl CO_2Me Cl + $:CF_2$ (5.76)

(76)

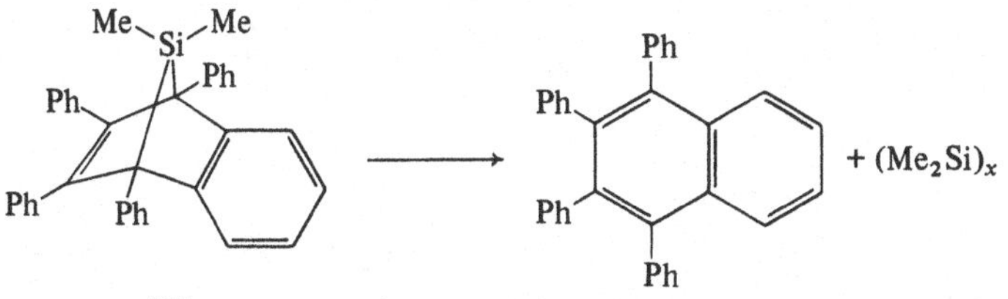

$$+ (Me_2Si)_x \qquad (5.77)$$

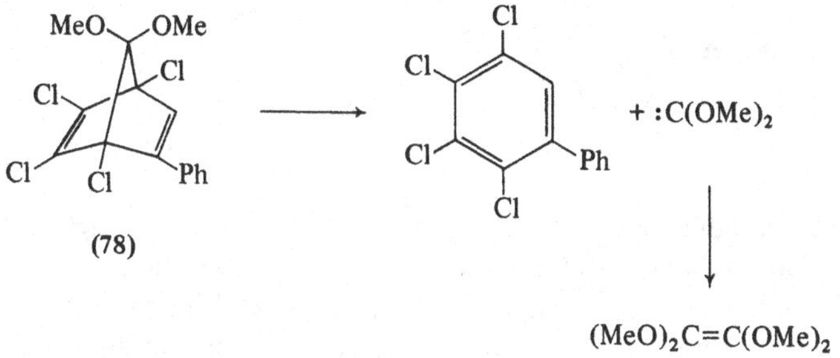

(77)

$$+ :C(OMe)_2$$

$$(MeO)_2C{=}C(OMe)_2 \qquad (5.78)$$

(78)

nones. Although no stereochemical studies have been carried out, the generality and ease of the reaction suggests that it is concerted.[98] The loss of carbon monoxide occurs particularly readily when an aromatic product is formed. Thus addition of cyclopentadienones to acetylenes (5.75) and arynes first gives norbornadienone derivatives, for example **(75)**, which lose carbon monoxide so rapidly that no primary adducts with the carbon monoxide bridge retained have yet been isolated.[99]

Carbon monoxide is extruded so readily since it is thermodynamically stable. The bridgehead groups in, for example, compounds **(76)** to **(78)** are also lost on strong heating (equations 5.76–5.78).[9,100]

Decomposition of the ketal **(78)** provided the first synthesis of the interesting electron-rich olefin, tetramethoxyethylene.[101]

Norbornadienone ketals of type **(79)** are fragmented as shown in (5.79). This may involve concerted or stepwise extrusion of the carbene **(80)** which would be fragmented by a retro-1,3-dipolar cycloaddition, or the whole process may be concerted.[16]

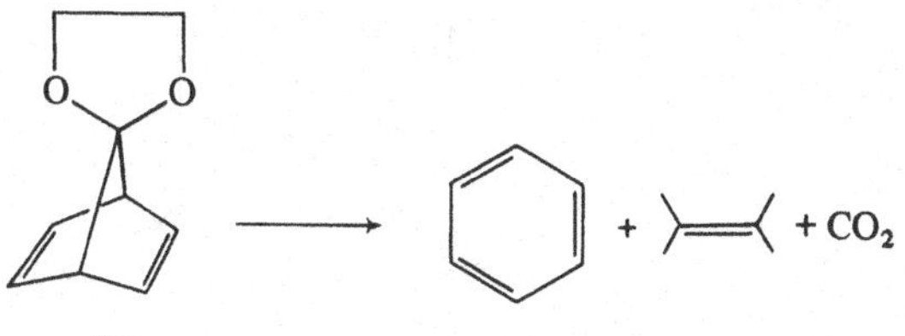

$$+ CO_2 \qquad (5.79)$$

(79)

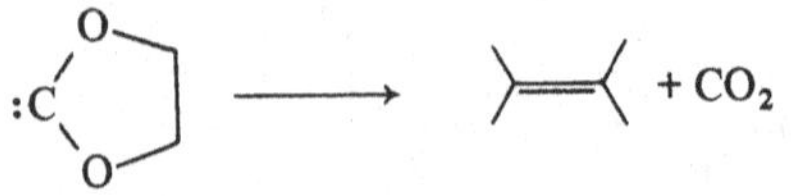

(80)

(5.80)

5.6. 2 + 2 + 2 Cycloadditions and eliminations (fig. 5.19)

Fig. 5.19

There are many examples of stepwise 2 + 2 + 2 additions proceeding through 1,4-dipolar intermediates (chapter 6). Concerted 2 + 2 + 2 cycloadditions are thermally allowed as ${}_{\pi}2_s + {}_{\pi}2_s + {}_{\pi}2_s$ or ${}_{\pi}2_s + {}_{\pi}2_a + {}_{\pi}2_a$ processes. The termolecular collisions necessary for these cycloadditions are very unlikely and the only examples known are those where at least two of the component π bonds are held together in one reactant.

Examples of such ${}_{\pi}2 + {}_{\pi}2 + {}_{\pi}2$ additions are the cycloadditions of tetracyanoethylene and other dienophiles to norbornadiene (5.81) and of tetracyanoethylene to 1,3,5-7-tetramethylenecyclo-octane (5.82).[102]

In both cases the geometry of the molecules favours the process. The first example is often called a homo-Diels–Alder reaction since a cyclopropane ring is formed rather than a π bond.[102]

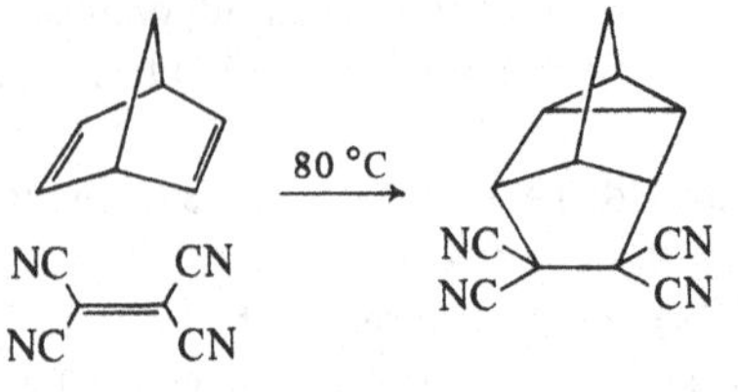

(5.81)

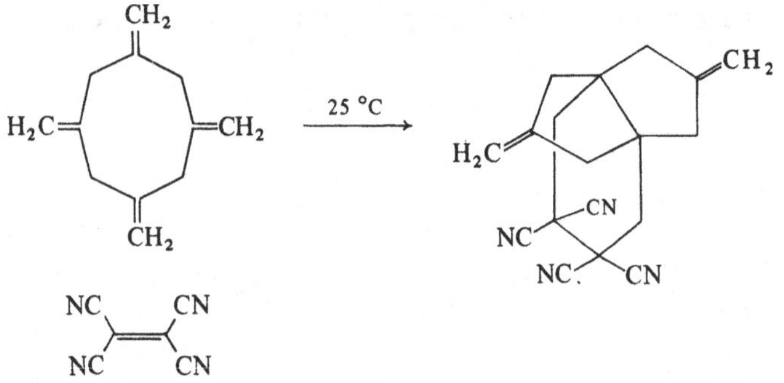

(5.82)

A possible 2 + 2 + 2 addition involving a cheletropic component is the addition of dihalocarbenes to norbornadiene (5.83).[103]

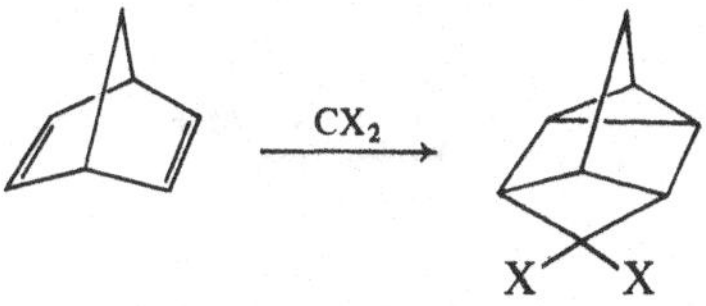

CX_2 X=F, Cl (5.83)

Bis-homo-2 + 2 + 2 cycloadditions are also known (5.84 and 5.85).[104,105] The second example is of interest as the first stage in the synthesis of the strained diazetidine discussed on p. 210. This is obtained from the adduct (**81**) by hydrolysis and decarboxylation followed by oxidation.[10]

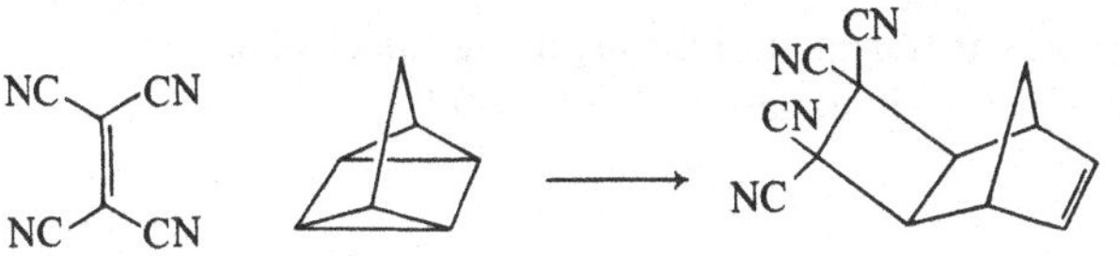

(5.84)

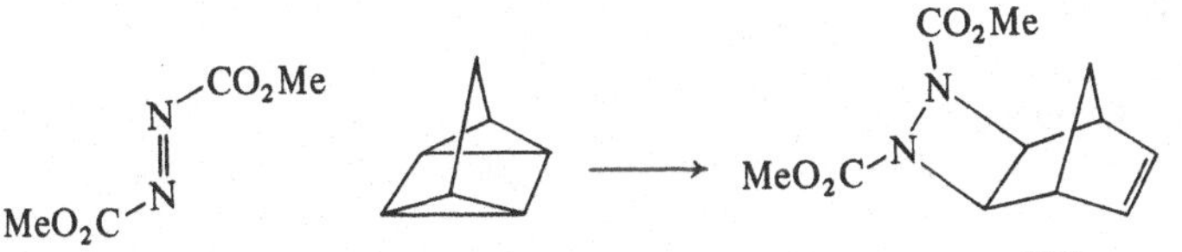

(5.85)

(**81**)

Concerted retro-2 + 2 + 2 cycloadditions ($_{\sigma}2_s + _{\sigma}2_s + _{\sigma}2_s$ or $_{\sigma}2_s + _{\sigma}2_a + _{\sigma}2_a$ fragmentations) do not suffer from the highly unfavourable entropy requirements of the forward reactions. The concerted retro-homo-Diels–Alder reaction (p. 131) can be considered as a retro-2 + 2 + 2 cycloaddition. Perhaps surprisingly, few simple retro-2 + 2 + 2 additions leading to three fragments have been reported. One example is the stereospecific fragmentation of the dihydro-oxadiazinone (**82**, equation 5.86).[106] Another example is the thermal fragmentation of the azo-lactone (**83**, equation 5.87) to diethylketen, cholestanone and nitrogen.[107] It is interesting to note that the photo-induced fragmentation of (**83**) takes an alternative course and is probably stepwise.

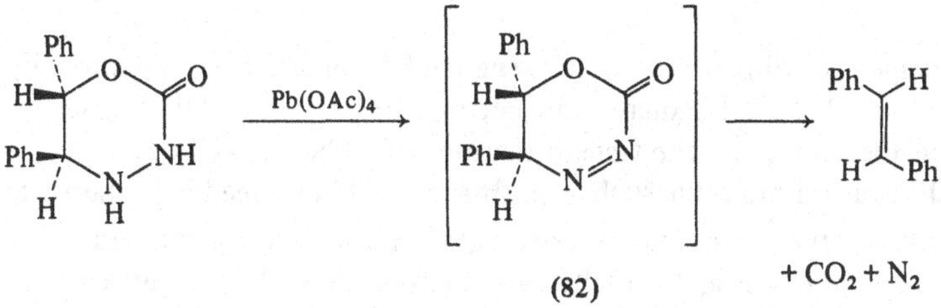

$Pb(OAc)_4$

(**82**) $+ CO_2 + N_2$ (5.86)

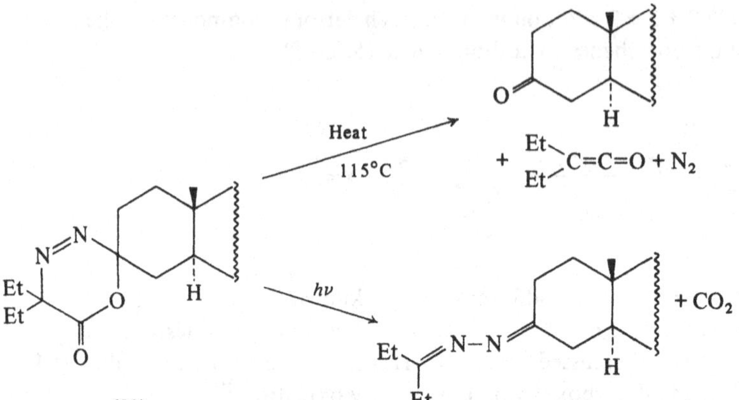

Heat 115°C

$+ Et_2C{=}C{=}O + N_2$

(5.87)

(83)

$h\nu$

$+ CO_2$

There are several examples of fragmentations involving cheletropic components, for example those shown in equations 5.88-5.90.[108]

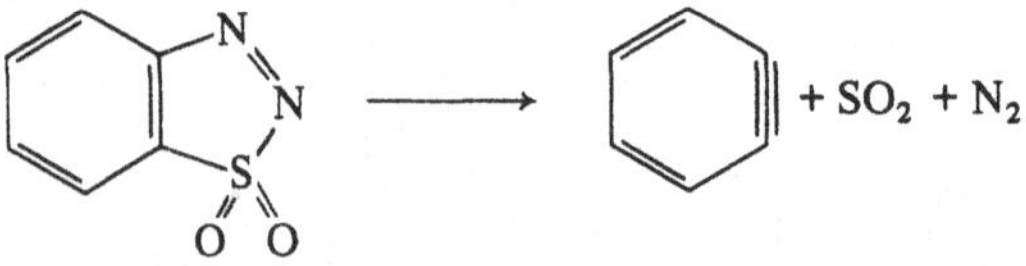

$\longrightarrow$ $+ SO_2 + N_2$ (5.88)

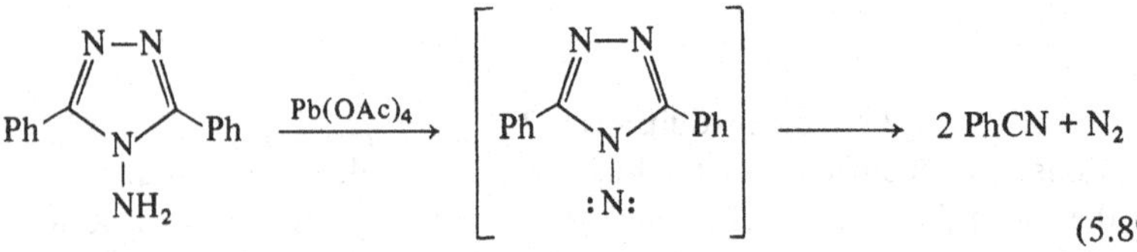

$\xrightarrow{Pb(OAc)_4}$ $\longrightarrow$ $2\ PhCN + N_2$ (5.89)

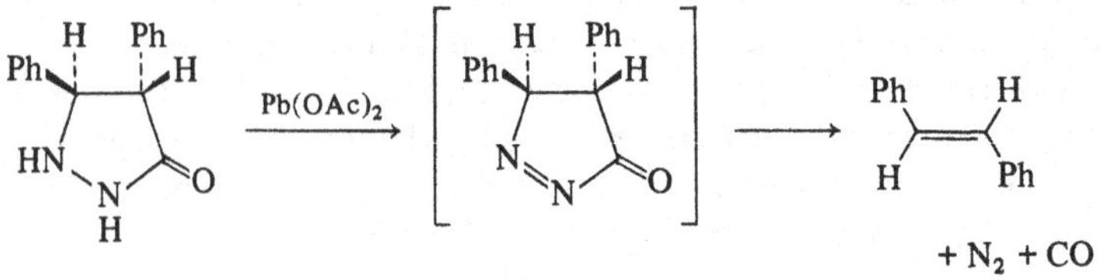

$\xrightarrow{Pb(OAc)_2}$ $\longrightarrow$ $+ N_2 + CO$ (5.90)

The rearrangement of prismane to benzene could geometrically take place by a concerted ${}_{\sigma}2_s + {}_{\sigma}2_s + {}_{\sigma}2_s$ fragmentation (retro-${}_{\pi}2_s + {}_{\pi}2_a + {}_{\pi}2_a$ addition). One might expect, therefore, that the transformation would be allowed and that a tremendously strained molecule such as prismane would very readily isomerise to benzene. However, the π components resulting from the cleavage are held together in a ring by σ bonds and actually form benzene orbitals. Two of the

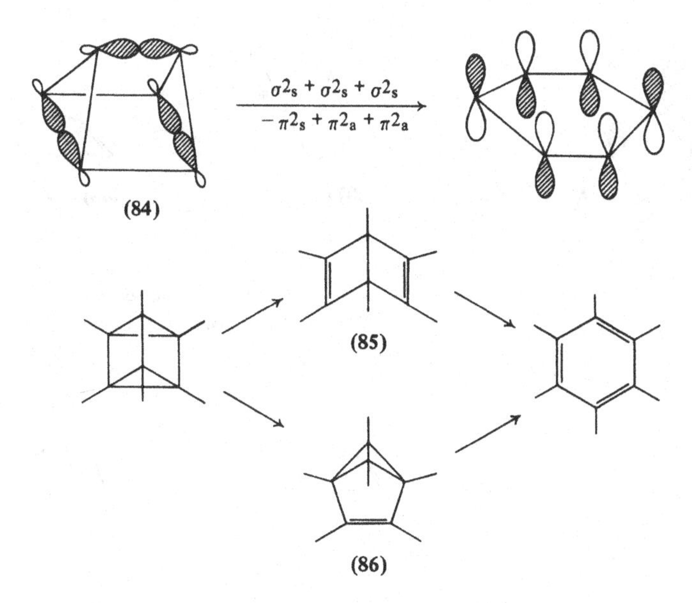

Fig. 5.20

prismane bonding orbitals correlate with bonding benzene orbitals but the fully symmetric orbital **(84)** actually correlates with a benzene antibonding orbital.[16] This explains why, although the process is exothermic by 381 kJ mol^{-1}, hexamethylprismane is only converted into hexamethylbenzene above 60 °C, and then not directly but via the valence isomers **(85)** and **(86)** (fig. 5.20).[109]

5.7. Other six-electron cycloadditions

This section covers some obscure but theoretically interesting reactions.

The $_{\pi}4_s + _{\pi}2_s$ additions of allyl anions have already been discussed. The symmetry of the HOMO of an allyl cation is such that it should undergo concerted $_{\pi}4_s + _{\pi}2_s$ addition to a diene.[110] This has been observed in the addition of the methylallyl cation (from Ag⊕-catalysed ionisation of 2-methylallyl iodide) to cyclopentadiene, cyclohexadiene and furan (fig. 5.21). Secondary orbital interactions indicate that *exo*-addition should be most favourable. This appears to be the case but the picture is complicated since the initially formed *exo*-cation **(87)** can 'flip' to the *endo*-form **(88)**. Under conditions where the adduct cation has a short lifetime and reacts rapidly with its counterion, only products derived from the *exo*-ion are formed. Other products result from proton loss from the two cations.[111]

Cyclopropanones undergo cycloadditions, apparently through the ring-opened form with which they are in equilibrium. This ring-opened form has

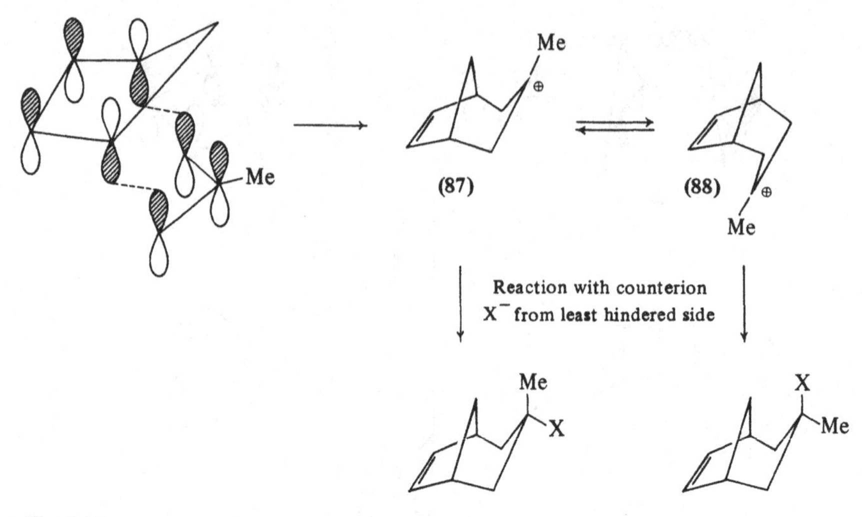

Fig. 5.21

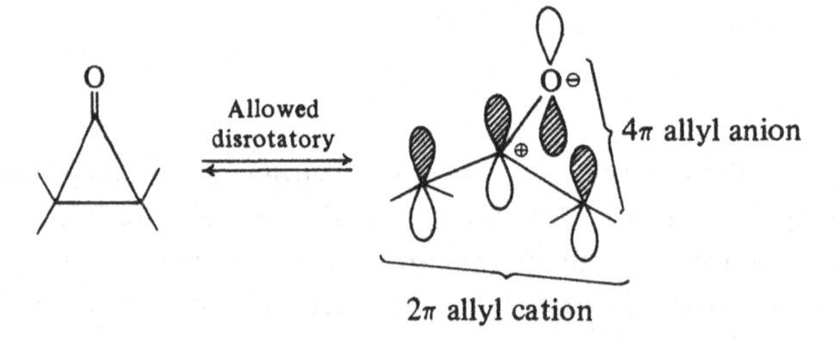

Fig. 5.22

Fig. 5.23

the orbital characteristics of both a 2π allyl cation and a 4π allyl anion (fig. 5.22). $4\pi + 2\pi$ Additions both of the allyl cation system to dienes and of the allyl anion system of the π bonds of trichloroacetaldehyde and sulphur dioxide have been observed (fig. 5.23).[112]

There is no clear evidence that these additions are concerted; in particular a stepwise mechanism seems likely for the additions of trichloroacetaldehyde and sulphur dioxide.

Yet a further way in which ${}_{\pi}4_s + {}_{\pi}2_s$ addition could arise is in the addition of 4π cations to a monoene. The HOMO of the cation **(89)** is as shown in **(90)**. Woodward has pointed out a reaction which can be fitted into this type of scheme (5.91).[16]

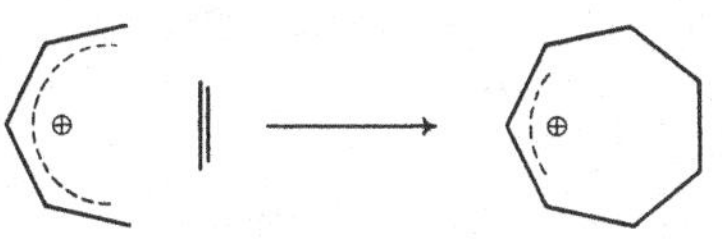

(89) 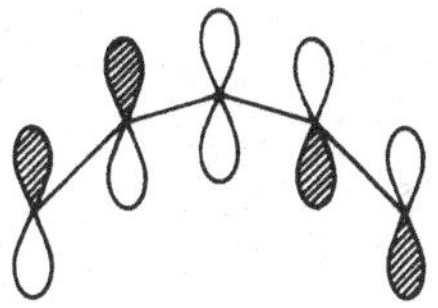 **(90)**

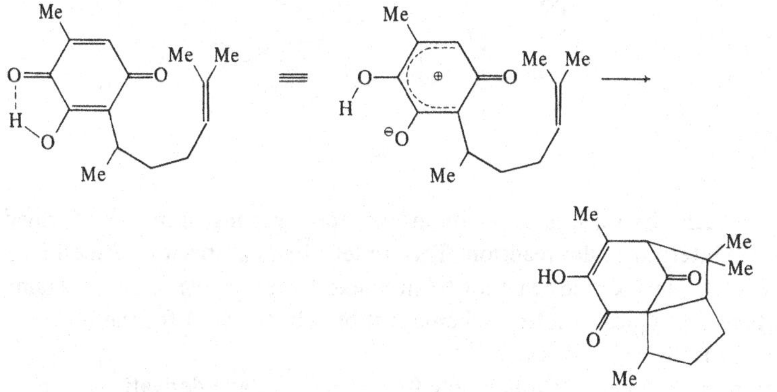

(5.91)

Photochemical cycloadditions[113]

Intermolecular cycloadditions involving six π electrons are rarely brought about photochemically, although some synthetically useful examples do exist as in the addition of *o*-quinones to alkenes (5.92). There is no firm evidence for the

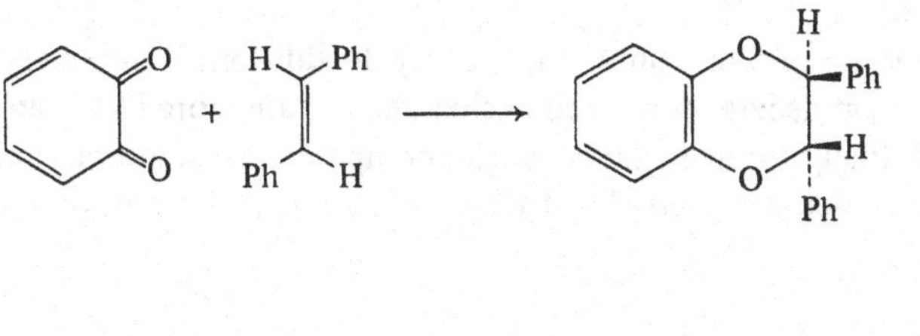

(5.92)

concerted $_{\pi}4_s + _{\pi}2_a$ or $_{\pi}4_a + _{\pi}2_s$ modes of addition which are predicted by the selection rules, assuming the simple photochemical excitation of an electron to the next highest orbital.

When mixtures of dienes or dienes and moneones are irradiated, some six-membered ring products are formed. However, these are usually minor products and are accompanied by considerable amounts of 2 + 2 adducts. The most reasonable explanation for this is that triplet excited reactants are involved. These

$$\underset{\text{Sensitiser}}{S} \xrightarrow{h\nu} {}^1S^* \longrightarrow {}^3S^* \xrightarrow{\text{Butadiene}} \underset{\textit{cisoid} \text{ and } \textit{transoid}}{S + {}^3(\text{butadiene})^*}$$

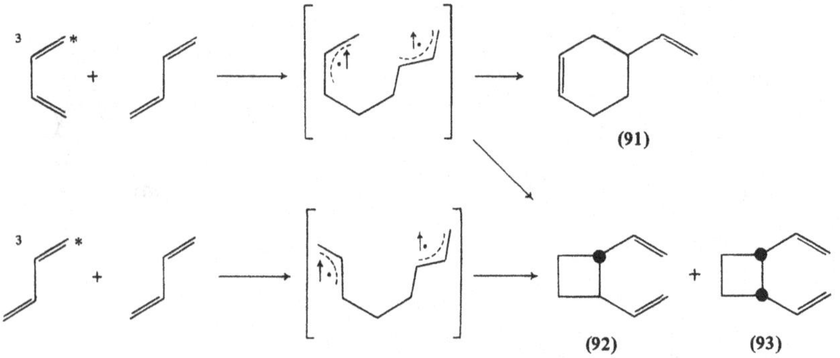

(91)

(92) (93)

Fig. 5.24

are formed directly (by sensitisers) or by intersystem crossing of initially formed singlets before intermolecular reaction. The triplet diradical species, formed by addition of the triplet excited reactant to an unexcited molecule, collapse mainly to cyclobutanes. A typical reaction scheme is shown in fig. 5.24 for the photosensitised dimerisation of butadiene.[114]

The ratio of cyclohexene (for example **91**) to cyclobutane derivatives (for example **92** and **93**) in the photosensitised dimerisation of butadiene and similar open-chain dienes, depends on the energy of the triplet sensitiser. *cisoid* Diene molecules have a lower excitation energy than *transoid* and so can be selectively excited by lower energy sensitisers. The diradicals formed by addition of these *cisoid* triplet diene molecules have a greater chance of closing to cyclohexene than those from *transoid* dienes, which lead exclusively to cyclobutanes (see chapter 6).

Although intermolecular photochemical cycloadditions are unlikely to be concerted, intramolecular concerted cycloadditions are more likely since in these cases, initially formed excited singlet components can interact prior to relaxation to triplet states. *cis*-Hexatrienes are converted into bicyclo[3,1,0]

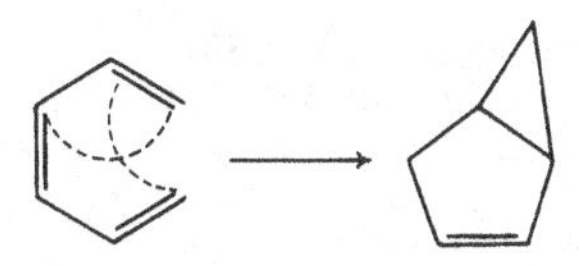

Fig. 5.25

hexenes on irradiation (fig. 5.25). This could involve concerted $_{\pi}4_s + _{\pi}2_a$ or $_{\pi}4_a + _{\pi}2_a$ processes but so far the reaction has not been carried out with sufficiently labelled reactants to clarify this.[115]

References

1. The scope and synthetic potential of the reaction have been the subject of many excellent reviews: (*a*) Huisgen, R., Grashey, R. and Sauer, J., in *The chemistry of alkenes*, ed. S. Patai, p. 739, Interscience, London, 1964; (*b*) Sauer, J., *Angew, Chem., Int. Edn Engl.*, **5**, 211 (1966); (*c*) Wassermann, A., *Diels–Alder reactions*. Elsevier, Amsterdam, 1965; (*d*) Hamer, J., *1,4-Cycloaddition reactions*, Academic Press, New York, 1967; and references in March, J., *Advanced organic chemistry*, 2nd Edn., McGraw-Hill, Tokyo 1977; (*e*) Schmidt, R. R., *Angew. Chem., Int. Edn Engl.*, **12**, 212 (1973). For a review of the stereochemistry, see Martin, J. G. and Hill, R. K., *Chem. Rev.*, **61**, 537 (1961). For reviews of the Diels–Alder reaction of heterodienes and dienophiles see Needleman, S. B. and Chang Kuo, M. C., *Chem. Rev.*, **62**, 405 (1962) and Desimoni, G. and Tacconi, G., *Chem. Rev.*, **75**, 651 (1975).
2. Oppolzer, W., *Angew. Chem., Int. Edn Engl.*, **16**, 10 (1977).
3. Klemm, L. H. and Gopinath, K. W., *Tetrahedron Lett.*, 1963, 1243; Klemm, L. H., Olson, D. R. and White, D. V., *J. org. Chem.*, **36**, 3740 (1971); Klemm, L. H., McGuire, T. M. and Gopinath, K. W., *J. org. Chem.*, **41**, 2571 (1976).
4. Stark, B. P. and Duke, A. J., *Extrusion reactions*, Chapter 11, Pergamon Press, Oxford, 1967.
5. Hoffmann, R. W., *Dehydrobenzene and cycloalkynes*, Academic Press, New York, 1967.
6. Gollnick, K., *Adv. Photochem.*, **6**, 1 (1968); Schaap, A. P. (ed.), *Singlet molecular oxygen*, Dowden, Hutchinson and Ross, Stroudsburg, 1976.
7. For a detailed analysis see Kearns, D. R., *J. Am. chem. Soc.*, **91**, 6554 (1969).
8. Reviews: (*a*) Sauer, J., *Angew. Chem., Int. Edn Engl.*, **6**, 16 (1967); (*b*) Seltzer, S., *Adv. alicyclic Chem.*, **2**, 1 (1968).
9. Review: Kwart, H. and King, K., *Chem. Rev.*, **68**, 415 (1968); Ripoll, J. L., Rouessac, A. and Rouessac, F., *Tetrahedron*, **34**, 19 (1978).
10. Lambert, J. B. and Roberts, J. D., *Tetrahedron Lett.*, 1965, 1457.
11. Van Sickle, D. E., and Rodin, J. O., *J. Am. chem. Soc.*, **86**, 3091 (1964).
12. Dolbier, W. R. and Dai, S.-H., *J. Am. chem. Soc.*, **90**, 5028 (1968); Brown, P. and Cookson, R. C., *Tetrahedron*, **21**, 1977, 1993 (1965).
13. Review: Kwart, H. and King, K., *Chem. Rev.*, **68**, 415 (1968); Seltzer, S., *J. Am. chem. Soc.*, **87**, 1534 (1965).
14. Grieger, R. A. and Eckert, C. A., *J. Am. chem. Soc.*, **92**, 7149 (1970); McCabe, J. R. and Eckert, C. A., *Acc. chem. Res.*, **7**, 251 (1974).
15. Dauben, W. G. and Krabbenhoft, H. O., *J. Am. chem. Soc.*, **98**, 1992 (1976).
16. Woodward, R. B. and Hoffmann, R., *Angew. Chem., Int. Edn Engl.*, **8**, 781 (1969).
17. Herndon, W. C. and Hall, L. H., *Tetrahedron Lett.*, 1967, 3095.
18. Williamson, K. L., Hsu, Y. F. L., Lacko, R. and Youn, C. H., *J. Am. chem. Soc.*, **91**, 6129 (1969); Houk, K. N., *Tetrahedron Lett.*, 1970, 2621.
19. For leading references see 1*c*, 8*a* and 18. See Salem, L., *J. Am. chem. Soc.*, **90**, 553 (1968) for an alternative view of which secondary interactions are important in favouring *endo*-addition.

20. Fleming, I., Gianni, F. L. and Mah, T., *Tetrahedron Lett.*, 1976, 881.
21. Houk, K. N. and Strozier, R. W., *J. Am. chem. Soc.*, **95**, 4094 (1973).
22. See, however, Thompson, H. W. and Melillo, D. G., *J. Am. chem. Soc.*, **92**, 3218 (1970) for a catalysed Diels–Alder reaction which appears to be stepwise.
23. Kiselev, V. D. and Miller, J. G., *J. Am. chem. Soc.*, **97**, 4036 (1975).
24. Little, J. C., *J. Am. chem. Soc.*, **87**, 4036 (1975).
25. Woodward, R. B. and Katz, T. J., *Tetrahedron*, **5**, 70 (1959).
26. Criegee, R. and Askani, R., *Angew. Chem., Int. Edn Engl.*, **7**, 537 (1968).
27. (*a*) Review: Gompper, R., *Angew. Chem., Int. Edn Engl.*, **8**, 312 (1969); (*b*) certain polar 1,4-cycloadditions are believed to be stepwise; see reference 1*e* and Bradsher, C. K., *Adv. Heterocyclic Chem.*, **16**, 289 (1974).
28. Fleming, I. and Karger, M. H., *J. chem. Soc. (C)*, 1967, 226.
29. Hershberg, E. B. and Ruhoff, J. R., *Org. Synth.*, Coll. vol. II, 102.
30. Moffett, R. B., *Org. Synth.*, Coll. vol. IV, 238.
31. Alder, K., Flock, F. H. and Beumling, H., *Chem. Ber.*, **93**, 1896 (1960).
32. Corey, E. J. and Mock, W. L., *J. Am. chem. Soc.*, **84**, 685 (1962).
33. Vogel, E., Grimme, W. and Korte, S., *Tetrahedron Lett.*, 1965, 3625.
34. Avram, M., Dinulescu, I., Elian, M., Farcasiu, M., Marica, E., Mateescu, G. and Nenitzescu, C. D., *Chem. Ber.*, **97**, 372 (1964).
35. Berson, J. A. and Olin, S. S., *J. Am. chem. Soc.*, **91**, 777 (1969).
36. Goldstein, M. J. and Thayer, G. L., *J. Am. chem. Soc.*, **87**, 1925, 1933 (1965).
37. Rieber, N., Alberts, J., Lipsky, J. A. and Lemal, D. M., *J. Am. chem. Soc.*, **91**, 5668 (1969).
38. Martin, M. and Roth, W. R., *Chem. Ber.*, **102**, 811 (1969).
39. Allred, E. L. and Hinshaw, J. C., *Chem. Communs*, 1969, 1021.
40. Allred, E. L., Hinshaw, J. C. and Johnson, A. L., *J. Am. chem. Soc.*, **91**, 3382 (1969).
41. Olsen, H. and Snyder, J. P., *J. Am. chem. Soc.*, **99**, 1524 (1977); Oth, J. F. M., Olsen, H. and Snyder, J. P., *J. Am. chem. Soc.*, **99**, 8505 (1977).
42. Reviews: (*a*) Huisgen, R., *Angew. Chem., Int. Edn Engl.*, **2**, 565 (1963); (*b*) Huisgen, R., *Angew. Chem., Int. Edn Engl.*, **2**, 633 (1963); (*c*) Bianchi, G., De Micheli, C. and Gandolfi, R., *The chemistry of double-bonded functional groups*, ed. S. Patai, p. 369, Interscience, London, 1977; (*d*) Huisgen, R., *J. org. Chem.*, **41**, 403 (1976). The concept was first recognised by Smith, L. Z., *Chem. Rev.*, **23**, 193 (1938).
43. Eidenschink, R. and Kauffmann, T., *Angew. Chem., Int. Edn Engl.*, **11**, 292 (1972).
44. Marino, J. P. and Mesbergen, W. B., *J. Am. chem. Soc.*, **96**, 4050 (1974).
45. Kauffmann, T., Berg, H. and Köppelmann, E., *Angew. Chem., Int. Edn Engl.*, **9**, 380 (1970).
46. Kauffmann, T. and Köppelmann, E., *Angew. Chem., Int. Edn Engl.*, **11**, 290 (1972).
47. Buchner, E., *Berichte*, **21**, 2637 (1888); **23**, 701 (1890); Michael, A., *J. prakt. Chem.*, **48**(2), 94 (1893).
48. Huisgen, R., Stangl, H., Sturm, H. J., Raab, R. and Bunge, K., *Chem. Ber.*, **105**, 1258 (1972).
49. Padwa, A., *Acc. chem. Res.*, **9**, 371 (1976).
50. Huisgen, R., Grashey, R. and Sauer, J., in *The chemistry of alkenes*, ed. S. Patai, p. 806, Interscience, London, 1964.
51. Grundmann, C. and Grünanger, P., *The nitrile oxides*, Springer-Verlag, Berlin, 1971.
52. Franz, J. E., Howe, R. K. and Pearl, H. K., *J. org. Chem.*, **41**, 620 (1976).
53. Bianchi, G., De Micheli, C. and Gandolfi, R., *The chemistry of double-bonded functional groups*, ed. S. Patai, p. 369, Interscience, London, 1977.
54. Sheradsky, T., *The chemistry of the azido group*, ed. S. Patai; p. 331, Interscience, London, 1971.
55. Stable adducts have not been isolated, but conjugated and electron-rich alkenes give products which are consistent with the intermediacy of such adducts; Bridson-Jones, F. S., Buckley, G. D., Cross, L. H. and Driver, A. P., *J. chem. Soc.*, 1951, 2999.
56. Oppolzer, W., *Angew. Chem,. Int. Edn Engl.*, **16**, 10 (1977).

57. Black, D. St. C., Crozier, R. F. and Davis, V. C., *Synthesis,* 1975, 205.
58. Challand, S. R., Gait, S. F., Rance, M. J., Rees, C. W. and Storr, R. C., *J. chem. Soc. (C),* 1975, 26.
59. Cycloadditions of azoxy compounds are rare. For an example see Challand, S. R., Rees, C. W. and Storr, R. C., *J. chem. Soc. Chem. Communs,* 1973, 837.
60. Chapman, O. L., *Organic photochemistry,* vol. 3, p. 116, Dekker, New York, 1973.
61. Criegee, R., *Angew. Chem., Int. Edn Engl.,* **14**, 745 (1975).
62. Kellogg, R. M., Noteboom, M. and Kaiser, J. K., *J. org. Chem.,* **40**, 2573 (1975).
63. Gotthardt, H., *Chem. Ber.,* **105**, 196 (1972).
64. These systems normally function as 2π components (dienophiles) in cycloadditions rather than as 1,3-dipoles. Kresze, G. and Wucherpfennig, W., *Angew. Chem., Int. Edn Engl.,* **6**, 149 (1976).
65. Products which are formally 1,3-dipolar adducts of some of these species have been isolated. However, there is some doubt as to whether these are the primary products. For examples, for ketocarbenes see Paulissen, R., Moniotte, P., Hubert, A. J. and Teyssié, P., *Tetrahedron Lett.,* 1974, 3311; vinyl carbenes, Kirmse, W., *Carbene chemistry,* 2nd edn, Wiley, New York, 1971; carbonyl nitrenes, Lwowsky, W. (ed.), *Nitrenes,* Wiley Interscience, New York, 1970.
66. Review: Ollis, W. D. and Ramsden, C. A., *Adv. heterocyclic Chem.,* **19**, 1 (1976).
67. Wagner-Jauregg, T., *Synthesis,* 1976, 349.
68. Padwa, A., *Angew. Chem., Int. Edn Engl.,* **16**, 123 (1976).
69. (*a*) Huisgen, R., *J. org. Chem.,* **33**, 2291 (1968), Huisgen, R., *J. org. Chem.,* **41**, 403 (1976); (*b*) Firestone, R. A., *J. org. Chem.,* **33**, 2285 (1968), Firestone, R. A., *J. org. Chem.,* **37**, 2181 (1972), Firestone, R. A., *Tetrahedron,* **33**, 3009 (1977). It is very interesting to compare the arguments put forward by Firestone and Huisgen in these papers.
70. Bayne, W. F. and Snyder, E. I., *Tetrahedron Lett.,* 1970, 2263; Dolbier, W. R. and Dai, S.-H., *Tetrahedron Lett.,* 1970, 4645.
71. See, however, Padwa, A. and Carlsen, P. H. J., *J. Am. chem. Soc.,* **98**, 2006 (1976).
72. Morrocchi, S., Ricca, A. and Zanarotti, A., *Tetrahedron Lett.,* 1970, 3215; see also reference 42*d*, p. 415.
73. Morrocchi, S., Ricca, A., Zanarotti, A., Gandolfi, R., Bianchi, G. and Gruenanger, P., *Tetrahedron Lett.,* 1969, 3329.
74. Houk, K. N., Sims, J., Duke, R. E., Strozier, R. W. and George, J. K., *J. Am. chem. Soc.,* **95**, 7287 (1973); Houk, K. N., Sims, J., Watts, C. R. and Luskus, L. J., *J. Am. chem. Soc.,* **95**, 7301 (1973).
75. Sustmann, R. and Trill, H., *Angew, Chem., Int. Edn Engl.,* **11**, 838 (1972).
76. Caramella, P., Gandour, R. W., Hall, J. A., Deville, C. G. and Houk, K. N., *J. Am. chem. Soc.,* **99**, 385 (1977).
77. Padwa, A. and Smolanoff, J., *J. chem. Soc. Chem. Communs,* 1973, 342.
78. See, for example, Grée, R. and Carrié, R., *Tetrahedron Lett.,* 1971, 4117.
79. A systematic attempt to observe an *endo*-effect in diazoalkane additions failed since these are HOMO controlled. Agosta. W. C. and Smith, A. B., *J. org. Chem.,* **35**, 3856 (1970).
80. Bunge, K., Huisgen, R., Raab, R. and Sturm, H. J., *Chem. Ber.,* **105**, 1307 (1972); Padwa, A., Dharan, M., Smolanoff, J. and Wetmore, S. I., *J. Am. chem. Soc.,* **95**, 1945 (1973).
81. Eberhard, P. and Huisgen, R., *J. Am. chem. Soc.,* **94**, 1345 (1972).
82. Berlin, K. D., Rathore, B. S. and Peterson, M., *J. org. Chem.,* **30**, 226 (1965).
83. Schönberg, A., Černik, D. and Urban, W., *Berichte,* **64**, 2577 (1931).
84. Kauffmann, T., Busch, A., Habersaat, K. and Scheerer, B., *Tetrahedron Lett.,* 1973, 4047.
85. Corey, E. J. and Winter, R. A. E., *J. Am. chem. Soc.,* **85**, 2677 (1963); Corey, E. J., Carey, F. A. and Winter, R. A. E., *J. Am. chem. Soc.,* **87**, 934 (1965); Corey, E. J. and Shulman, J. I., *Tetrahedron Lett.,* 1968, 3655.
86. Huisgen, R. and Gotthardt, H., *Chem. Ber.,* **101**, 552, 839, 1059 (1968); Huisgen, R., Gotthardt, H. and Grashey, R., *Chem. Ber.,* **101**, 536, 829 (1968); Gotthardt, H., Huisgen, R. and Knorr, R., *Chem. Ber.,* **101**, 1056 (1968).

87. Baldwin, J. E. and Hong, S. Y., *Chem. Communs,* 1967, 1136; Baldwin, J. E. and Hong, S. Y., *Tetrahedron,* **24**, 3787 (1968). For other examples of retro-1,3-dipolar cycloadditions see Schneider, M. and Csacsko, B., *Angew. Chem., Int. Edn Engl.,* **16**, 867 (1977) and Sinbandhit, S. and Hamelin, J., *J. chem. Soc. Chem. Communs,* 1977, 768.
88. Review: Stark, B. P. and Duke, A. J., *Extrusion reactions,* Pergamon Press, Oxford, 1967.
89. Dodson, R. M. and Sauers, R. F., *Chem. Communs,* 1967, 1189.
90. Mock, W. L. and McCausland, J. H., *Tetrahedron Lett.*, 1968, 391.
91. See Hamer, J., *1,4-Cycloadditions reactions,* Chapter 13, Academic Press, New York, 1967.
92. Zwanenberg, B., Thijs, L., Broens, J. B. and Strating, J., *Recl. Trav. chim. Pays-Bas Belg.*, **91**, 443 (1972).
93. Mock, W. L., *J. Am. chem. Soc.*, **88**, 2857 (1966); McGregor, S. D. and Lemal, D. M., *J. Am. chem. Soc.*, **88**, 2858 (1966).
94. Saltiel, J. and Metts, L., *J. Am. chem. Soc.*, **89**, 2232 (1967).
95. Jung, F., Molin, M., Van Den Elzen, R. and Durst, T., *J. Am. chem. Soc.*, **96**, 935 (1974).
96. Heldeweg, R. F. and Hogeveen, H., *J. Am. chem. Soc.*, **98**, 2341 (1976).
97. Lemal, D. M. and McGregor, S. D., *J. Am. chem. Soc.*, **88**, 1335 (1966); Carpino, L. A., *Chem. Communs,* 1966, 494.
98. Baldwin, J. E., *Can. J. Chem.*, **44**, 2051 (1966).
99. Yankelevich, S. and Fuchs, B., *Tetrahedron, Lett.*, 1967, 4945, and references therein.
100. Hoffmann, R. W., *Angew. Chem., Int. Edn Engl.*, **10**, 529, 537 (1971). It is not clear whether these carbene extrusions are concerted; see, for example, Hoffmann, R. W. and Hirsch, R., *Tetrahedron Lett.*, 1970, 4819 and Ranken, P. F. and Battiste, M. A., *J. org. Chem.*, **36**, 1996 (1971). See also Griffin, G. W., *Angew. Chem., Int. Edn Engl.*, **10**, 537 (1971).
101. Hoffmann, R. W. and Hauser, H., *Tetrahedron,* **21**, 891 (1965).
102. Blomquist, A. T. and Meinwald, Y. C., *J. Am. chem. Soc.*, **81**, 667 (1959); Williams, Fickes, G. N. and Metz, T. E., *J. org. Chem.*, **43**, 4057 (1978).
103. Jefford, C. W., Heros, V. and Burger, U., *Tetrahedron Lett.*, 1976, 703.
104. Smith, C. D., *J. Am. chem. Soc.*, **88**, 4273 (1966).
105. Rieber, N., Alberts, J., Lipsky, J. M. and Lemal, D. M., *J. Am. chem. Soc.*, **91**, 5668 (1969).
106. Rosenblum, M., Longroy, A., Neveu, M. and Steel, C., *J. Am. chem. Soc.*, **87**, 5716 (1965).
107. Barton, D. H. R. and Willis, B. J., *Chem. Communs,* 1970, 1225.
108. (*a*) Wittig, G. and Hoffmann, R. W., *Chem. Ber.*, **95**, 2718, (1962); (*b*) Sakai, K. and Anselme, J. P., *Tetrahedron Lett.*, 1970, 3851; (*c*) Rees, C. W. and Yelland, M., personal communication.
109. Oth, J. F. M., *Angew. Chem., Int. Edn Engl.*, **7**, 646 (1968); Oth, J. F. M., *Recl. Trav. chim. Pays-Bas, Belg.*, **87**, 1185 (1968).
110. Review: Hoffmann, H. M. R., *Angew. Chem., Int. Edn Engl.*, **12**, 819 (1973).
111. Hoffmann, H. M. R., Joy, D. R. and Suter, A. K., *J. chem. Soc. (B)*, 1968, 57; Hoffmann, H. M. R. and Joy, D. R., *J. chem. Soc. (B)*, 1968, 1182.
112. Edelson, S. S. and Turro, N. J., *J. Am. chem. Soc.*, **92**, 2770 (1970); Turro, N. J., *Acc. chem. Res.*, **2**, 25 (1969).
113. Reviews: Lamola, A. A. and Turro, N. J., *Techniques of organic chemistry,* vol. xiv, ed. A. Weissberger, Interscience, New York, 1969; Chapman, O. L. and Lenz, G. in *Organic photochemistry,* vol. 1, p. 283, ed. O. L. Chapman, Arnold, London, 1967; Barltrop, J. A. and Coyle, J. D., *Excited states in organic chemistry,* Wiley, London, 1975.
114. Liu, R. S. H., Turro, N. J. and Hammond, G. S., *J. Am. chem. Soc.*, **87**, 3406 (1965).
115. The reaction is stereoselective at at least one centre: Chow, Y. L., Joseph, T. C., Quon, H. H. and Tam, J. N. S., *Can. J. Chem.*, **48**, 3045 (1970).

6 Other cycloadditions

Most cycloadditions, other than those with six electrons participating, involve four electrons. Considerations of orbital symmetry indicate that thermal ${}_{\pi}2_s + {}_{\pi}2_s$ cycloaddition is not allowed but that concerted ${}_{\pi}2_s + {}_{\pi}2_a$ addition (fig. 6.1) is.

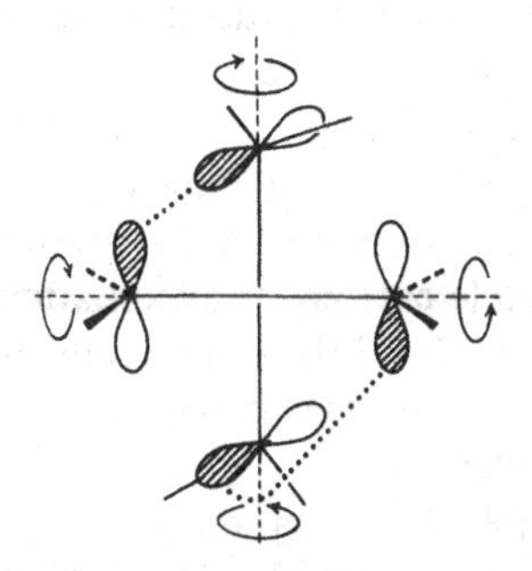

Fig. 6.1

For ${}_{\pi}2_s + {}_{\pi}2_a$ addition the reactants must approach as shown in fig. 6.1. This involves inefficient orbital overlap and considerable twisting of the π bonds as the overlap is increased. Also the approach of the reactants is hindered by non-bonding steric interactions between substituents on the reactants. These effects combine to make the allowed ${}_{\pi}2_s + {}_{\pi}2_a$ mode of addition very unfavourable and so the simple picture has emerged that 2 + 2 cycloadditions only occur by stepwise mechanisms involving diradical or zwitterionic intermediates. This model fits the facts very well. Thus 2 + 2 cycloadditions are largely restricted to those involving highly strained alkenes, polyhaloalkenes, conjugated dienes, highly nucleophilic and electrophilic alkenes and cumulenes, all systems where such intermediates would be highly stabilised. As we shall see (§6.4) the orbital structure of cumulenes renders a low-energy concerted pathway possible so this important class of 2 + 2 additions is dealt with separately. However, the straightforward stepwise mechanism has also been questioned for some of the other

2 + 2 cycloadditions, particularly those between highly polar reactants. This arises because although the allowed ${}_{\pi}2_s + {}_{\pi}2_a$ mode of addition can never involve favourable geometrical overlap it could be favoured electronically by substitutents which close the HOMO-LUMO energy separation. Moreover, closing the HOMO-LUMO gap can make the contribution of charge transfer configurations to the transition state important and render the concerted but formally disallowed ${}_{\pi}2_s + {}_{\pi}2_s$ mode energetically the most favourable. The problem is a subtle one because the type of substituent which will narrow the HOMO-LUMO energy difference is of course the type which will favour formation of an intermediate.

6.1. Non-polar 2 + 2 cycloadditions[1] (those involving diradical intermediates)

These are typified by the dimerisation of ethylene and involve reactants without substituents which are strongly electron withdrawing or releasing. For such 2 + 2 additions involving non-polar reactants, configuration interaction is irrelevant and the concerted ${}_{\pi}2_s + {}_{\pi}2_a$ mode derives no electronic stabilisation so that a stepwise mechanism via a diradical intermediate is the most reasonable.

The thermal dimerisation of ethylene to cyclobutane is extremely unfavourable (E_A = 184 kJ mol^{-1}) although the reaction enthalpy is favourable (ΔH° = 79.4 kJ mol^{-1}) and the equilibrium concentration of cyclobutane at 25 °C and 1 atm. is 99.8 mole %. Raising the temperature to overcome the activation barrier does not help because of the unfavourable entropy of dimerisation (ΔS° = -192 J mol^{-1} K^{-1}) which means that at 500 °C the equilibrium concentration of cyclobutane is only 0.01 mole %.[2] Thus cyclobutane cannot be prepared conveniently by dimerisation of ethylene.

In order for 2 + 2 cycloaddition to be feasible the activation energy must be reduced. This can be brought about by raising the energy of the reactant π bonds or lowering the energy of the diradical intermediate and hence the transition state leading to it, or by effecting both of these changes. Thus twisting will raise the energy of the π bond and this accounts for the reactivity of strained cyclic alkenes towards dimerisation. Fluorine destabilises a π bond and chlorine stabilises a radical centre making polychlorofluoroalkenes exceptionally reactive in this type of addition. They dimerise or react with other activated alkenes. Their often almost exclusive 2 + 2 addition to dienes is also a striking feature; for example, dichlorodifluoroethylene reacts with butadiene to give a vinylcyclobutane (equation 6.4) whereas ethylene and most other olefins give Diels-Alder adducts.

Some dienes also dimerise to give divinylcyclobutanes via allylic stabilised diradicals in preference to, or in competition with, concerted Diels-Alder modes of dimerisation.

Equations 6.1 to 6.4 are examples of the type of system which undergoes 2 + 2 cycloadditions via diradical intermediates. Forcing conditions are generally required; typically these cycloadditions are carried out at 100–200 °C under pressure.

Evidence that these are radical additions comes from the observed orientations, which are always those which would arise from the most stable diradical intermediate. Thus the dimerisation of acrylonitrile (6.1) leads to 1,2-dicyanocyclobutane since the diradical intermediate (**1**), in which both radical centres are stabilised by α-cyano groups, is more stable than the alternative diradical (**2**). This orientation is not consistent with an ionic mechanism since the zwitterion (**3**), in which the positive charge would be destabilised by the strongly electron-withdrawing cyanide, should be less stable than the zwitterion (**4**) which would give 1,3-dicyanocyclobutane.[1c] Other examples which illustrate the orientation are shown in equations 6.2–6.4. An α-chlorine is better able to stabilise a radical centre than an α-fluorine; thus the addition of dichlorodifluoroethylene to butadiene proceeds as shown in equation 6.4. The diradical (**5**) is estimated to be approximately 33 kJ mol^{-1} lower in energy than (**6**) and more than 88 kJ mol^{-1} lower than (**7**). Such energy differences are sufficient to ensure that the orientation is virtually all as shown.[1d]

$$2\ CH_2{=}CH{-}CN \longrightarrow \underset{(\mathbf{1})}{\dot{C}H(CN){-}CH_2{-}CH_2{-}\dot{C}H(CN)} \longrightarrow \text{1,2-dicyanocyclobutane} \qquad (6.1)$$

(**2**) $NC{-}CH{-}\dot{C}H_2$ / $CH_2{-}\dot{C}H{-}CN$ (**3**) $CH_2{-}\overset{\ominus}{C}H{-}CN$ / $CH_2{-}\overset{\oplus}{C}H{-}CN$ (**4**) $CH_2{-}\overset{\ominus}{C}H{-}CN$ / $NC{-}CH{-}\overset{\oplus}{C}H_2$

$$2\ CCl_2{=}CF_2 \longrightarrow \dot{C}Cl_2{-}CF_2{-}CF_2{-}\dot{C}Cl_2 \longrightarrow \text{1,1,2,2-tetrachloro-3,3,4,4-tetrafluorocyclobutane} \qquad (6.2)$$

$$PhCH{=}CH_2 + CCl_2{=}CF_2 \longrightarrow Ph\dot{C}H{-}CH_2{-}CF_2{-}\dot{C}Cl_2 \longrightarrow \text{1,1-dichloro-2,2-difluoro-4-phenylcyclobutane} \qquad (6.3)$$

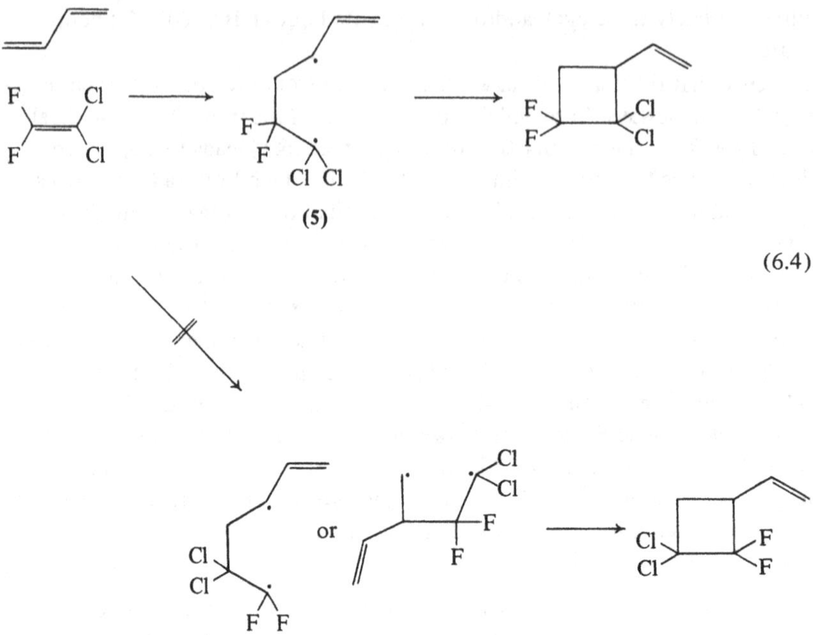

(5)

(6.4)

or

(6) (7)

These additions are also independent of solvent polarity, often proceeding well in non-polar solvents or even in the gas phase. The rate of addition of 1,1-dichloro-2,2-difluoroethylene to butadiene increases by a factor of less than three as the solvent is changed from hexane to methanol.[1b] Since the reactions are not radical chain reactions they are not initiated or inhibited by the usual free radical initiators or inhibitors.

These non-polar 2 + 2 cycloadditions also show the lack of stereoselectivity expected from an intermediate diradical in which free rotation about σ bonds can occur. The diradical intermediate for a thermal addition should be formed as a singlet so that there is no spin inversion barrier to ring closure. However, the conformation in which the diradical is initially formed is unlikely to be right for immediate ring closure. Thus, bond rotations can occur in the time taken for the molecule to attain the required conformation for closure of the second bond. For example, in the addition of 1,1-dichloro-2,2-difluoroethylene and tetrafluoroethylene to *trans, trans*-hexa-2,4-diene and the isomers of 1,4-dichlorobutadiene, loss of configuration occurs at the double bond which becomes part of the cyclobutane ring, but, as expected, configuration at the other double bond is largely or completely retained because of the barrier to rotation in the allyl radical residue (fig. 6.2).[1b,3]

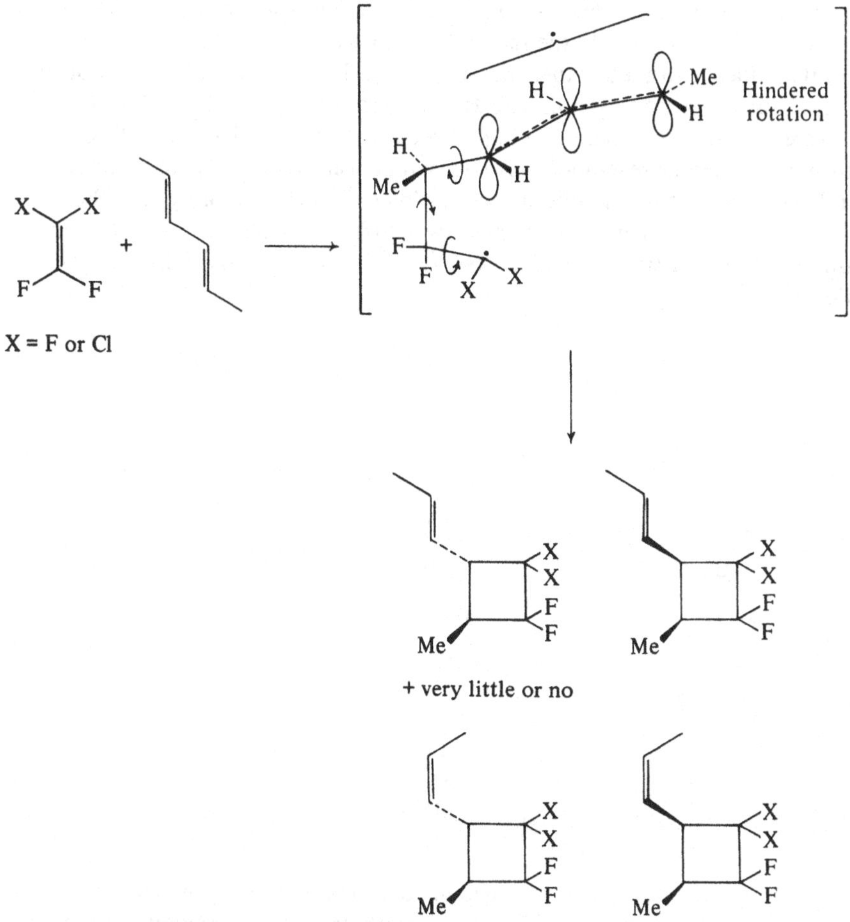

X = F or Cl

Hindered rotation

+ very little or no

Fig. 6.2

The intermediate diradicals can revert to reactants rather than cyclise. This is shown by the recovery of isomerised dienes from the reaction shown in fig. 6.2. Since the dienes do not themselves isomerise under the reaction conditions, the isomerised dienes must result from cleavage of the diradicals after rotation has occurred.[4]

It has been estimated that in the addition shown in fig. 6.2 bond rotation is about ten times faster than cyclisation which is in turn about four times faster than cleavage.[3,5] In the addition of tetrafluoroethylene to *cis*-2-butene, rotation in the diradical was estimated to be four times faster than ring closure.[6]

Diradicals are also formed in photochemical cycloadditions, but here they are often triplet diradicals. In this case, in addition to the diradical having to attain the right conformation, there is a further barrier to ring closure, namely spin inversion. It is not clear how the rate of spin inversion in this type of triplet diradical compares with the rate of bond rotation. There is some experimental evidence to suggest that triplet diradicals have a longer lifetime prior to ring closure than the singlet diradicals produced in the thermal reaction. There is a greater loss of configuration in photosensitised triplet diradical additions than in the thermal reaction.[1b] Also when diradicals (**8**) and (**9**) are generated in the singlet state by thermal decomposition or by direct photolysis of the azo compounds (**10**) and (**11**), they show a different pattern of products from the corresponding triplet diradicals generated by photosensitised decomposition.[1b]

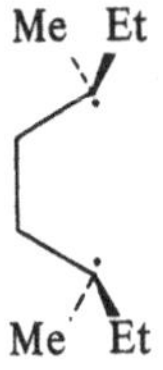
Me Et
Me Et

(8)

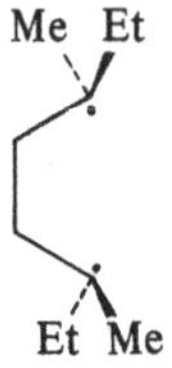
Me Et
Et Me

(9)

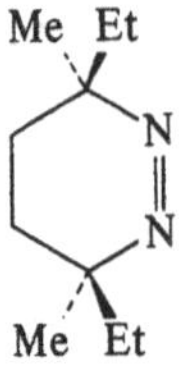
Me Et
N
N
Me Et

(10)

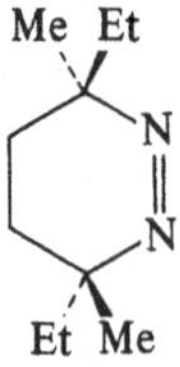
Me Et
N
N
Et Me

(11)

Both types of diradical are formed in conformations which are ideally disposed for ring closure, but cleavage to methylbut-1-ene represents a large proportion of the reaction. Significantly, cyclobutane formed from the singlet diradical almost completely retains the configuration of the azo compound, but the triplet diradical gives cyclobutane with appreciable loss of configuration. A greater proportion of cleavage also occurs for the triplet diradical, in line with it having a greater barrier to ring closure.

A possible alternative explanation of these stereochemical results is based on the conformational flexibility of the intermediate rather than on spin. Two ground state olefins cannot correlate directly with ground state cyclobutane in any geometrically accessible way: as the first bond is formed, there is a symmetry-imposed barrier to concerted formation of the second bond. An intermediate is produced which has at least a conformational barrier to ring closure. This species, identifiable with the 'singlet' of the previous treatment, can therefore undergo loss of configuration before closure: the cycloaddition may show partial, but not complete, stereoselectivity. An olefin in an excited electronic configuration will react with a ground state olefin to produce an excited configuration of the intermediate; this may well have different barriers to conformational change - it may be more 'floppy' - and may therefore close less stereoselectively than the ground state species. This type of argument, backed up by molecular orbital calculations, has been applied successfully to the similar additions of singlet and triplet carbenes (see §6.6).

Concerted versus stepwise radical additions to dienes[1b,d]

Generally the concerted 4 + 2 Diels-Alder reaction is the most favourable mode of addition of a monoene to a diene. However, if this mode of addition is particularly disfavoured or if formation of a diradical is particularly favoured, a diradical addition can compete. In some cases the diradical formation may be sufficiently favourable compared with the concerted Diels-Alder reaction for it to occur almost to the exclusion of the latter. The addition of 1,1-dichloro-2,2-difluoroethylene to butadiene is just such an example. It can be seen from table 6.1 that the addition gives mainly cyclobutane but a trace of cyclohexene is also formed. This is to be expected from the diradical mechanism. The butadiene exists in *cisoid* or *transoid* conformations which should have essentially the same reactivity for diradical formation. The diradical is allyl stabilised

Table 6.1. Addition of olefins to butadiene

	Percentage composition of product	
Olefin	2 + 2 Adduct	2 + 4 Adduct
$CF_2{=}CCl_2$	99	1
$CHF{=}CF_2$	85.8	14.2
$CH_2{=}CF_2$	35	65
$CH_2{=}C(OAc)(CN)$	14	86
$CH_2{=}CH_2$	0.02	99.98

and therefore bond rotations within the allyl residue will be hindered compared with rotations about the other bonds. The diradical formed from a *transoid* diene cannot close to a cyclohexene without rotations within this allyl portion of the molecule. Such diradicals would therefore be expected to close exclusively to cyclobutanes. On the other hand, diradicals formed from *cisoid* butadiene can close to both cyclobutane or cyclohexene without rotations within the allyl system. Since butadiene and most other open-chain dienes exist predominantly in the *transoid* form, the small amount of cyclohexene produced can be explained solely in terms of diradical formation (fig. 6.3).

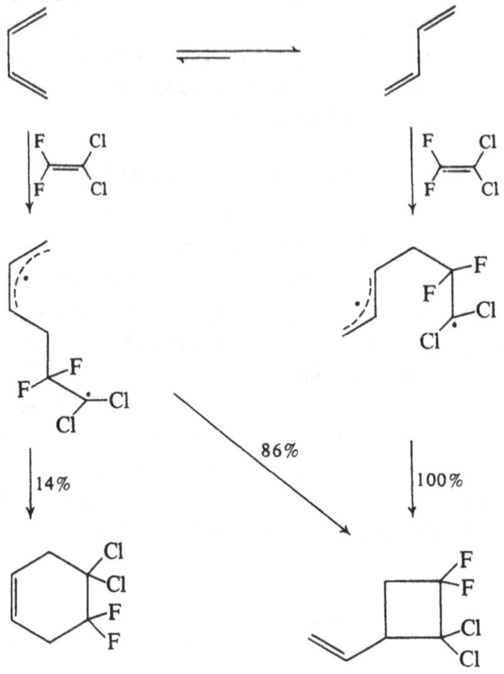

Fig. 6.3

It can be seen that cyclobutane formation is favoured over cyclohexene formation even for the *cisoid* diene since conformations of diradical which lead to the cyclohexenes are less favourable than those which lead to cyclobutanes.

In this type of radical addition, where rotation within the allylic system is disfavoured because of delocalisation, the proportion of cyclohexene should parallel, but never exceed, the proportion of the diene which is in a *cisoid* conformation (provided, of course, that the rates of diradical formation from *cisoid* and *transoid* dienes are comparable). This is found to be the case. The fraction of 2 + 4 adduct from 1,1-dichloro-2,2-difluoroethylene and butadiene increases with temperature in the same way as the fraction of butadiene having the *cisoid* configuration.[7] For 2-substituted butadienes, the proportion of 2 + 4 adduct increases with the size of the 2-substituent, since the larger the 2-substituent the greater the proportion of *cisoid* diene.[8]

Although the small amount of cyclohexene derivative formed in the addition of 1,1-dichloro-2,2-difluoroethylene to butadiene can be accommodated by the diradical mechanism, it is also possible that some or all of this 2 + 4 adduct arises by a competing Diels–Alder reaction which would be equally favoured by an increasing fraction of *cisoid* diene.

Where the proportion of 2 + 4 adduct rises above the proportion of *cisoid* diene, it seems certain that the 2 + 4 adducts arise largely by the concerted Diels–Alder mechanism and that the 2 + 2 adducts arise by a competing stepwise reaction. For example, the thermal addition of trifluoroethylene to butadiene at 215 °C gives 85.8 % 2 + 2 and 14.2 % 2 + 4 adducts.[1b] Taking the photosensitised addition of trifluoroethylene to butadiene as a model for the reaction proceeding entirely through a diradical intermediate, it can be estimated that only 2 % of 2 + 4 adduct in the thermal reaction would arise from the radical mechanism. The remainder therefore results from a concerted Diels–Alder reaction. The additions of 1,1-difluoroethylene and trifluoroethylene (table 6.1) are therefore examples where diradical formation and concerted Diels–Alder reaction are energetically comparable. This is reasonable because fluorine is less able to stabilise a radical than chlorine, so these olefins would give less stable diradicals than that from 1,1-dichloro-2,2-difluoroethylene, which reacts almost entirely by the diradical mechanism. As expected, diradical formation by trifluoroethylene is less favourable than concerted Diels–Alder reaction when *cis* fusion of the diene favours the latter. Thus the thermal addition of trifluoroethylene to cyclopentadiene gives almost exclusively 2 + 4 adduct with less than 0.1 % 2 + 2 adduct. When the addition is forced to proceed by the stepwise radical mechanism, as in the photosensitised reaction, the product contains 87 % 2 + 2 and 13 % 2 + 4 adduct.[1b]

In principle, a simple way to determine whether 2 + 2 or 2 + 4 adducts arise by competing stepwise and concerted modes would be to study the stereoselectivity of the reaction. In practice there are few systems which give both 2 + 2 and 2 + 4 adducts simultaneously where this can be studied. The only example reported so far is the addition of *cis*- and *trans*-1,2-dichloro-1,2-difluoroethylene to cyclopentadiene. This gives more than 95 % of 2 + 4 adduct stereospecifically. The minor portion of 2 + 2 adduct is formed non-stereospecifically, in line with concerted 2 + 4 and stepwise 2 + 2 mechanisms.[9]

α-Acetoxyacrylonitrile gives, with butadiene, a ratio of 2 + 2 to 2 + 4 addition that is little affected by variation of solvent or temperature. This was originally taken as evidence that both types of adduct were formed through a common diradical intermediate, since it seemed unlikely that competing mechanisms would depend so similarly on conditions.[10] This is now better interpreted as a competition between concerted 2 + 4 and stepwise 2 + 2 additions, because the related addition of *trans, trans*-hexadiene to α-acetoxyacrylonitrile gives only 2 + 4 adduct stereospecifically and therefore almost certainly by a concerted mechanism.[11] Introduction of the two electron-releasing methyl groups into butadiene favours the Diels–Alder reaction (p. 112) and hinders radical addition (initial attack on a secondary carbon is less favourable than on a primary carbon) so that any competing radical process falls below the level of detection.

Careful investigation of the addition of ethylene to butadiene discloses a trace of vinylcyclobutane.[12] This can be accounted for by a radical reaction which competes unfavourably with the concerted Diels–Alder process. It is obvious that a negligible amount of the 2 + 4 adduct can be formed by the radical route.

6.2. The possibility of concerted non-polar 2 + 2 cycloadditions

The situation where a concerted ${}_{\pi}2_s + {}_{\pi}2_a$ addition is most likely to be encountered is in the dimerisation of strained alkenes. This is because twisting of the π bonds is necessary to maximise overlap in the ${}_{\pi}2_s + {}_{\pi}2_a$ transition state and in a strained alkene the π bond is already twisted in the right sense. Significantly strained alkenes such as adamantene (**12**)[13] and *cis,trans*-1,5-cyclo-octadiene (**13**) do dimerise spontaneously.[14] Woodward and Hoffmann[15]

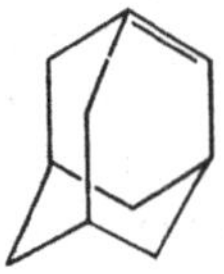

(12)

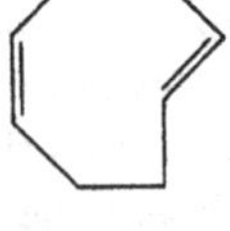

(13)

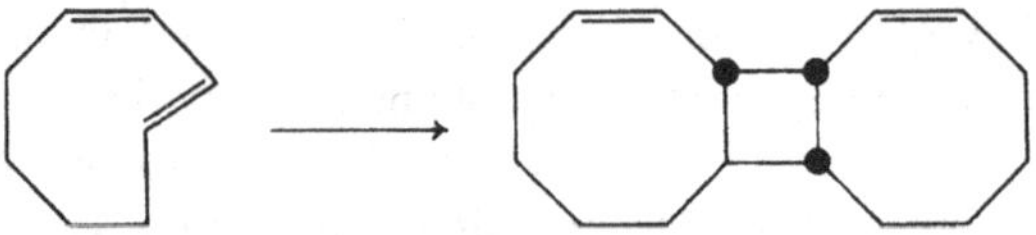

(14)

(6.5)

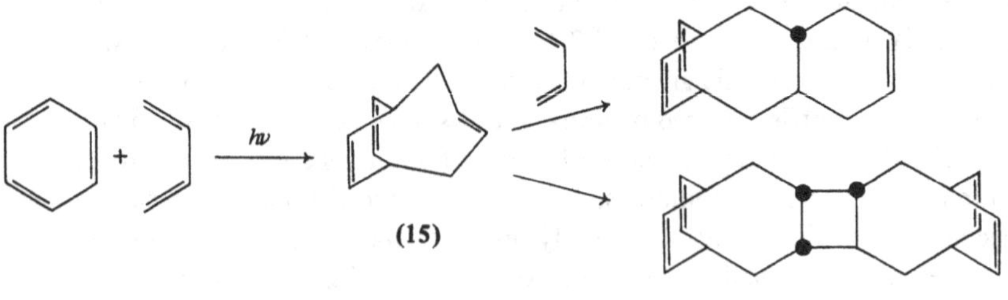

$h\nu$

(15)

(16)

(6.6)

have suggested that one of the products (**16**) obtained when benzene and butadiene are irradiated is the result of spontaneous concerted ${}_{\pi}2_s + {}_{\pi}2_a$ dimerisation of the initially formed photoadduct (**15**). The *trans* double bond in this photoadduct is necessarily twisted because it is constrained in an eight-membered ring. The Diels–Alder adduct of (**15**) and butadiene is also formed (6.6).[16]

However, in the case of the octadiene (**14**) although the major product does have the $_{\pi}2_s + _{\pi}2_a$ stereochemistry (6.5), *trans, trans* and *cis, cis* fused products are also formed and a diradical intermediate seems to be involved in the formation of all the dimers since they are formed in a constant ratio over a wide range of temperatures and in the presence of dienes and dienophiles which divert a major portion of the reaction.[17]

In the addition of 1,2-dichloroethylenes to *trans*-cyclo-octene, loss of configuration is observed for the former but not the latter. This is not therefore consistent with $_{\pi}2_s + _{\pi}2_a$ modes but is better explained by the intervention of a diradical intermediate in which rotation in the cyclic portion is hindered by transannular interactions.[18]

Identical product mixtures are obtained in the 2 + 2 addition of both *cis*- and *trans*-dideuterioethylene to tetrafluoroethylene. This is again more easily accommodated by a stepwise mechanism involving a diradical intermediate.[19] The HOMO and LUMO energies of $CF_2{=}CF_2$ are very similar to those of $CH_2{=}CH_2$ and so no electronic stabilisation of the concerted $_{\pi}2_s + _{\pi}2_a$ mode over that for simple ethylene dimerisation is to be expected.

However, in the case of the halogenated alkenes and dienes some narrowing of the HOMO–LUMO separation is to be expected and such systems may be described as semipolar 2 + 2 cycloadditions. In view of this, Epiotis[20] has suggested that the stereochemistry of these additions may be explained by competing *concerted* $_{\pi}2_s + _{\pi}2_a$ and *concerted* $_{\pi}2_s + _{\pi}2_s$ modes, the latter being stabilised by configuration interaction. Detailed analysis of the steric effect of substituents on the ability of the reactants to overlap in the $_{\pi}2_s + _{\pi}2_a$ sense and balancing this against the $_{\pi}2_s + _{\pi}2_s$ mode in which there will be less electronic stabilisation but better overlap does lead to a rationalisation of the detailed stereochemical data available.

In conclusion there is general agreement that non-polar 2 + 2 cycloadditions proceed via diradical intermediates and even in the case of the semipolar 2 + 2 cycloadditions the diradical model is still highly satisfactory and the most widely accepted.

6.3. Polar 2 + 2 cycloadditions[1a, b, 21]

2 + 2 Cycloadditions of electron-rich to electron-deficient alkenes often proceed in high yield under mild conditions. All of these additions can be explained by the formation of a stabilised zwitterionic intermediate. In some cases the additions are reversible, the dipolar intermediate being in equilibrium with both products and reactants. Normally, formation of the intermediate is rate determining. In the event of the zwitterion being exceptionally well stabilised, ring closure of the zwitterion may be rate determining.

Typical alkenes participating in such polar 2 + 2 cycloadditions are shown below. Some of these cycloadditions are shown in equations 6.7–6.9. Orienta-

Electron-deficient component	*Electron-rich component*
$(CN)_2C{=}C(CN)_2$	$H_2C{=}CHOR$
$(CN)_2C{=}C(CN)Cl$	$H_2C{=}CHSR$
$(CF_3)_2C{=}C(CN)_2$	$H_2C{=}CHNR_2$
$CF_3(CN)C{=}C(CN)CF_3$	
$H_2C{=}CHCO_2R$	
$H_2C{=}CHSO_2R$	
$H_2C{=}CHNO_2$	

tion is always that resulting from the most stabilised zwitterion as illustrated in (6.7).

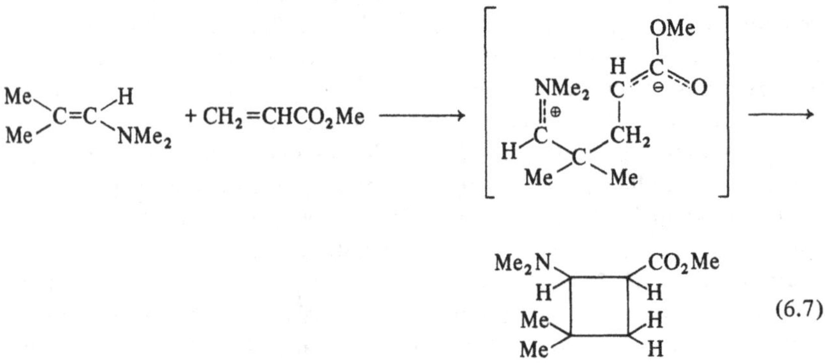

Me, Me C=C H, NMe_2 + $CH_2{=}CHCO_2Me$ ⟶ [$NMe_2^{\oplus}$=CH–C(Me)(Me)–CH_2–CH=C(OMe)–$O^{\ominus}$] ⟶ cyclobutane (Me_2N, H; Me, Me; CO_2Me, H; H, H) (6.7)

MeO–C_6H_4–CH=CH_2 + (NC)(NC)C=C(CN)(CN) ⟶ cyclobutane (H, H; NC, NC; 4-MeOC_6H_4, H; CN, CN) (6.8)

$CH_2{=}C$(H)(S–CMe_3) + F_3C(NC)C=C(CF_3)(CN) ⟶ cyclobutane (F_3C, NC; H, H; CF_3, CN; $SCMe_3$, H) + cyclobutane (NC, F_3C; H, H; CN, CF_3; $SCMe_3$, H) (6.9)

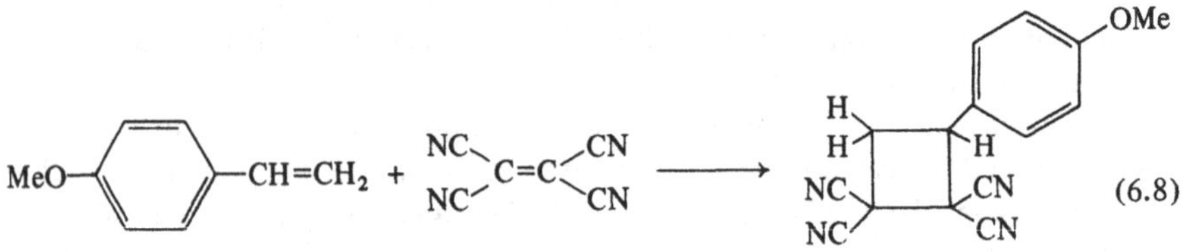

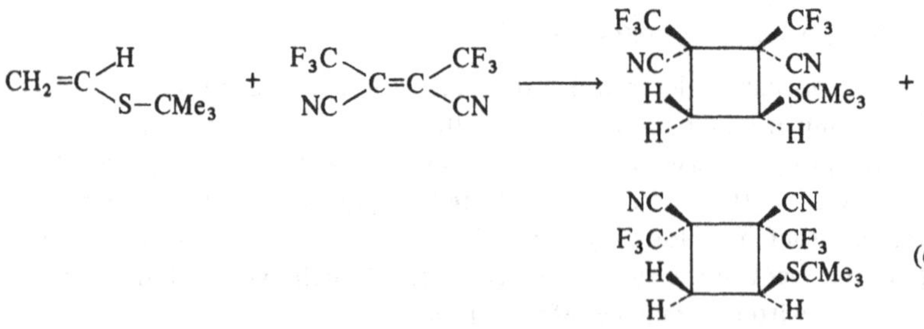

On mixing, these olefins form deeply coloured solutions. The colours, which are attributed to π complexes formed between the electron-rich and electron-deficient olefins, are discharged as the reaction proceeds.

The characteristics of these reactions are consistent with a stepwise process involving a zwitterionic intermediate. The rates frequently show great dependence on solvent polarity, being much greater in more polar solvents; for example, the addition shown in equation 6.8 is 6.3×10^4 times faster in acetonitrile than in cyclohexane.[21a, 22] In some cases, however, the additions are only moderately sensitive to solvent polarity (see p. 15).

The additions are extremely sensitive to substituent effects. For example, addition of styrenes to tetracyanoethylene only occurs when the styrenes have electron-releasing *para*-substituents (e.g. *p*-OMe, equation 6.8) and the rate of these additions varies greatly with the electron-releasing ability of the *para*-substituent: $k_{p\text{-OMe}}/k_{p\text{-cyclopropyl}} = 500$.[1b]

As would be expected in a two-step reaction it is only substituents located at the points of developing positive and negative charge which affect the rate of reaction profoundly. Thus in the addition of enol ethers (17) to tetracyanoethylene, R^1 and R^2 have a large effect on the rate whereas R^3 and R^4 are much less significant.[21c] Also, 1,1-di-, 1,1,2-tri- and tetracyanoethylene all

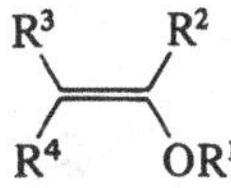

R^3 R^2

R^4 OR^1

(17)

undergo reaction with isobutenyl methyl ether at much the same rate since in all three the developing negative charge is localised on a carbon bearing two cyano groups.

Cyano- and *trans*-1,2-dicyanoethylenes react immeasurably slowly under the same conditions since only a single cyano group is available for stabilisation. It is very revealing to compare these substituent effects with those of an allowed, concerted reaction. In going from cyano- to tetracyanoethylene the rates of addition to cyclopentadiene and 9,10-dimethylanthracene increase by factors of 4×10^7 and 15×10^9, respectively and introduction of CN at either carbon has a dramatic effect (table 6.2).[21c] Clearly the concerted and stepwise mechanisms respond quite differently to substituents and this provides a criterion for distinguishing them. The rates of the Diels–Alder reactions correlate well with frontier orbital energy separations since the concerted reaction has an early transition state resembling reactants. The transition state for the stepwise reaction is a late one and resembles the intermediate zwitterion and so PMO theory is not applicable. Significantly the rates of the allowed 4 + 2 addition are many orders of magnitude greater than that of the forbidden 2 + 2 addition.

Table 6.2. Rate constants ($10^5 k_2$ litre^{-1} mol^{-1} s^{-1}) for 2 + 2 and 2 + 4 cycloadditions of cyanoethylenes

		CN	NC, CN	CN, CN	NC, CN	NC, CN, NC, CN
Me, Me, OMe	*a*	0	0	31.6	2.39	3.97
	b	1.04	81	4.55×10^4	4.80×10^5	4.30×10^7
Me, Me	*b*	0.89	139	1.27×10^5	5.90×10^6	1.30×10^{10}

[a] Benzene 25 °C. [b] Dioxane 20 °C.

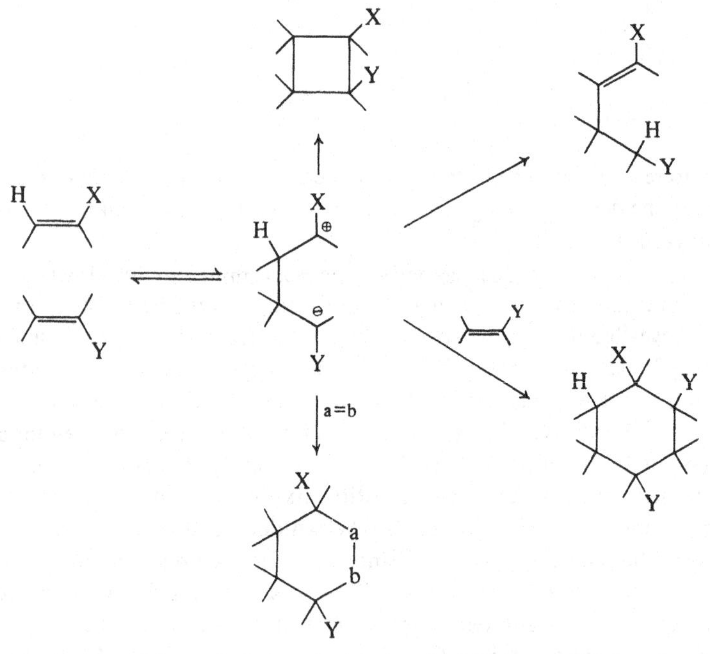

Fig. 6.4. Reactions of electron-rich with electron-deficient olefins. x = electron releasing; Y = electron withdrawing.

Further strong evidence for the stepwise nature of these cycloadditions comes from the observation of alternative reactions of the 1,4-dipolar intermediates. These are shown schematically in fig. 6.4.

Proton transfer can compete with ring closure when the electron-rich component has a β-hydrogen atom. For example, the enamine (**18**) gives cyclobutane (**19**) with nitro-olefins in non-polar solvents but gives nitroenamines (**20**) in more polar solvents (6.10). It is assumed that the intermediate zwitterion is better stabilised in the more polar solvents and is therefore more selective so that proton transfer competes with ring closure.[23] Such competing reactions are general for enamines with β-hydrogen atoms.

$CH_2{=}CH \cdot NO_2$ + (**18**) → zwitterion; Non-polar solvent → (**19**); Polar solvent → (**20**) (6.10)

In some cases the 1,4-dipole has been intercepted with nucleophiles such as alcohols or by cycloaddition to another reactant to give a six-membered ring.[24] This reactant may be the solvent, for example acetonitrile or acetone, or a second molecule of the electron-deficient olefin. The enamine (21) and β-nitrostyrene normally give the cyclobutane but in the presence of excess β-nitrostyrene the 1 : 2-adduct (**22**) is isolated (6.11).[23]

The enamine (**23**) forms cyclobutanes with acrylic acid derivatives (equation 6.7) but with methylenemalonic ester, the 1,4-dipole, which is even better stabilised, reacts with another molecule of ester to give the six-membered ring adduct (**24**) rather than closing to form the cyclobutane (6.12).[25]

This type of six-membered ring formation from a 1,4-dipole and a 1,4-dipolarophile is known as 1,4-dipolar cycloaddition.[26] It is quite general and promises great synthetic application. 1,4-Dipoles are fundamentally different from 1,3-dipoles (§5.3). The latter have four π electron systems extending over

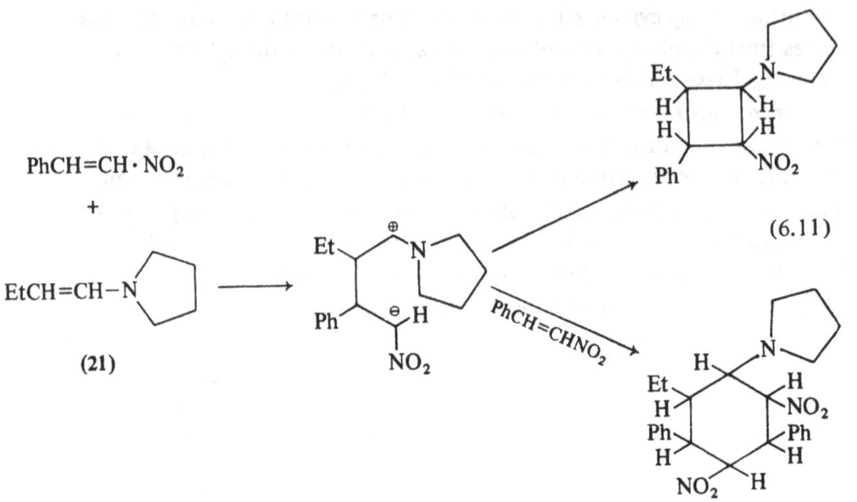

(6.11)

(21)

(22)

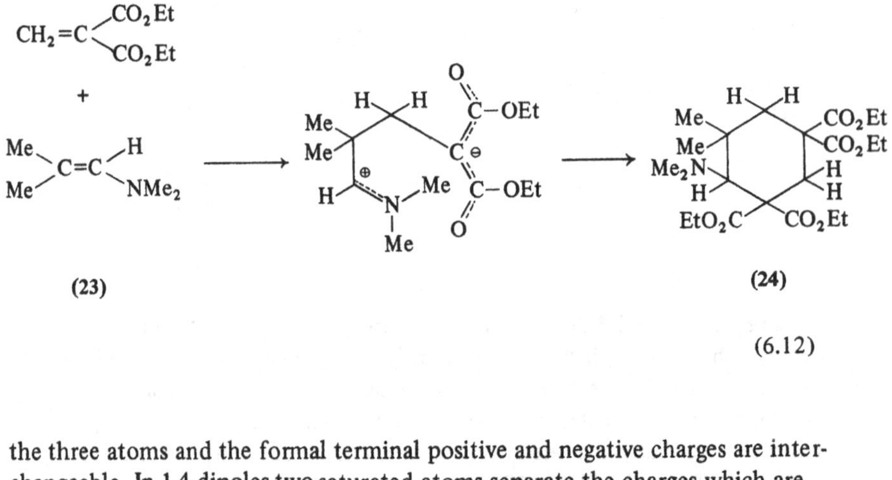

(23)

(24)

(6.12)

the three atoms and the formal terminal positive and negative charges are interchangeable. In 1,4-dipoles two saturated atoms separate the charges which are not interchangeable. The 1,4-dipole may also be produced by cleavage of one σ bond of a suitably substituted cyclobutane (the first step of a polar retro-2 + 2 addition).

In certain cases 1,4-dipolar intermediates can actually be isolated; for example, the exceptionally well-stabilised dipoles (**27**) in equation 6.13.[21a]

The yellow to orange crystalline 1,4-dipoles are formed at low temperatures but at higher temperatures they close to cyclobutanes which then undergo elimination to give cyclobutenes. Electrocyclic ring opening of the cyclobutenes gives butadienes, the observed products. The 1,4-dipoles from (**25**, X = SMe) and *p*-nitrobenzylidene malononitrile (**26**, $R^4 = p\text{-}NO_2C_6H_4$, R^5 = H) or 1,1-dicyano-2,2-bistrifluoromethylethylene (**26**, $R^4 = R^5 = CF_3$) are particularly stable and only rearrange to butadienes on heating to about 80 °C.

(**25**) + (**26**) → (**27**) → [cyclobutane → (− HX) cyclobutene] → butadiene (6.13)

X = SMe, NMe_2
R^1 = H, Me, Et, Pr^i, Ph
R^4 = Ph, $p\text{-}NO_2C_6H_4$, CF_3, CN
$R^2 + R^3 = (CH_2)_{2\text{-}4}$
R^5 = H, CF_3, CN

Lack of stereospecificity has been observed in ionic 2 + 2 cycloadditions, in line with the presence of an intermediate in which bond rotation can occur before closure of the second bond.[1b, 21a, c, 27] In general, however, ionic additions are highly stereoselective; this contrasts with the much lower stereoselectivity observed for cycloadditions with diradical intermediates. Ionic 2 + 2 additions are much more stereoselective because the electrostatic attraction between the ends of the 1,4-dipole favours its formation in a configuration approaching (**28**); the second bond may therefore be formed without rotation of other σ bonds. In the diradical intermediate no such electrostatic attraction is present and steric factors tend to make an open configuration (**29**) more likely.[28] This is particularly so since the carbon atoms bearing the unpaired electrons tend to have the bulky substituents which stabilise the radical centres.[1b] In the time taken for bond rotation to bring the diradical into a suitable conformation for ring closure, greater loss of configuration can occur.

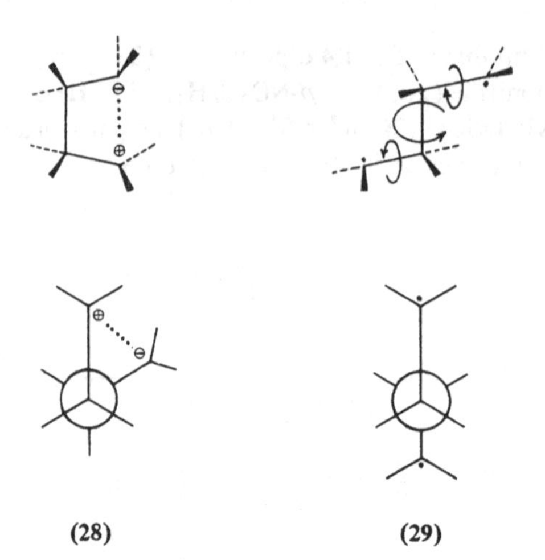

(28) (29)

The high stereoselectivity is illustrated by the addition of ethyl but-1-enyl ether to tetracyanoethylene which is in fact one of the least selective polar 2 + 2 additions. In benzene, the *cis* enol ether gives only 2 % of the adduct resulting from rotation about the enol ether component. The proportion of this 'wrong' adduct increases with more polar solvents (7 % CH_2Cl_2, 10 % CH_3CO_2Et, 18 % MeCN) as would be expected because the zwitterion is better solvated and the Coulombic interaction between the charged centres is reduced, so giving greater opportunity for bond rotation to compete with ring closure. The *trans*-isomer behaves similarly and for this system it is estimated that in acetonitrile, ring closure is five times faster than bond rotation, the difference being even greater in less polar solvents (in contrast, rotation can be up to ten times faster than closure for the diradical). It is also estimated that the zwitterion reverts to reactants at virtually the same rate as it closes, this being revealed by isomerisation of unreacted enol ether resulting from cleavage of the zwitterion after rotation.[21c]

The very high stereoselectivity of many 2 + 2 cycloadditions has raised the question of whether they should be considered as concerted. Epiotis[29] has suggested that they are indeed concerted $_{\pi}2_s + _{\pi}2_s$ additions in which the polar substituents cause a strong donor–acceptor relationship between the reactants. Configuration interaction can then favour the $_{\pi}2_s + _{\pi}2_s$ pericyclic mode and render it 'allowed'. Where the reactions are not completely $_{\pi}2_s + _{\pi}2_s$-stereoselective it is possible that the $_{\pi}2_s + _{\pi}2_s$ and $_{\pi}2_s + _{\pi}2_a$ modes are finely balanced. In a $_{\pi}2_s + _{\pi}2_a$ process the donor molecule will be the $_{\pi}2_s$ component and the acceptor the $_{\pi}2_a$, and there are cases where rotation is observed to occur only in the acceptor.[27, 30] It is not easy to dismiss the concerted $_{\pi}2_s + _{\pi}2_s$ mechanism; however, if these reactions were truly pericyclic then one would not expect the different response to substituents that is observed in the 2 + 2 and concerted

4 + 2 additions discussed above. Also, as solvent polarity is increased one would anticipate that the contribution of charge transfer configurations to the transition state would become more important and consequently that the $_{\pi}2_s + _{\pi}2_s$ stereoselectivity would increase. While an example which supports this prediction can be found,[27, 30] the more typical results with enol ethers[21c] and substituted styrenes clearly contradict it.[1b] It has also been pointed[31] out that while the two-step mechanism via a zwitterion correctly rationalises the observed 'head-to-head' regioselectivity in all cases the prediction based on the Epiotis model is less clear. Thus in the $_{\pi}2_s + _{\pi}2_s$ addition favoured by configuration interaction the important frontier orbital interactions are between HOMO and HOMO and LUMO and LUMO. The coefficients of these orbitals for a typical donor–acceptor pair are as shown in fig. 6.5. The former interaction predicts 'head-to-head' regioselectivity, the latter 'head-to-tail'.

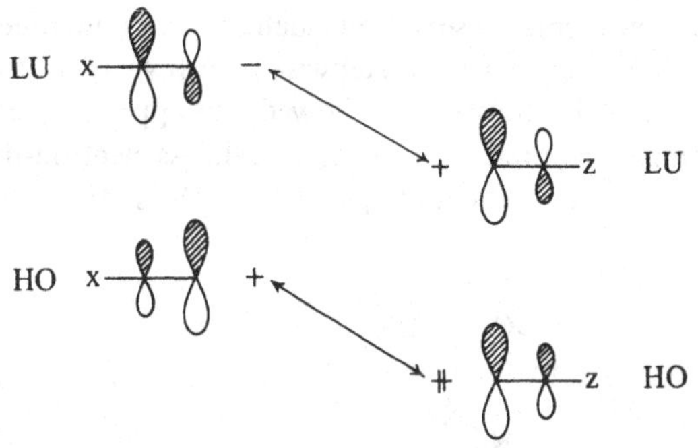

Fig. 6.5

In summary, the stepwise mechanism involving a zwitterionic intermediate in which bond rotation is substantially or completely suppressed offers the best general model for these 2 + 2 additions. However, the concerted mechanism offers an interesting alternative and will no doubt continue to stimulate further mechanistic investigation.

Four-membered ring formation by an ionic cycloaddition is not limited to cyclobutanes. Imines dimerise to give 1,3-diazetidines (6.14). These are frequently unstable and revert to imines which may or may not be the same as the starting imines.[21b]

Azodicarboxylic esters, in addition to being reactive dienophiles, also undergo 2 + 2 additions with electron-rich olefins to give 1,2-diazetidines (6.15).[32, 33]

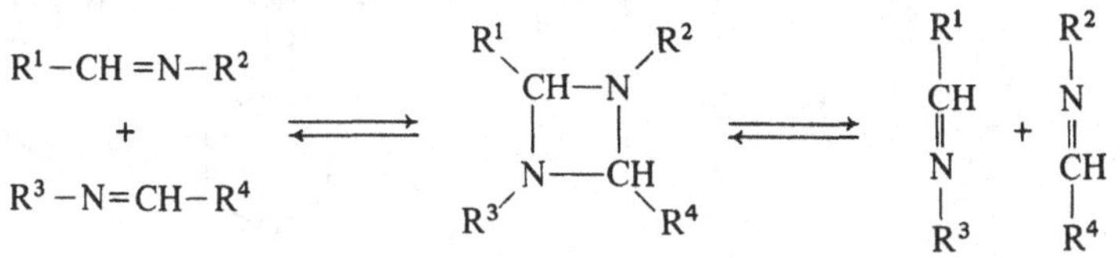

$$R^1{-}CH{=}N{-}R^2 + R^3{-}N{=}CH{-}R^4 \rightleftharpoons \text{(1,3-diazetidine: } R^1CH{-}NR^2{-}CHR^4{-}NR^3\text{)} \rightleftharpoons R^1{-}CH{=}N{-}R^3 + R^2{-}N{=}CH{-}R^4 \qquad (6.14)$$

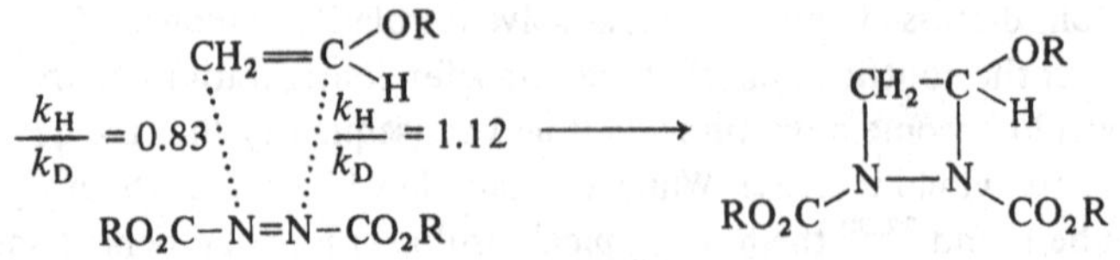

$$\frac{k_H}{k_D} = 0.83 \quad \frac{k_H}{k_D} = 1.12 \qquad CH_2{=}C(OR)H + RO_2C{-}N{=}N{-}CO_2R \longrightarrow \text{azetidine adduct} \qquad (6.15)$$

With vinyl ethers, the addition is stereospecific and almost solvent independent; the values of $\Delta H^{\ddagger}$ and $\Delta S^{\ddagger}$ also support a concerted addition. The secondary H/D isotope effects are shown. They are remarkably similar to those observed for the concerted addition of keten to styrene and suggest that the two bond-forming processes occur simultaneously but are very different at each terminus. The workers prefer to explain these different isotope effects by a stepwise reaction through a dipolar intermediate, an idea supported by the trapping of a dipolar intermediate in the 2 + 2 addition of indene to 4-phenyl-1,2,4-triazoline-3,5-dione.[33] Experimental observations such as these underline the need for caution in deciding on concerted or stepwise pathways. Obviously, further subtle investigation of this formally disallowed, but apparently concerted, 2 + 2 cycloaddition is required. With symmetrical, less electron-rich olefins, azodicarboxylic esters act as dienes to give formal Diels-Alder adducts (6.16).[33,34]

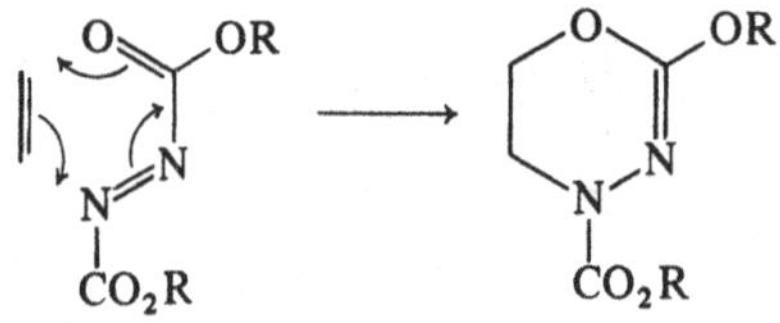

(6.16)

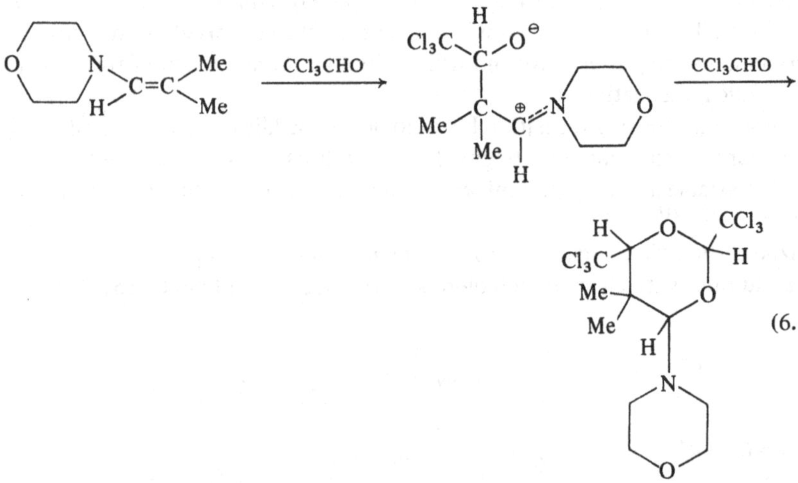

(6.17)

Many other cycloadditions are known in which hetero-1,4-dipoles are formed as intermediates.[21a, 26] These give four-membered ring heterocycles by ring closure or six-membered ring heterocycles by addition to another component, as, for example, in equation 6.17.[35] Many of these involve cycloadditions of heterocumulenes, which are discussed later.

The highly electrophilic acetylenedicarboxylic esters are worth a special mention since these have been used extensively for annelation of electron-rich heterocyclic systems with a cyclobutene ring. In some cases the cyclobutenes can be isolated but usually ring opening occurs so providing a useful method of ring expansion (6.18).[21d, 36]

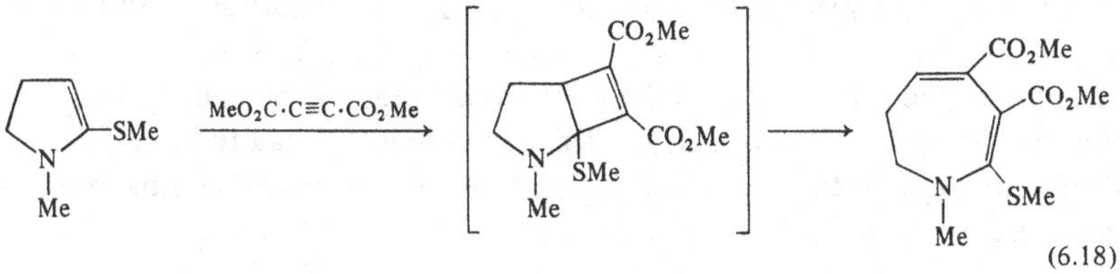
SMe
N
Me
$MeO_2C{\cdot}C{\equiv}C{\cdot}CO_2Me$
CO_2Me
CO_2Me
N
SMe
Me
CO_2Me
CO_2Me
N
SMe
Me

(6.18)

Electron-rich acetylenes give similar reactions with heterocycles containing electron-deficient C=C or C=N bonds.

6.4. 2 + 2 Cycloadditions of cumulenes[37]

Cumulenes have at least two orthogonal π bonds, with a central atom common to both of the π bonds. Allene (**30**) is a cumulene with an all-carbon skeleton. Heterocumulenes can be considered as being derived from allene by replacement of one, two or three carbon atoms by heteroatoms; keten (**31**) is an example.

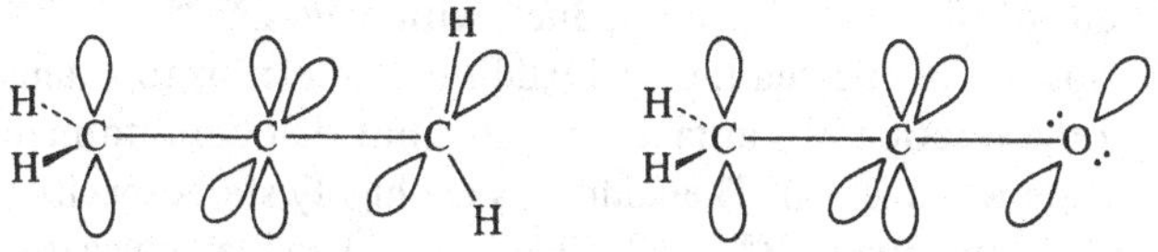
H
H
C
H
C
C
H
H
H
C
C
O

(30) **(31)**

Table 6.3 shows the types of systems which are known. Some 1,3-dipolar species can be written with cumulenic resonance forms, for example

$$R{-}C{\equiv}\overset{\oplus}{N}{-}\overset{\ominus}{O} \leftrightarrow R{-}\overset{\ominus}{C}{=}\overset{\oplus}{N}{=}O$$

but only those systems where the main uncharged resonance form has a cumulene structure will be considered.

The reactivity and stability of heterocumulenes vary considerably. Sulphur dioxide, carbon disulphide and carbon dioxide are very stable, relatively unreactive compounds. Ketens are of intermediate reactivity; some diaryl ketens, though reactive, are isolable, but other ketens have to be generated *in situ*. At the other extreme, sulphenes have so far only been detected and trapped as reactive intermediates.

The separate π bonds in cumulenes can act as dienophiles or dipolarophiles in Diels–Alder or 1,3-dipolar cycloadditions, and molecules in which a double bond is conjugated with the cumulene can function as dienes in Diels–Alder additions (fig. 6.6). In general, however, the notable and characteristic feature of the cycloaddition reactions of cumulenes is the tendency to form four-membered rings. Such reactions are formally $2\pi + 2\pi$ cycloadditions, and were therefore originally thought to go by stepwise mechanisms. This seemed reasonable because the sort of intermediates involved would, in general, be well-stabilised zwitterions or allyl diradicals. This general picture had to be re-examined, however, when the cycloadditions of ketens to certain olefins were shown to be concerted.

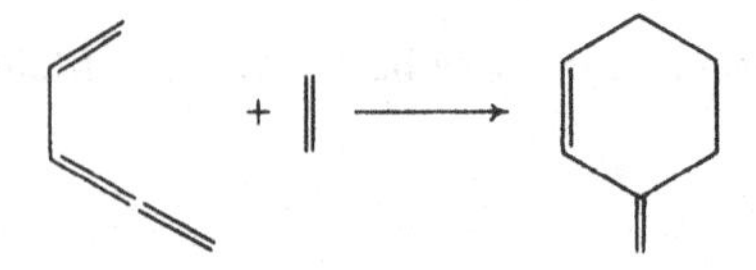

Fig. 6. 6

Heterocumulene cycloadditions. Mechanism

Since concerted thermal ${}_{\pi}2_s + {}_{\pi}2_s$ cycloadditions are not allowed, considerable interest was aroused by the stereospecific additions of olefins to ketens.[38, 39] It seemed even more remarkable that ketens form only the disallowed 2 + 2 adduct, and none of the allowed 2 + 4 adduct, with dienes.[38, 40]

Huisgen[41] closely investigated the cycloadditions of olefins to ketens and concluded that the reaction was in fact concerted but that bond formation in the transition state was unequal. The additions are highly stereospecific, with low enthalpies of activation ($\Delta H^{\ddagger} \triangleq$ 38–42 kJ mol^{-1}) and large negative entropies of activation ($\Delta S^{\ddagger} \triangleq$ -167 J mol^{-1} K^{-1}). The effect of solvent polarity on the rate of reaction is small but consistent with some charge build-up in the transition state. Thus, for example, a 50-fold change of rate is observed in going from cyclohexane to acetonitrile, compared with 63 000 for the stepwise ionic addition of tetracyanoethylene to *p*-methoxystyrene. This small charge buildup in the transition state explains the orientation of the additions which is such that substituents are best able to stabilise the developing charges (fig. 6.7). (See also p. 197).

Table 6.3. Heterocumulenes

Formula	Name	Class
$R_2C{=}C{=}O$	Ketens	One heteroatom
$O{=}C{=}C{=}C{=}O$	Carbon suboxide	
$R_2C{=}C{=}NR$	Ketenimines	
$R_2C{=}C{=}S$	Thioketens	
$R_2C{=}S{=}CR_2$	Sulphilidines	
$O{=}C{=}O$	Carbon dioxide	Two heteroatoms
$S{=}C{=}S$	Carbon disulphide	
$RN{=}C{=}NR$	Carbodiimides	
$R_2C{=}S{=}O$	Sulphines	
$R_2C{=}S({=}O){=}O$	Sulphenes	
$S{=}C{=}O$	Carbonyl sulphide	
$RN{=}C{=}O$	Isocyanates	
$RN{=}C{=}S$	Isothiocyanates	
$O{=}N{=}O$	Nitrogen dioxide	Three heteroatoms
$O{=}S{=}O$	Sulphur dioxide	
$RN{=}S{=}NR$	Sulphur diimines	
$RN{=}S{=}O$	Sulphinyl amines	
$RN{=}S({=}O){=}O$	Sulphuryl amines	

Fig. 6.7

The addition of ketens to norbornene proceeds without any of the skeletal rearrangement which might be expected from a stepwise reaction involving a zwitterion with a norbornyl cation component. Finally, Baldwin has shown that for the addition of styrene to diphenylketen, secondary deuterium isotope effects occur at each terminus. The magnitudes of these effects are interesting (fig. 6.8): that for bond *a* is consistent with a change of hybridisation from sp^2 to sp^3 in the rate-determining step. However, that for bond *b* is not consistent with a simple change in hybridisation but it does show that some form of interaction occurs in the rate-determining step.[42]

Ph
CH=CH_2
$\frac{k_H}{k_D} = 1.23$ (bond *b*) $\frac{k_H}{k_D} = 0.91$ (bond *a*)
Ph_2C=C=O

Fig. 6.8

For the concerted keten additions to fit in with orbital symmetry predictions, they must be ${}_{\pi}2_s + {}_{\pi}2_a$ processes in which the olefin is always found to act as the ${}_{\pi}2_s$ component and the keten as the ${}_{\pi}2_a$ component, and which for some reason are much more favourable than other ${}_{\pi}2_s + {}_{\pi}2_a$ cycloadditions.

Woodward and Hoffmann have suggested that there is a favourable secondary interaction between the highest occupied olefin orbital and the vacant orthogonal π^* antibonding orbital of the keten molecule. This sufficiently stabilises the transition state for the allowed ${}_{\pi}2_s + {}_{\pi}2_a$ concerted mode of addition (fig. 6.9) so that this process occurs in preference to a stepwise alternative.[15,43]

Consistent with this type of crossed transition state it is found that *cis*-1,2-disubstituted alkenes are more reactive than the *trans*-isomers. The latter must add through a transition state in which one of the alkene substituents interferes with the keten substituents whereas with the former both alkene substituents can be oriented away from the keten substituent.

Another way of looking at keten cycloadditions is to consider them as 2 + 2 + 2 cycloadditions, formally either ${}_{\pi}2_s + {}_{\pi}2_s + {}_{\pi}2_s$ or ${}_{\pi}2_s + {}_{\pi}2_a + {}_{\pi}2_a$ and therefore allowed, in which both keten π bonds are involved (fig. 6.10). This

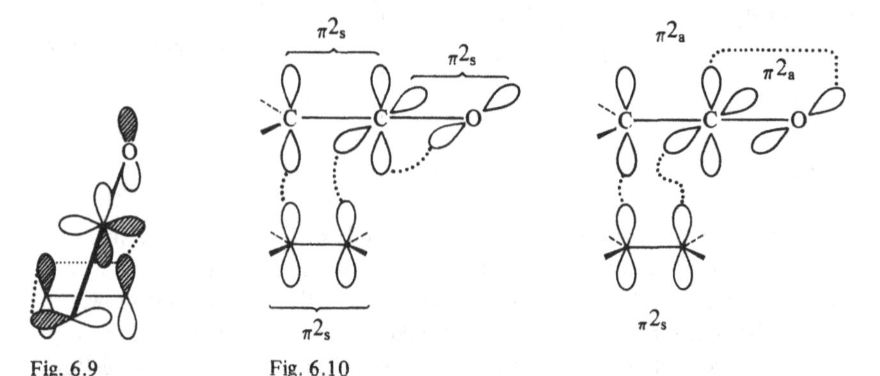

Fig. 6.9

Fig. 6.10

would require a transition state similar to that in fig. 6.9 but slightly less tight. It would also be a more realistic transition state because overlap between the ketenophile HOMO and the keten π^* carbonyl orbital orthogonal to the keten C=C bond is likely to be the most important frontier orbital interaction. This follows because this carbonyl π^* orbital is in fact the keten LUMO (fig. 6.11).[44] This LUMO also has a large coefficient on the central carbon so the observed regioselectivity in keten addition (fig. 6.7) is readily explained by frontier orbital theory, assuming the dominant interaction is between the ketenophile HOMO and the keten LUMO (the larger ketenophile HOMO coefficient will be on the unsubstituted carbon). This frontier orbital view of keten additions therefore corresponds very closely to that proposed by Huisgen.[41]

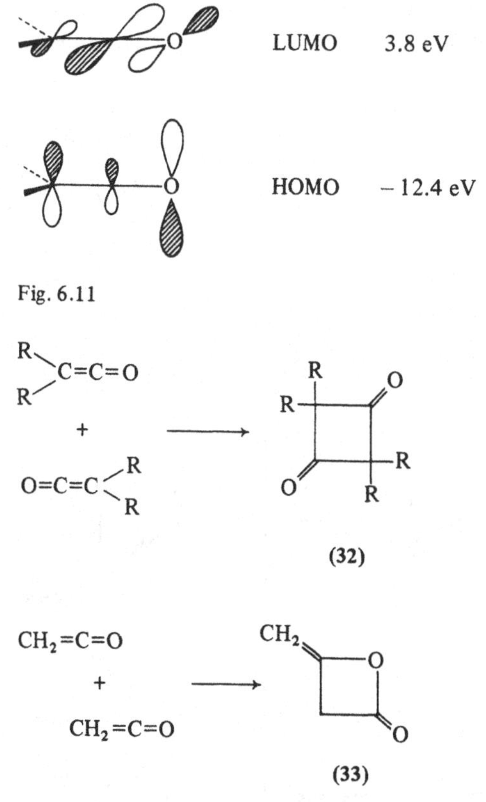

Fig. 6.11

Fig. 6.12

Dimerisation of substituted ketens to give 1,3-cyclobutanediones **(32)** is also a concerted process; again frontier orbital theory explains the observed head-to-tail regioselectivity.[45] Keten itself dimerises differently to give the lactone **(33)** (fig. 6.12).

Since keten dimerisation and cycloaddition to olefins can be concerted, it is tempting to speculate that all cumulenic systems may undergo concerted cycloaddition since they have an orthogonal vacant π^* antibonding orbital. Such cycloadditions are not all concerted, however.[44a] The ability of the π^* orbital to interact depends on its energy relative to the π orbital of the olefin. In ketens the π^* orbital is exceptionally low-lying so that efficient mixing with the olefin π orbital can occur. Even so, keten additions can be stepwise where the zwitterionic intermediates in the alternative non-concerted mode are particularly well stabilised. Thus, although the additions of vinyl ethers are always concerted, the additions of enamines can be stepwise.[46] Huisgen has shown that both concerted and stepwise additions through a zwitterion can occur simultaneously.[47] As expected, the stepwise process becomes more important in more polar solvents which can better stabilise the zwitterionic intermediate. It is estimated that the addition of dimethylketen to *N*-isobutenylpyrrolidine

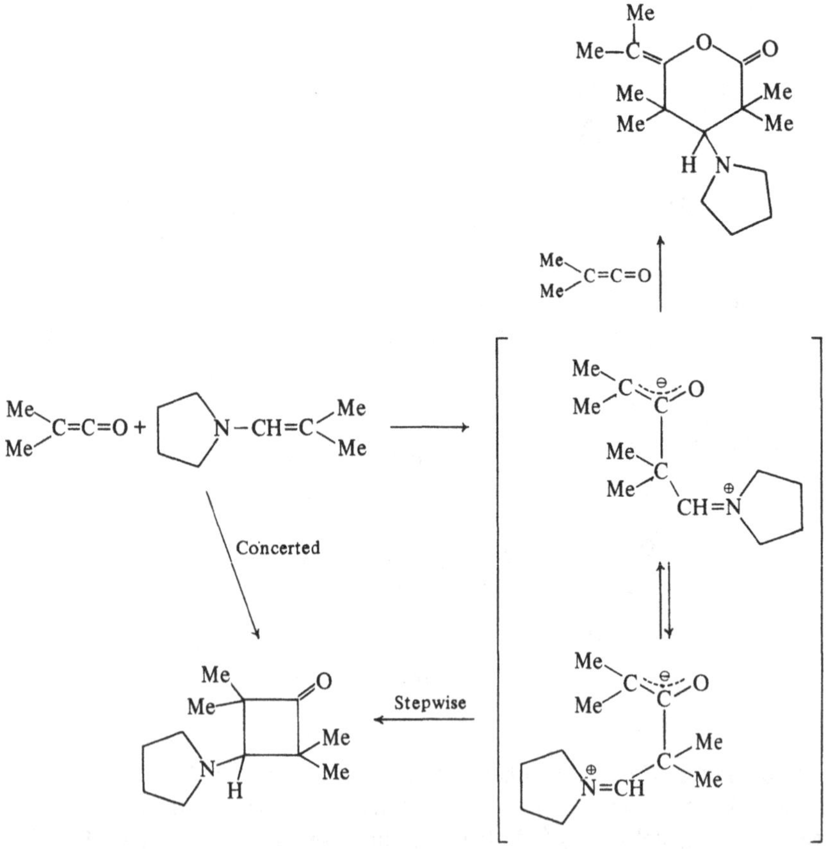

Fig. 6.13

is 57 % stepwise in acetonitrile but only 8 % in cyclohexane. That enamines undergo stepwise addition whereas addition to vinyl ethers is concerted is reasonable since the less electronegative nitrogen atom is better able to accommodate the positive charge in a zwitterionic intermediate than is oxygen. Evidence for the stepwise addition comes from the isolation of 2 : 1 six-membered ring adducts, as shown in fig. 6.13.[47]

The addition of ketens to Schiff bases is also stepwise; 2 : 1 adducts have been isolated[48] (fig. 6.14). The intermediate zwitterions in such reactions have also been trapped by water and methanol[49] (fig. 6.15) and by sulphur dioxide.[50]

Very little systematic investigation into the mechanism of the cycloadditions of other heterocumulenes has been undertaken. It is possible that other hetero-

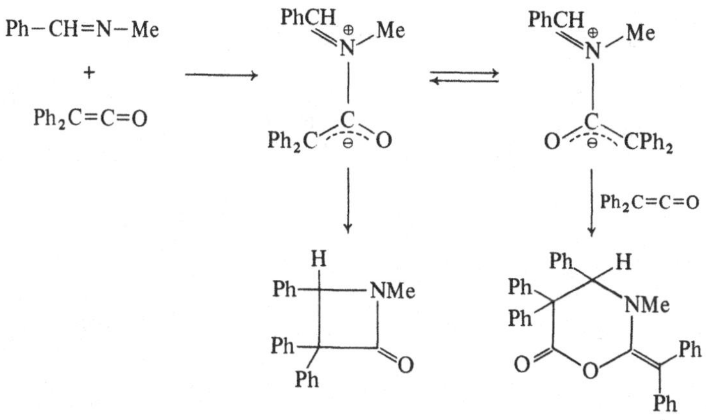

Ph–CH=N–Me + $Ph_2C=C=O$ → PhCH=N⊕(Me)–C(⊖)(O)=CPh_2 ⇌ ...; Ph, H, NMe, Ph, Ph, O; $Ph_2C=C=O$; Ph, H, Ph, Ph, NMe, O, O, Ph, Ph

Fig. 6.14

Ph–CH=N–Ph + $Ph_2C=C=O$ → PhCH=N⊕(Ph)–C(⊖)(O)=CPh_2 → H, Ph, N, Ph, Ph, Ph, O; ROH → OR, PhCH–N–Ph, C=O, Ph_2CH

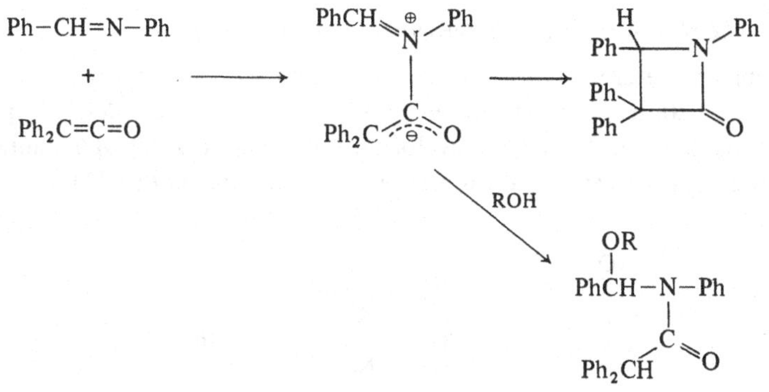

Fig. 6.15

cumulenes can undergo cycloadditions by the allowed concerted $_{\pi}2_s + _{\pi}2_a$ mode. On the other hand, since all heterocumulenes have an electrophilic central atom and nucleophilic terminal atoms, stepwise mechanisms through 1,4-dipolar intermediates are also possible.[21a] Frequently the ionic intermediates are highly stabilised and in many cases the ionic mechanism almost certainly applies. Evidence for this is the trapping of the 1,4-dipolar intermediates, as, for example, in the reaction of ketens with carbodiimides to give azetidinones **(34)**. If water is added the intermediate dipole **(35)** is trapped as **(36)**. Control experiments show that the azetidinones are not converted into **(36)** by reaction with water.[51]

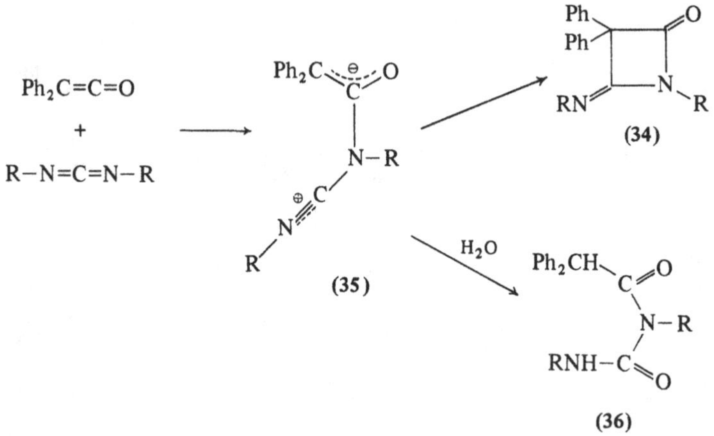

$Ph_2C{=}C{=}O$ + $R{-}N{=}C{=}N{-}R$ → **(35)** → **(34)**; **(35)** + H_2O → **(36)**

(35) **(34)** **(36)**

In summary, concerted $_{\pi}2_s + _{\pi}2_a$ additions are feasible for cumulenes where the π^* orbital is of low energy and where the reactants are such that formation of intermediates is not too favourable.

Other examples of heterocumulene cycloaddition

Examples of heterocumulenes acting as dienophiles are shown in equations 6.19 and 6.20. Because they are such unsymmetrical dienophiles, it is quite likely that many of these 4 + 2 cycloadditions occur in a stepwise manner. Heterocumulenes can also form part of a diene component (6.21).

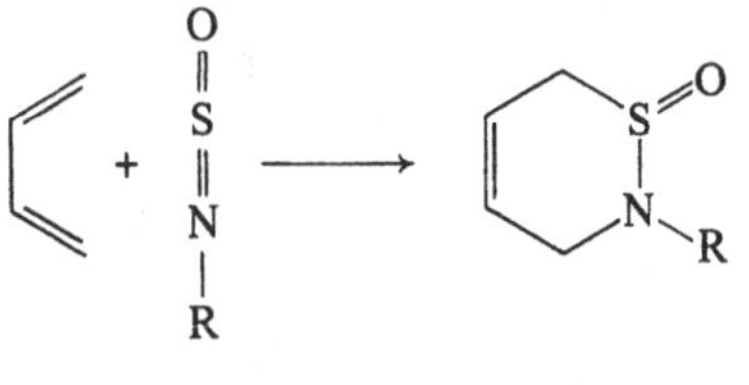

(6.19)

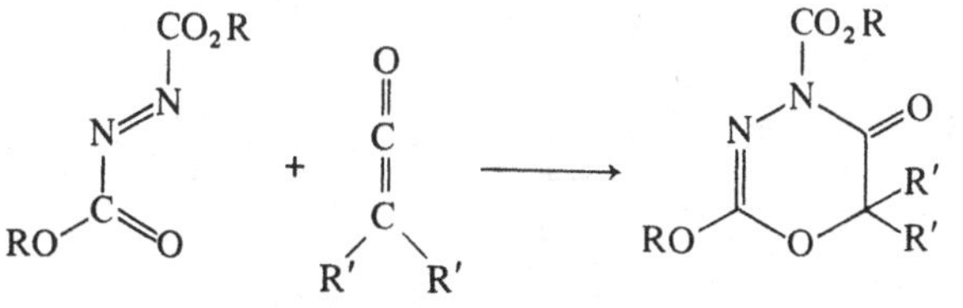

(6.20)

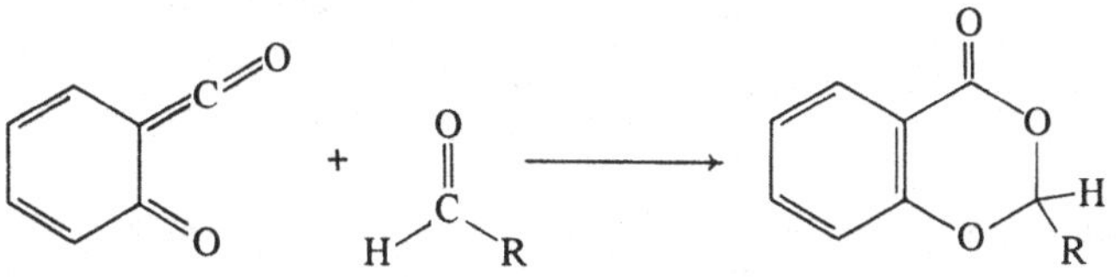

(6.21)

Heterocumulenes are usually efficient 1,3-dipolarophiles. Whether these cycloadditions are concerted is very much open to question since ionic intermediates would often be highly stabilised. Because of the possible combinations of heteroatoms in 1,3-dipoles and in heterocumulenes the synthetic potential of 1,3-dipolar additions to heterocumulenes is immense. Examples of some available ring systems are shown in equations 6.22-6.25.

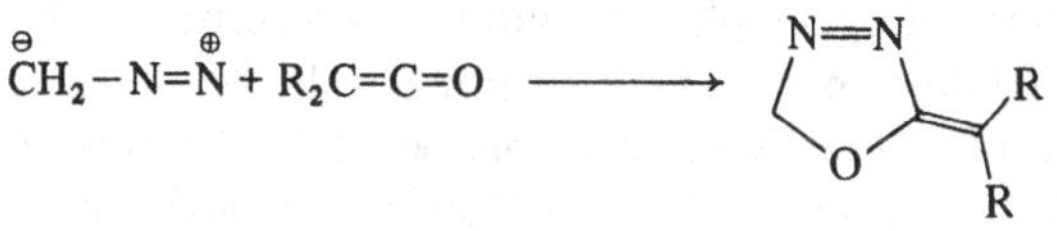

$$\overset{\ominus}{C}H_2-N=\overset{\oplus}{N} + R_2C=C=O \longrightarrow \qquad (6.22)$$

$$R_2\overset{\ominus}{C}-\underset{}{\overset{R}{N}}-\overset{\oplus}{N}-R + R-N=C=O \longrightarrow \qquad (6.23)$$

$$\overset{\ominus}{C}H_2-N=\overset{\oplus}{N} + R-N=C=N-R \longrightarrow \qquad (6.24)$$

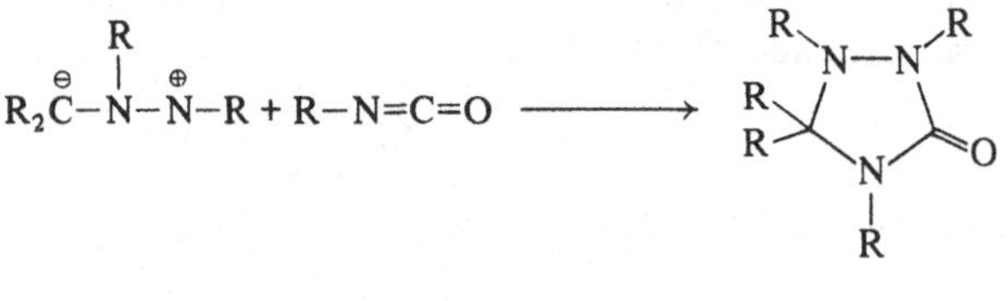

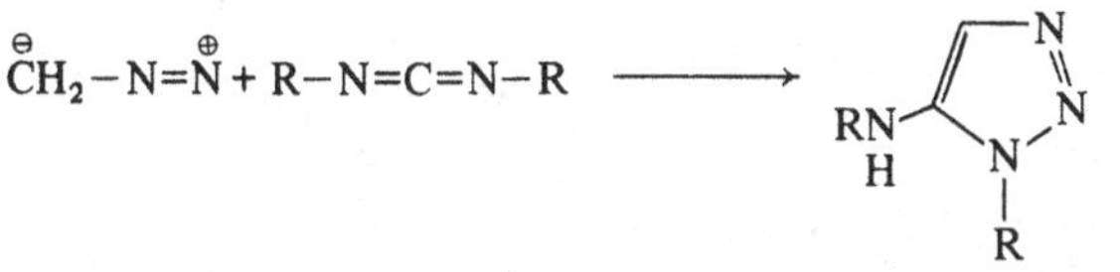

$$R-\overset{\ominus}{C}=N-\overset{\oplus}{O} + R-N=S=O \longrightarrow \qquad (6.25)$$

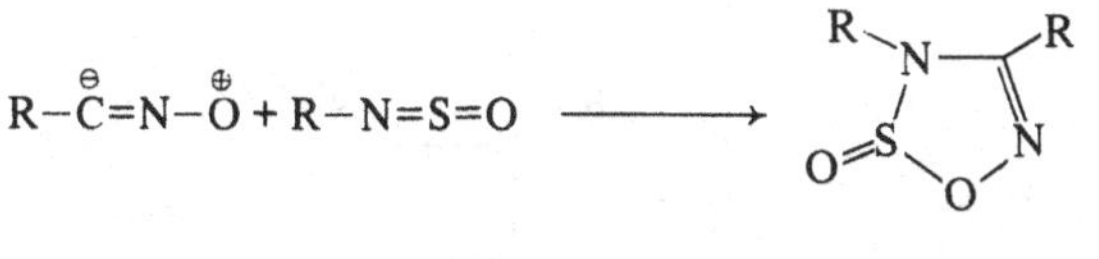

Heterocumulenes readily undergo 2 + 2 cycloadditions to form a variety of four-membered heterocycles. The more reactive heterocumulenes tend to dimerise readily and this often leads to difficulty in their isolation (6.26-6.28).

$$R_2C{=}C{=}O \longrightarrow \text{(cyclobutane-1,3-dione)} + \text{(4-alkylidene-2-oxetanone)} \qquad (6.26)$$

Less commonly

$$R_2C{=}C{=}S \longrightarrow \text{(2,4-dialkylidene-1,3-dithietane)} \qquad (6.27)$$

$$R{-}N{=}C{=}N{-}R \longrightarrow \text{(1,3-diazetidine-2,4-diimine)} \qquad (6.28)$$

Cycloadditions of one heterocumulene to another also occur readily (for example, 6.29 and 6.30). In many cases the resulting cycloadducts are unstable and fragment so that overall interchange reactions occur. By selective removal of one component this type of reaction can be made synthetically useful; an example is the formation of unsymmetrical carbodiimides from isocyanates and symmetrical carbodiimides (6.31).[37]

$$R{-}N{=}C{=}O + R{-}N{=}S{=}N{-}R \longrightarrow \text{(cycloadduct)} \qquad (6.29)$$

$$R_2C{=}C{=}O + R{-}N{=}C{=}O \longrightarrow \text{(cycloadduct)} \qquad (6.30)$$

$$R^1{-}N{=}C{=}O + R^2{-}N{=}C{=}N{-}R^2 \rightleftharpoons \text{(cycloadduct)} \rightleftharpoons R^2{-}N{=}C{=}O + R^1{-}N{=}C{=}N{-}R^2 \qquad (6.31)$$

2 + 2 Cycloadditions to a great variety of other multiple bonds are known. Additions of heterocumulenes to C=C, C≡C, C=O, C=N, C=S, N=O, N=N, N=S, S=O, P=C, P=O, P=N and P=S have been reported (6.32-6.35),[37] and some of these are very useful in synthesis. For example, 2 + 2 additions of chlorosulphonyl isocyanate have been extensively exploited as a route to β-lactams (6.32).

$R_2C{=}C{=}O + R_2C{=}N{-}R$ → β-lactam (ring: R₂C–C(=O)–N(R)–CR₂) ← $R{-}N{=}C{=}O + R_2C{=}CR_2$ (6.32)

$R_2C{=}C{=}O + R{-}N{=}O \longrightarrow$ (ring: R₂C–C(=O)–N(R)–O) (6.33)

$R{-}N{=}C{=}O + R_2C{=}N{-}R \longrightarrow$ (ring: R–N–C(=O)–N(R)–CR₂) (6.34)

$R_2C{=}C{=}O +$ R–N=N–R $\longrightarrow$ (ring: R₂C–C(=O)–N(R)–N(R)) (6.35)

cis more reactive than *trans*

Cycloadditions of allenes[1a,c]

On heating, allenes dimerise and also undergo cycloadditions to activated olefins, giving methylenecyclobutanes. In the past these reactions have been assumed to involve diradical intermediates, a view which fitted in well with initial orbital symmetry predictions that 2 + 2 cycloadditions must be step-wise. The orientations of the dimerisations and cycloadditions to olefins are consistent with processes which proceed through the most stabilised diradical intermediate. In allene cycloadditions, initial bond formation is generally assumed to occur to the central allene carbon since this leads to an allyl stabilised radical (after 90° rotation of one of the methylene groups). However, because such rotation must precede appreciable stabilisation, it seems unlikely that this effect alone determines the site of initial bond formation.

Head-to-head dimers (**37**) are always formed (only allene itself forms a minor amount of head-to-tail dimer) and the additions in fig. 6.16 are clearly those resulting from the most stabilised diradical.[52] Other characteristics of the reactions are similar to those of other 2 + 2 radical cycloadditions (§6.3).

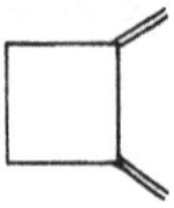

(**37**)

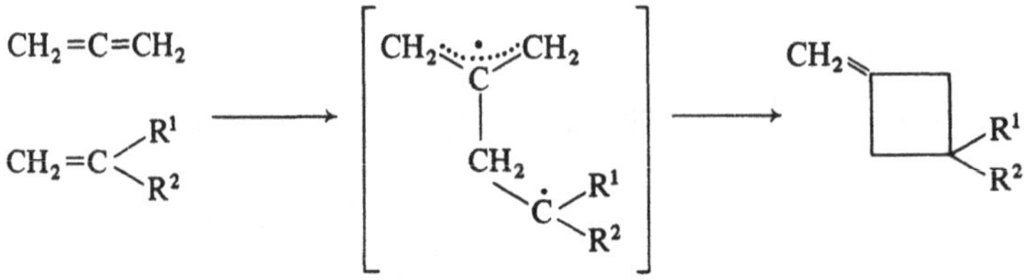

R^1	CN	CN	CO_2Me	CO_2H	CHO	CH_2CO_2Et	CO_2Et
R^2	Me	OAc	H	H	Me	CO_2Et	CO_2Et

Fig. 6.16

However, the demonstration of and rationalisation of concerted keten additions and the observation of high stereospecificity in certain allene cycloadditions have led to speculation that the additions may be concerted $_{\pi}2_s + _{\pi}2_a$ processes. Diethyl maleate and fumarate add stereospecifically to 1,1-dimethyl-

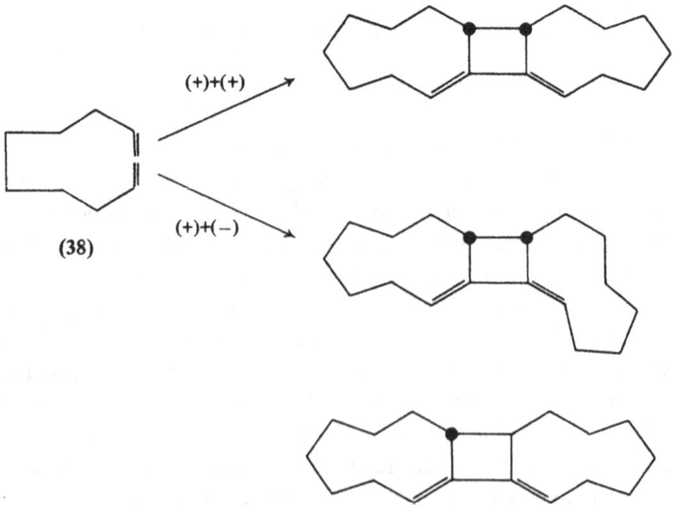

(**38**)

allene without any of the rotation expected from a diradical intermediate. The related addition to acrylonitrile is virtually independent of solvent polarity, thus ruling out a highly stereoselective stepwise reaction through a zwitterion.[53] Optically active adducts are formed in the addition of optically active allenes to olefins, indicating stereospecificity in the allene component also.[54]

Optically active and racemic cyclo-1,2-nonadienes **(38)** give dimers whose stereochemistries are precisely those expected for the allowed $_{\pi}2_s + _{\pi}2_a$ addition.[55]

Although these results appear to be consistent with a concerted $_{\pi}2_s + _{\pi}2_a$ mechanism for allene cycloadditions, some workers maintain that the reactions are stepwise, the stereospecificity of the second bond formation being attributed to other factors. For example, an intermediate such as **(39)** may be so sterically crowded that bond rotations are unable to compete with ring closure.

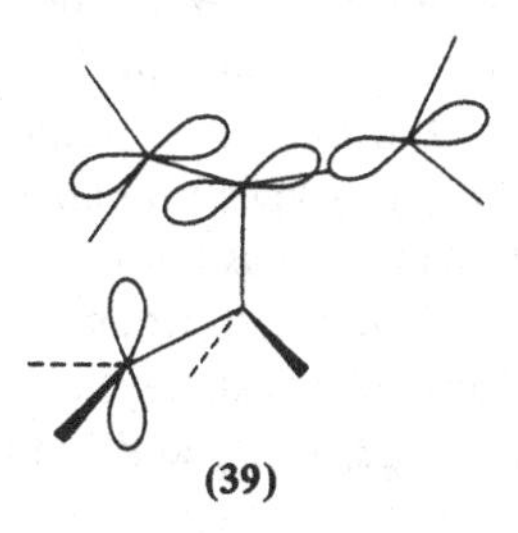

(39)

A concerted addition is likely to be less favourable than it is with ketens since the π^* orbital of allenes is energetically less accessible for efficient secondary interaction.

The main evidence for an intermediate comes from isotope studies.[56] The dimerisation of tetradeuterioallene occurs at the same rate as the dimerisation of undeuterated allene, suggesting that only bond formation between the central carbon atoms is important in the transition state. Very small or negligible intermolecular secondary deuterium isotope effects are also observed in the addition of 1,1-dideuterioallene to 1,1-dichloro-2,2-difluoroethylene and to acrylonitrile. On the other hand, the dimerisation of 1,1-dideuterioallene shows an intramolecular kinetic isotope effect, $k_H/k_D = 1.14$. Deuterium is found predominantly in the methylene groups rather than in the cyclobutane ring, indicating that ring closure to CH_2 is faster than that to CD_2. Thus the reaction must involve an intermediate which differentiates between two modes of ring closure in a step which is not overall rate determining.

Also the dimerisation of 1,1-dimethylallene gives the same three products, **(40)** to **(42)**, as are formed from the bis allylic radical **(44)** generated by thermolysis or photolysis of the azo compound **(43)**.[57]

Thus while the mechanism of allene cycloaddition is not finally settled a stepwise process seems most likely.

(40) (41) (42) (43)

(44)

Cycloadditions of vinyl cations

Vinyl cations have the structure shown in fig. 6.17. They are formed by addition of electrophiles to acetylenes or allenes. The vacant p orbital orthogonal to the olefin π bond is ideally suited for stabilisation of the transition state for concerted ${}_{\pi}2_s + {}_{\pi}2_a$ addition by overlap with the HOMO of the other component (fig. 6.18). They can therefore be considered as the prototypes for concerted cumulene additions; the vacant p orbital plays the same role as the π^* orbital of the cumulene.[25] In addition to the stabilisation discussed above, since vinyl cations have an sp-hybridised carbon, these additions are less susceptible to the unfavourable steric interactions which inhibit the ${}_{\pi}2_s + {}_{\pi}2_a$ addition of one olefin to another.

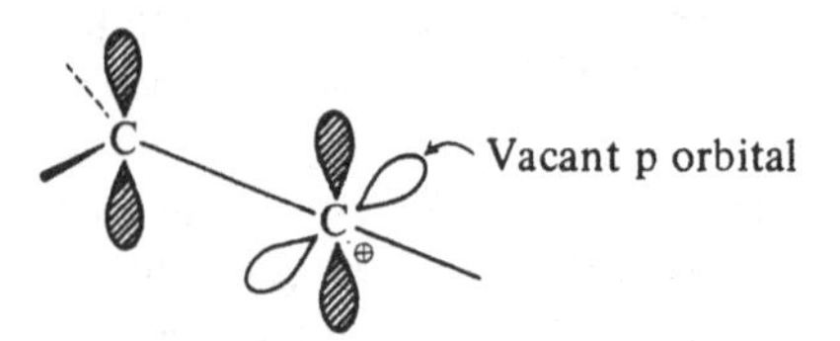

Fig. 6.17

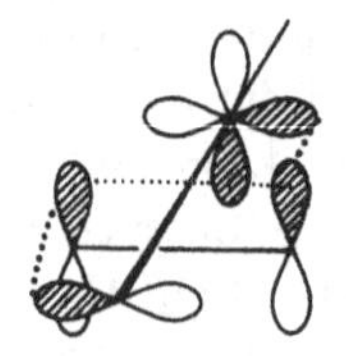

Fig. 6.18

Reactions which can be rationalised by the facile 2 + 2 addition of vinyl cation intermediates[58] are the formation of the cyclobutane (**45**) from allene and HCl (6.36),[59] and of the cyclobutene (**46**) from 2-butyne and chlorine (6.37).[60]

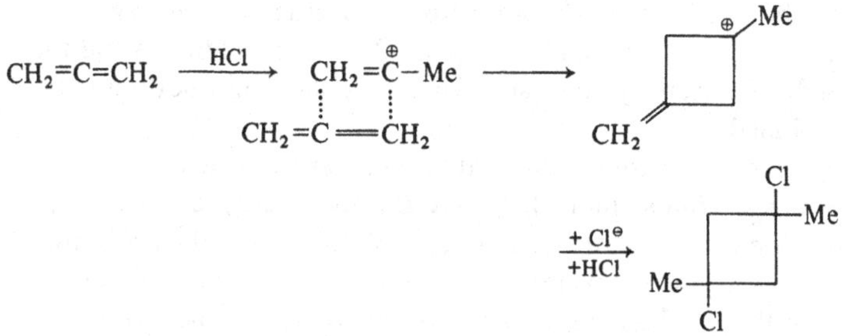

$$CH_2{=}C{=}CH_2 \xrightarrow{HCl} \begin{matrix} CH_2{=}\overset{\oplus}{C}{-}Me \\ \vdots \quad \vdots \\ CH_2{=}C{=}CH_2 \end{matrix} \longrightarrow \text{(3-methylenecyclobutyl cation, } \overset{\oplus}{C}{-}Me\text{)} \xrightarrow[+HCl]{+Cl^{\ominus}} \text{1,3-dichloro-1,3-dimethylcyclobutane} \quad (6.36)$$

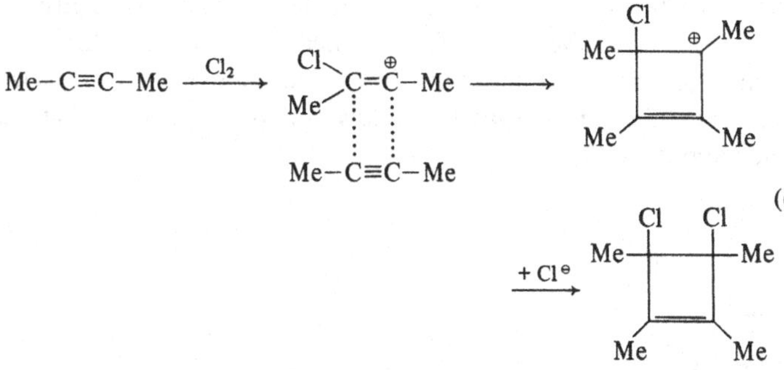

(45)

$$Me{-}C{\equiv}C{-}Me \xrightarrow{Cl_2} \begin{matrix} Cl(Me)C{=}\overset{\oplus}{C}{-}Me \\ \vdots \quad \vdots \\ Me{-}C{\equiv}C{-}Me \end{matrix} \longrightarrow \text{(Cl, Me, Me, Me, } \overset{\oplus}{}Me\text{ cyclobutenyl cation)} \xrightarrow{+Cl^{\ominus}} \text{3,4-dichloro-1,2,3,4-tetramethylcyclobutene} \quad (6.37)$$

(46)

6.5. Retro-2 + 2 additions (fig. 6.19)[21d]

Fig. 6.19

Concerted, retro-2 + 2 cycloadditions would have to be ${}_{\sigma}2_s + {}_{\sigma}2_a$ processes and this would involve considerable twisting of the four-membered ring (fig. 6.20). The concerted reaction would be expected to be more energy demanding than a stepwise reaction involving cleavage of one σ bond to give a 1,4-diradical. In line with this, cleavage of cyclobutanes requires high temperatures and shows the characteristics of stepwise processes. In some cases, however, there is evidence that the unfavourable ${}_{\sigma}2_s + {}_{\sigma}2_a$ mode does compete.

Fig. 6.20

Where the cyclobutane bears highly polar groups, cleavage to give a zwitterion may occur readily; the reversibility of polar 2 + 2 cycloadditions and the generation of 1,4-dipoles from cyclobutanes for 1,4-dipolar cycloaddition has already been mentioned.

Retro-2 + 2 cycloadditions for systems with non-polar substituents have high activation energies ($E_a \simeq 250$ kJ mol^{-1}).[61] They also show fairly large positive entropies of activation indicative of a large release of ordering in the transition state. This is expected for the breaking of one σ bond to give a 'floppy' intermediate but not for the simultaneous breaking of two σ bonds, where the potential fragments are still highly ordered with respect to each other in the transition state. For example, concerted retro-Diels–Alder reactions generally show $\Delta S^{\ddagger} \sim 0$. Cleavage of substituted cyclobutanes proceeds to give mainly fragments derived from the most stable diradical (fig. 6.21) and those cyclobutanes which give more stabilised diradicals cleave more easily. Thus isopropenylcyclobutane (**47**) decomposes with lower activation energy than does isopropylcyclobutane (**48**).[61]

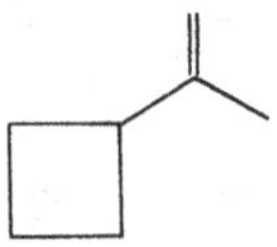

(47)

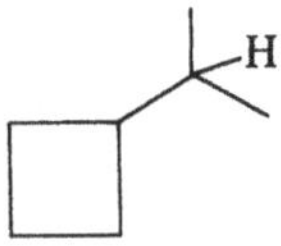

(48)

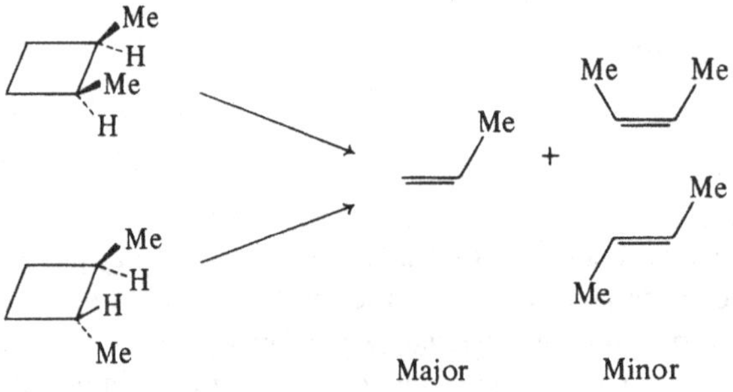

Fig. 6.21

The formation of mixtures of butenes from both *cis*- and *trans*-dimethylcyclobutanes (fig. 6.21) shows that some loss of configuration has occurred.[62] It could be argued that since a concerted ${}_{\sigma}2_s + {}_{\sigma}2_a$ process involves inversion of one fragment, such stereochemical results are also consistent with concerted cleavage in which inversion either of the butene or of the ethylene fragment occurs (fig. 6.22).

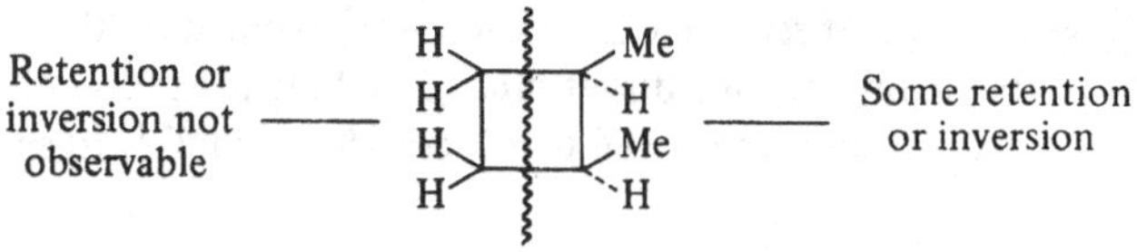

Retention or inversion not observable — H H H H | Me H Me H — Some retention or inversion

Fig. 6.22

However, systems have been chosen in which the configuration of one of the olefins produced is fixed as *cis* by virtue of it being in a small ring. The stereochemistry of the other fragment is then studied. For a concerted ${}_{\sigma}2_s + {}_{\sigma}2_a$ reaction its configuration should be completely inverted. The *trans*-dimethylbicycloheptane **(49)** gives 2-butene which is 75 % *trans* (6.38). The *cis*-isomer **(50)** gives equal amounts of *cis*- and *trans*-butene. This is just as expected for a diradical intermediate in which cleavage of the second σ bond competes with bond rotation; the driving force for rotation is greater in the diradical in which two methyl groups are *cis*.[63]

The bicyclo-octane **(51)**, however, gives cyclohexene and dideuterioethylene which is 57–62 % *trans* (6.39). A diradical intermediate in which bond rotation reached equilibrium before cleavage of the second bond, should give 50 % *cis* and 50 % *trans*; if cleavage occurred before complete equilibration then *cis* should predominate.[64] These results suggest, therefore, that the concerted ${}_{\sigma}2_s + {}_{\sigma}2_a$ process is competing to some extent with the stepwise diradical route. (Note that in this case the diradical would be less stabilised.)

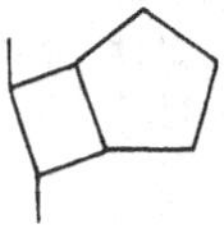

(49) 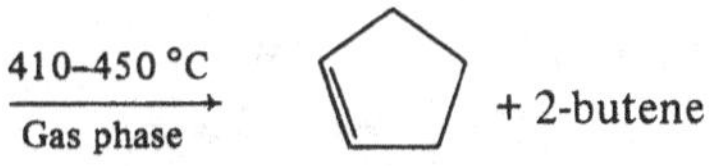 —410–450 °C, Gas phase→ cyclopentene + 2-butene (6.38)

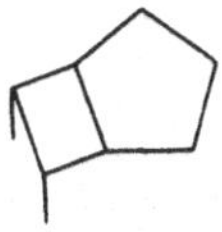

(50)

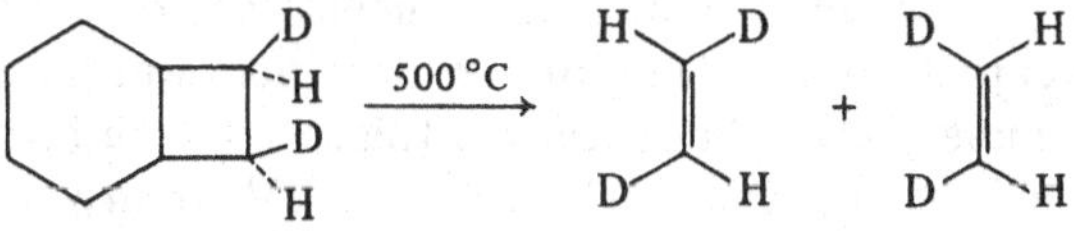

(51) —500 °C→ HDC=CDH (57–62%) + DHC=CDH (6.39)

The disallowed nature of retro-2 + 2 addition is underlined by the remarkable stability of the highly strained diazetidine **(52)**. Fragmentation of **(52)** to norbornene and nitrogen is estimated to release 250 kJ mol^{-1}. In spite of

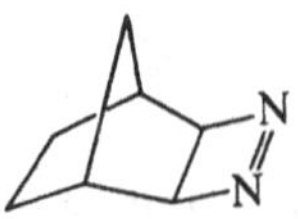

(52) 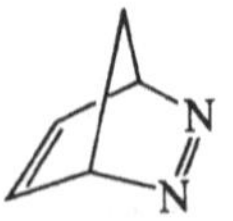 **(53)**

this, decomposition only occurs on heating and the activation parameters are $\Delta H^{\ddagger} \triangleq 140$ kJ mol^{-1}, $\Delta S^{\ddagger} \triangleq 12$ J mol^{-1} K^{-1}. This is in striking contrast to compound **(53)** which is far less strained but which loses nitrogen as fast as it is formed, even at –78 °C. The rates of nitrogen loss from **(52)** and **(53)** differ by a factor of about 10^{22} and a difference of activation enthalpy of about 75 kJ mol^{-1} is estimated for the two processes.[65] This reflects very well the difference between a symmetry-allowed concerted retro-Diels–Alder reaction and a reaction for which either the geometrically favoured mode is disallowed (${}_{\sigma}2_s + {}_{\sigma}2_s$), or the allowed mode is geometrically very unfavourable (${}_{\sigma}2_s + {}_{\sigma}2_a$), and which therefore most likely proceeds by diradical stepwise loss of nitrogen.

Bicyclobutanes are formed by irradiation of dienes. In spite of the strain in such molecules they are remarkably resistant to isomerisation to the much more stable butadienes. This high-temperature isomerisation, however, appears to be a sterically unfavourable but concerted ${}_{\sigma}2_s + {}_{\sigma}2_a$ process rather than a stepwise reaction with a diradical intermediate. The ${}_{\sigma}2_s + {}_{\sigma}2_a$ stereochemistry is illustrated by the examples shown in equations 6.40 and 6.41.[66]

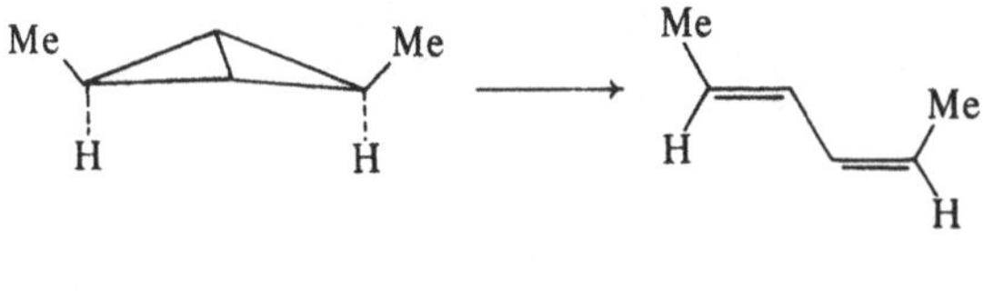

Me Me H H → Me H Me H (6.40)

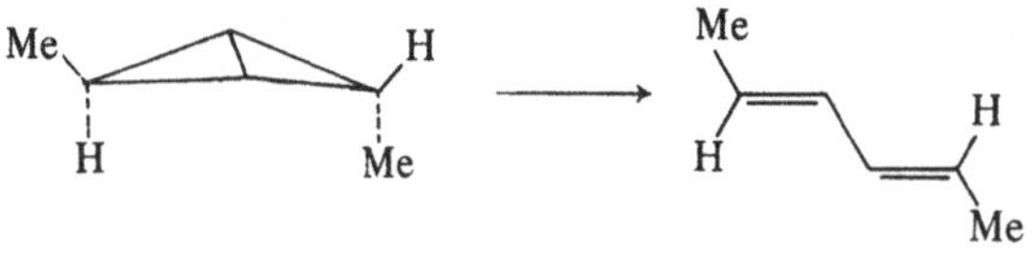

Me H H Me → Me H H Me (6.41)

Since concerted ${}_{\pi}2_s + {}_{\pi}2_a$ addition of certain cumulenes to olefins occurs readily, it might be expected that those retro-2 + 2 additions which lead to a cumulene would be more likely to be concerted. This seems to be borne out by the often ready and stereospecific loss of carbon dioxide from β-lactones (6.42).[67]

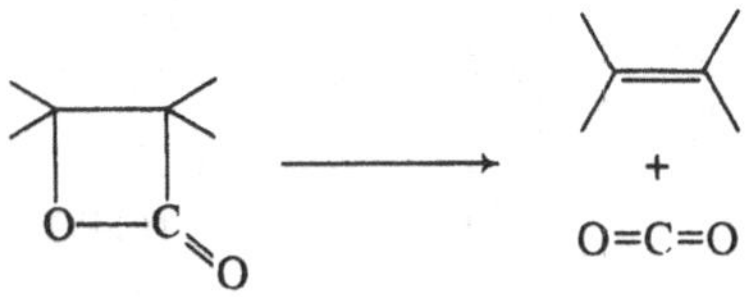

(6.42)

β-Lactams also lose HN=C=O stereospecifically on strong heating (6.43), whereas in both the *cis*- and *trans-O*-alkylated lactams (**54**), under similar conditions, only one bond is cleaved to give the same three products (6.44).[68]

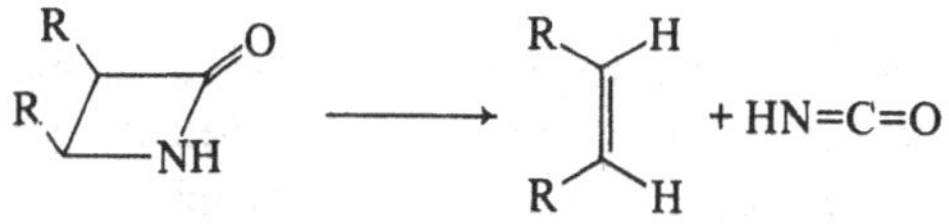

+ HN=C=O (6.43)

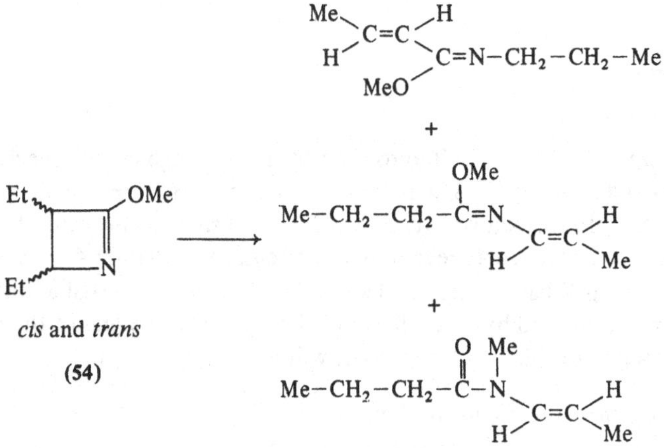

cis and *trans*

(**54**)

(6.44)

A particularly interesting retro-2 + 2 addition is the cleavage of 1,2-dioxetanes. If this is a concerted retro-$_{\pi}2_s + _{\pi}2_s$ reaction, then one of the keto groups must be formed in an excited state. A case where this occurs is the thermal decomposition of the dioxetane (**55**) which initiates 'photochemical reactions' of other substrates present (6.45). There are several such examples of photochemistry without light.[69]

The electronically excited species which are responsible for certain types of chemiluminescence and bioluminescence also probably arise by this type of dioxetane cleavage.[70]

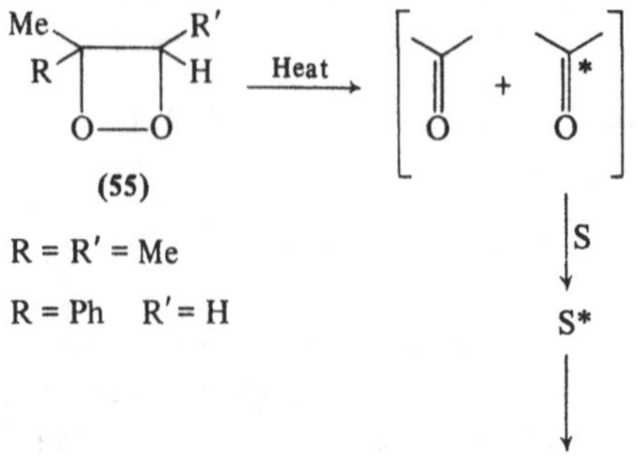

Me R' R H O—O **(55)** Heat → [O + O*] → S → S* → Photo products (6.45)

R = R' = Me

R = Ph R' = H

6.6. Cheletropic reactions

The cycloaddition of an atom or group X to an olefin to form a three-membered ring and the reverse process constitute a further type of four-electron cycloaddition or elimination. If concerted, such reactions are examples of cheletropic processes (fig. 6.23).

X ⇌ X (three-membered ring)

Fig. 6.23

The forward reaction is limited to the additions of high-energy species such as carbenes and nitrenes.[21b] The reverse reaction occurs more widely and a variety of extrusions is known where X is usually a small stable molecule.[71]

The theory of cheletropic reactions was discussed in chapter 4. It was shown that when X has a vacant and a filled orbital the concerted addition or extrusion must proceed by a non-linear cheletropic route. Some of the better known examples of this type of reaction will now be discussed.

Addition of carbenes and nitrenes[21b, 72]

Carbenes ($:CR_2$) and nitrenes ($:\dot{N}R$) are short-lived reactive intermediates which are electron deficient; they contain a carbon or nitrogen atom with two non-bonding orbitals between which are distributed two electrons. Both the electrons can be in the same orbital, **(56)**, or one electron may be in each, **(57)** and **(58)** (fig. 6.24).

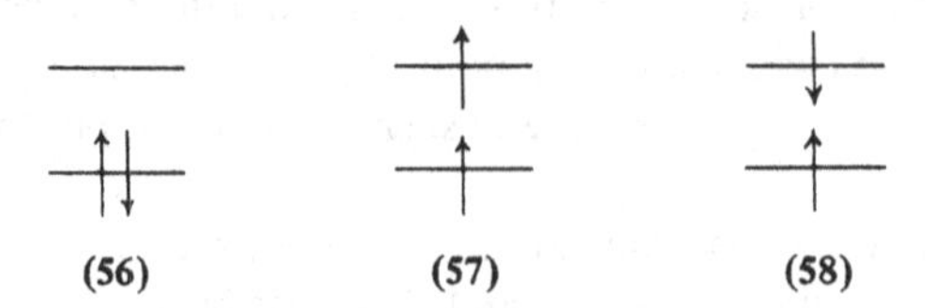

Fig. 6.24

Possible non-linear (**59**) and linear (**60**) structures for a carbene $:CR_2$ are shown. The bent structure, which accentuates the difference in energy of the non-bonding orbitals, is likely to be favoured for the singlet ground configuration (**56**), whereas the excited configurations (**57**) and (**58**) are likely in a linear or nearly linear structure such as (**60**), where the non-bonding orbitals are close or equivalent in energy.

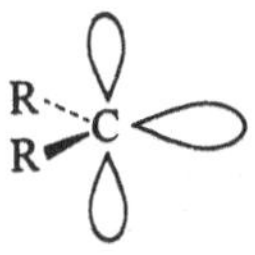

(**59**)  (**60**)

Depending on the nature of the substituents, either configuration (**56**) or configuration (**57**) can be the actual ground state (the lowest energy state) of the carbene or nitrene. Most ground states are triplets, of type (**57**), but the singlet configuration (**56**) is normally the ground state when both substituents have a lone pair (F, Cl, Br, OR, NR_2, etc.). This can be envisaged, in valence bond terms, as being due to resonance stabilisation of configuration (**56**) by the lone pair on the heteroatom (fig. 6.25).

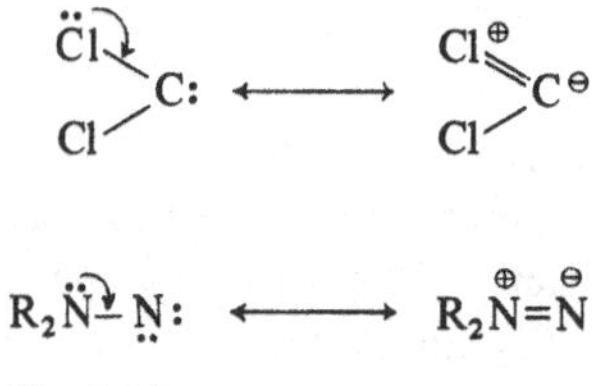

$Cl_2C: \longleftrightarrow Cl^{\oplus}{=}C^{\ominus}(Cl)$

$R_2\ddot{N}-\ddot{N}: \longleftrightarrow R_2\overset{\oplus}{N}=\overset{\ominus}{N}$

Fig. 6.25

Carbenes and nitrenes add to olefins, to give cyclopropanes and aziridines, respectively. Such reactions are synthetically very useful. These additions are not given by all carbenes and nitrenes: many preferentially undergo rearrangement, fragmentation, insertion, or hydrogen abstraction. Sometimes these reactions can be avoided by using metal catalysis in the decomposition of the carbene precursors, or by using organometallic sources of the intermediates. For example, thermal uncatalysed decomposition of diazoketones gives mainly products of the Wolff rearrangement, but in the presence of a copper catalyst, cycloaddition to an olefin can compete (6.46). In such cases, the cycloadditions probably do not involve free carbenes, but 'carbenoids' – a term loosely used to describe complexed carbenes or carbene-like intermediates. Intermolecular cycloadditions by free, uncomplexed carbenes and nitrenes are limited to relatively few types; the major ones are shown in table 6.4.

$$\mathrm{RCOCHN_2} \xrightarrow{\text{Heat}} \mathrm{R{-}CH{=}C{=}O + N_2} \tag{6.46}$$

Cu + (tetramethylethylene), Heat → R–C(=O)(H)–tetramethylcyclopropane

Table 6.4 Carbenes and nitrenes which commonly add to olefins

Carbenes	
methylene	$:CH_2$
dihalogenocarbenes	$:CX_2$ (X = F,Cl,Br)
alkoxycarbonylcarbenes	$:CHCO_2R$
vinylidene carbenes	$:C{=}CR_2$
diarylcarbenes	$:CAr_2$
arylhalogenocarbenes	:CArX
atomic carbon	:C:
Nitrenes	
alkoxycarbonylnitrenes	$:\ddot{N}CO_2R$
cyanonitrene	$:\ddot{N}CN$
aminonitrenes	$:\ddot{N}NR_2$

In some cases, the addition is stereospecific, and in others it is not. Stereospecific addition to olefins is observed with carbenoids, and for carbenes which have been shown independently to have singlet ground states of type **(56)**; examples are dihalogenocarbenes, such as $:CCl_2$ and $:CF_2$, and aminonitrenes (diazenes), $:\ddot{N}NR_2$. With species for which the ground state is probably a triplet, **(57)**, the pattern is more complex. These intermediates may be generated in configuration **(56)**, and if they react as such before decay to the ground state, their addition is stereospecific. If they are deactivated by collision before addition, the reaction is non-stereospecific.

This pattern is in accord with the orbital symmetry predictions. Only a species of configuration **(56)**, with a filled and a vacant orbital, can participate in the concerted, non-linear cheletropic mode of addition described in chapter 4. The cycloaddition of species **(57)** or **(58)** must be stepwise and is therefore likely to be non-stereospecific. In the particular case of the addition of methylene, $:CH_2$, to ethylene, Hoffmann has carried out molecular orbital calculations which show that the excited configurations **(57)** and **(58)** of methylene react with ethylene to produce an excited configuration of cyclopropane (6.47), which is an open three-carbon intermediate (trimethylene) with no barriers to rotation.[73]

$$H_2\dot{C}\cdot + \begin{matrix} CH_2 \\ \| \\ CH_2 \end{matrix} \longrightarrow H_2\dot{C}-CH_2-\dot{C}H_2 \qquad (6.47)$$

This excited configuration of trimethylene has no particular driving force for ring closure and will therefore presumably undergo rotation about the C—C bonds before collapsing to cyclopropane.

The stereospecificity of the addition therefore depends on whether the species reacts in the electronic configuration **(56)**, with both electrons in one orbital, or in the configuration **(57)** or **(58)**, with one electron in each. It is *not* primarily the spin state which determines the stereospecificity, but the distribution of the electrons.

An earlier, widely accepted explanation for the stereospecificity of the additions of the different types of carbene was based on the principle of the conservation of spin. It was assumed that only spin-paired singlet carbenes could add in a concerted manner; triplet carbenes would necessarily produce a triplet diradical intermediate which could only ring close after spin inversion. Since this was assumed to be slow compared to bond rotation, addition of a triplet carbene would be non-stereospecific. This theory, known as the *Skell hypothesis*, has proved to be a very useful practical guide.[74]

Extrusions

The overall picture for extrusion reactions is by no means clear, and general conclusions as to the mechanism cannot be made. Extrusions of some fragments, for example sulphur dioxide and nitrous oxide, are stereospecific, whereas others, for example sulphur monoxide, are not. It is tempting to

$$MeCH_2SO_2CHClMe \rightleftharpoons Me\overset{\ominus}{C}HSO_2CHClMe \xrightarrow{\text{slow}}$$

$$\underset{\textbf{(61)}}{\overset{}{MeCH-CHMe \atop \diagdown S O_2 \diagup}} \xrightarrow{\text{heat}} \underset{\text{stereospecific}}{MeCH{=}CHMe + SO_2} \qquad (6.48)$$

speculate that those extrusions which are stereospecific are concerted and are examples of non-linear cheletropic processes, but this may be an oversimplification.

Fragmentation of episulphones and episulphoxides[75]

Episulphones **(61)** are intermediates in the formation of alkenes from α-chlorosulphones and bases (6.48). They can also be synthesised independently; for example, from a sulphonyl chloride, triethylamine and diazomethane (6.49).

$$RCH_2SO_2Cl + CH_2N_2 + Et_3N \longrightarrow \text{(R-episulphone, } SO_2\text{)} + N_2 + Et_3\overset{\oplus}{N}HCl^{\ominus} \qquad (6.49)$$

When they are heated, alone or in solution, episulphones are fragmented stereospecifically to sulphur dioxide and alkenes. This can be rationalised by a concerted, non-linear extrusion of sulphur dioxide. However, it must be emphasised that this is a rationalisation of the observed stereospecificity, not a proven mechanism. Alternative stepwise routes, involving, for example, initial ring expansion of an episulphone to a four-membered ring isomer, have also been proposed.[76] It may well be that different mechanisms operate in different systems. In contrast, the extrusion of sulphur monoxide from episulphoxides is non-stereospecific and therefore most probably occurs in a stepwise manner (6.50).[77]

$$\text{RCH–CHR (S=O bridged)} \longrightarrow RCH{=}CHR + SO \qquad (6.50)$$

non-stereospecific

Fragmentation of aziridine derivatives

N-Nitrosoaziridines are unstable compounds formed from aziridines and nitrosyl chloride at low temperatures (6.51). At room temperature they decompose stereospecifically giving nitrous oxide and the olefin.[78] Again the result can be rationalised by the 'non-linear extrusion' mechanism.

$$\text{aziridine (N–H; } R^1, R^2, R^3, R^4) \xrightarrow[\text{low temperature}]{NOCl} \text{N-nitrosoaziridine (N–NO; } R^1, R^2, R^3, R^4) \xrightarrow{\text{room temperature}} R^1R^2C{=}CR^3R^4 + N_2O \qquad (6.51)$$

The extrusion of nitrogen from the diazenes (**62**) is more confusing. The deamination of aziridines (**63**) and (**64**, R = Me) by difluoroamine, which proceeds through the diazene, is stereospecific.[79] However, the diazenes (**62**, R = Ph), generated by oxidation of the *N*-aminoaziridines (**65**), fragment non-stereospecifically. The *trans*-isomer gives 100 % *trans*-stilbene; the *cis* gives 85 % *trans* and 15 % *cis*. It is not yet clear whether the extrusion is non-stereospecific or whether the lack of stereospecificity results from isomerisation of the diazene prior to fragmentation.[80]

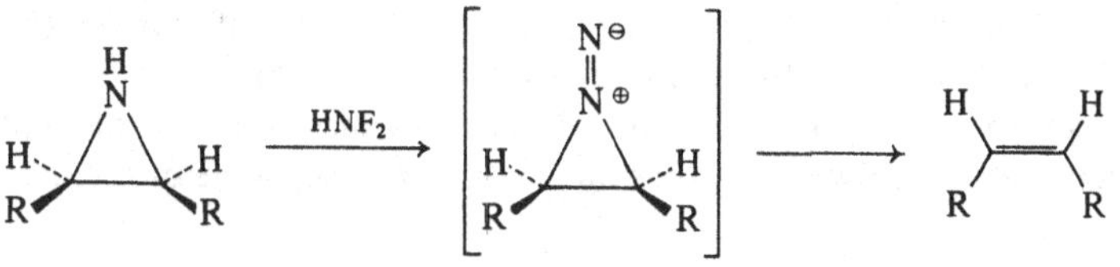

(63) (62)

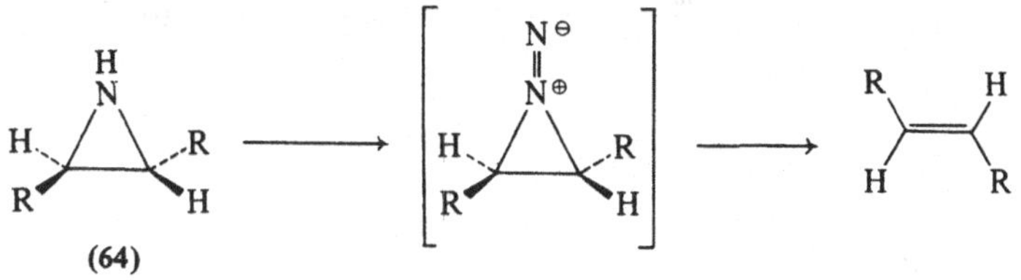

(64)

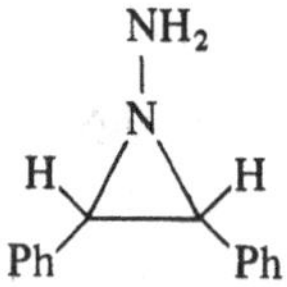

cis and trans

(65)

The *N*-aminoaziridines **(65)** are surprisingly unstable and are fragmented to give an olefin together with a species of stoichiometry H_2N_2, not yet identified.[80]

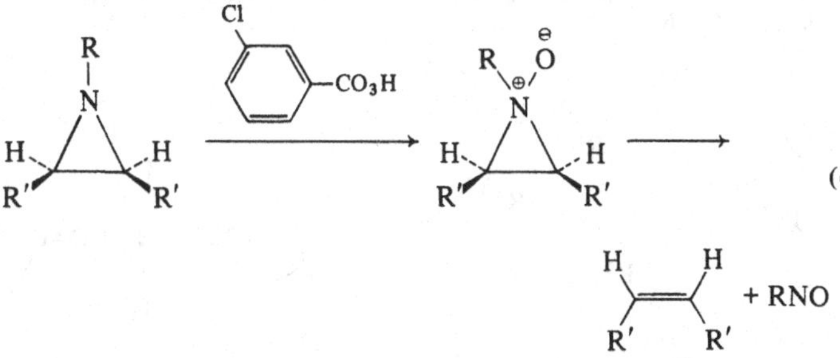

(6.52)

N-Alkylaziridines can be converted stereospecifically to olefins by treatment with *m*-chloroperbenzoic acid. This presumably involves extrusion of a nitroso compound from an intermediate aziridine *N*-oxide (6.52).[81]

Miscellaneous fragmentations

There are several examples of the fragmentation of cyclopropanes to give carbenes (6.53).[72] This is generally a photochemical reaction. Carbenes have also been produced by the photochemical fragmentation of oxiranes (6.54), and nitrenes have been obtained similarly from oxaziridines (6.55). It is not yet possible to draw general conclusions as to the mechanism of these photofragmentations, although there is evidence that at least some of them are concerted.

$$\text{(1,2-diphenylcyclopropane)} \xrightarrow{h\nu} :CH_2 + PhCH{=}CH_2 \qquad (6.53)$$

$$\text{(2,3-diphenyloxirane)} \xrightarrow{h\nu} Ph\ddot{C}H + PhCHO \qquad (6.54)$$

$$\text{(2,2,3-triphenyloxaziridine)} \longrightarrow Ph\ddot{N} + Ph_2CO \qquad (6.55)$$

Cyclopropyl and related carbenes and nitrenes fragment to give acetylenes and nitriles (6.56).[82] This type of extrusion is the basis of a preparatively useful fragmentation of *N*-aziridinylhydrazones (6.57).[83]

$$\text{(cyclopropyl)}{-}\ddot{C}{-}R \longrightarrow \| + R{-}C{\equiv}C{-}R \qquad (6.56)$$

$$\text{(aziridinyl)}N{-}N{=}C(R^1)\text{(epoxide with } R^2, R^3, R^4) \xrightarrow{\text{heat}} \text{(alkene)} + N_2 + R^1C{\equiv}CR^2 + O{=}C(R^3)R^4 \qquad (6.57)$$

6.7. Photochemical 2 + 2 cycloadditions[1a, 21b, d, 84]

Photochemical 2 + 2 cycloadditions have been known for some considerable time but it is only since the 1950s that the synthetic scope of the reaction has been fully exploited.

The situation regarding the mechanism of photochemical 2 + 2 cycloaddition is quite complicated. The first step involves formation of a photo-excited reactant which then interacts with a second reactant in its ground state. Complexes between the excited and ground state reactants, known as exciplexes or excimers, can be detected in some cases.

A variety of excited states is possible. Excitation may involve a $\pi \rightarrow \pi^*$ transition in which an electron is transferred from the π bond (HOMO) to the π^* antibond (LUMO). In systems containing heteroatoms, such as carbonyl compounds, $n \rightarrow \pi^*$ excitation can occur with transfer of an electron from a lone pair orbital to the π^* orbital. Such transitions are generally of lower energy than $\pi \rightarrow \pi^*$ transitions. Initial excitation gives rise to a singlet state but these have a very short lifetime (approximately 10^{-9} s). Intersystem crossing to a lower energy and longer-lived triplet state is frequently faster than intermolecular reactions so that many additions proceed through the triplet state. Photochemical additions are often achieved by the use of sensitisers which absorb the light energy and then transfer it to the reactant, so exciting it indirectly. Such sensitisation normally leads to a triplet excited reactant.

Orbital symmetry considerations indicate that ${}_{\pi}2_s + {}_{\pi}2_s$ addition is allowed if it involves a singlet excited reactant. However, the inability of the reactants to achieve the critical alignment necessary for a concerted reaction within the very short lifetime of the excited singlet state will tend to make such reactions uncommon. It is also possible that the singlet excited reactant interacts with the ground state reactant in a stepwise fashion via a singlet diradical intermediate. In the case of the triplet excited molecule there is a spin inversion barrier to concerted addition, and photochemical additions involving triplet states can best be considered as proceeding through a diradical intermediate. In fact most photochemical additions appear to be stepwise.

As would be expected 2 + 2 photoadditions involving singlet states show higher stereoselectivity than those involving triplet states where loss of configuration is commonly found. Exciplex formation and excitation of charge transfer complexes of ground state reactants have been held responsible in certain cases for determining stereo- and regioselectivity. However, frontier orbital theory again provides the simplest overall rationalisation of regioselectivity.[85] In some solid state 2 + 2 photodimerisations the stereo- and regioselectivity is controlled by crystal structure.[86]

Examples most likely to involve photo-induced ${}_{\pi}2_s + {}_{\pi}2_s$ addition are the direct irradiation of pure tetramethylethylene[87] and of *cis*- and *trans*-butenes. The latter proceeds stereospecifically as shown in fig. 6.26.[88] The photochemical retro- 2 + 2 additions in fig. 6.27 also proceed stereospecifically, suggesting that they are concerted.[89]

There are many examples of non-concerted photochemical 2 + 2 additions. Cyclic alkenes can be photodimerised using high energy triplet sensitisers, such as acetone. Intramolecular photosensitised addition has found wide application in the synthesis of strained systems (e.g. 6.58, 6.59).

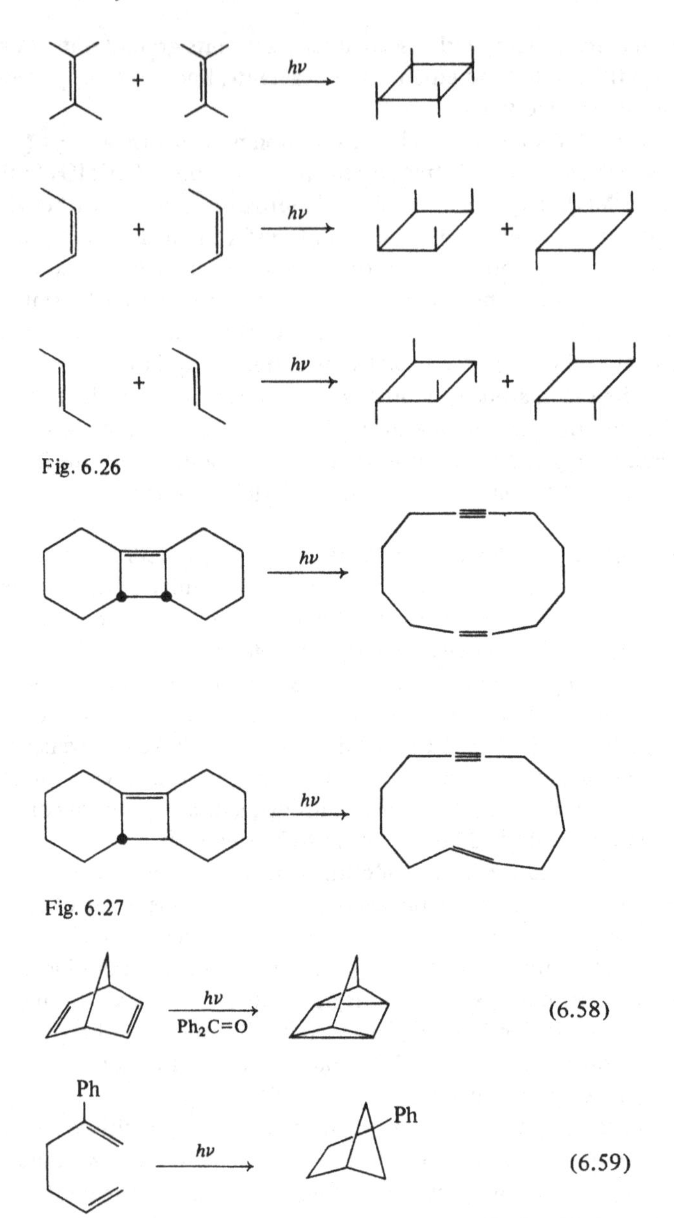

Fig. 6.26

Fig. 6.27

Triplet-sensitised photodimerisation of dienes and photoaddition of dienes to monoenes leads generally to 2 + 2 adducts although some 2 + 4 addition is also observed.[84a, e]

Bicyclobutane (**66**) formation accompanies cyclobutene formation in the direct irradiation of conjugated dienes (6.60). The diene molecules with the *transoid* configuration lead to the bicyclobutane in an intramolecular 2 + 2 addition via the diradical **(67)**.[84e]

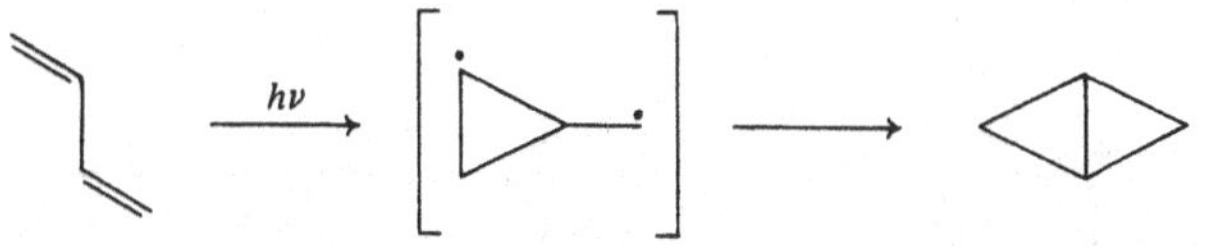

(6.60)

(67) **(66)**

Photoadditions of α,β-unsaturated carbonyl compounds are particularly important.

On irradiation, cyclopentenones and cyclohexenones dimerise or add to olefins, acetylenes and allenes (6.61 and 6.62).[89, 84] These photoadditions are

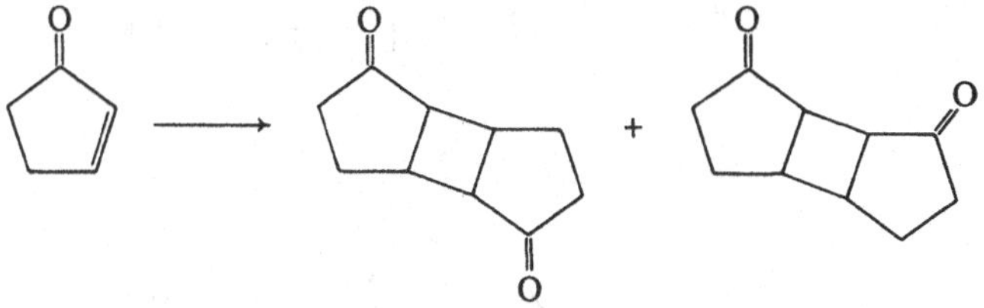

(6.61)

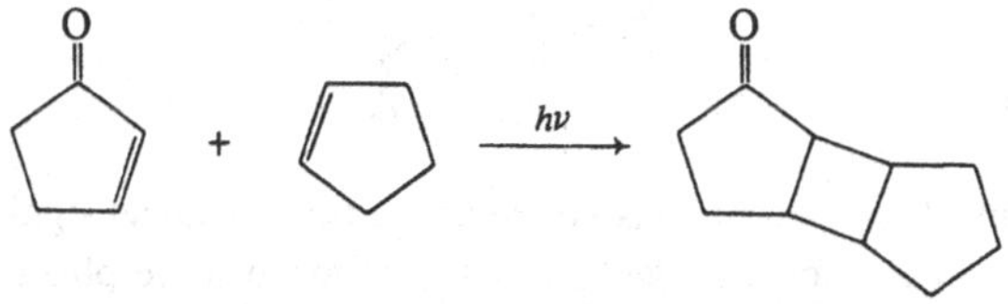

(6.62)

believed to be stepwise and to involve triplet excited enones. They appear to be limited to cyclic enones but otherwise they are quite general and have proved valuable in organic synthesis. A range of novel cage compounds has been

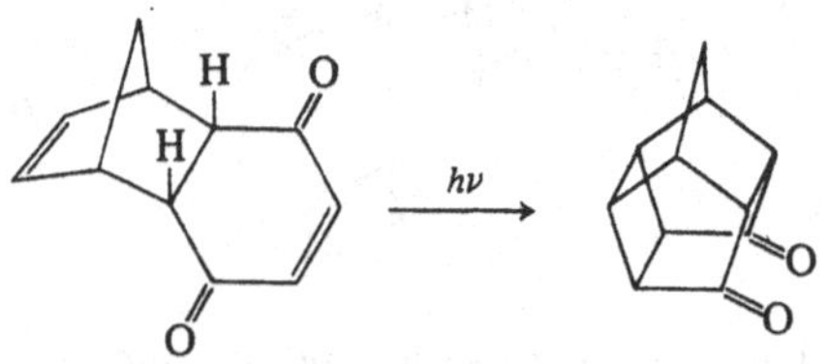

(6.63)

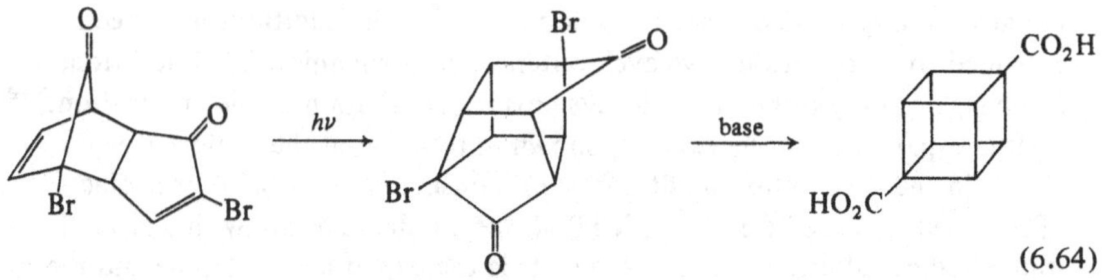

(6.64)

produced by intramolecular reactions of this type (6.63, 6.64). Such photo-additions are key steps in Corey's elegant syntheses of caryophyllene (**68**, 6.65) and α-caryophyllene alcohol (**69**, 6.66).[90]

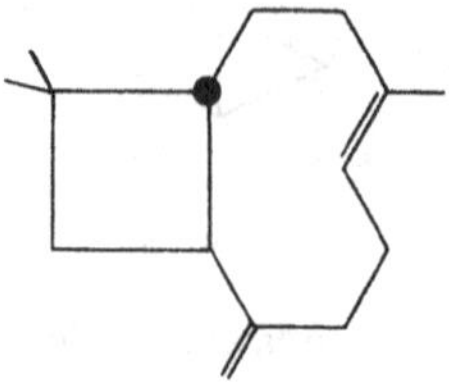

(**68**) 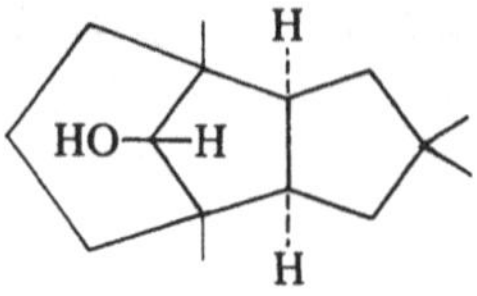 (**69**)

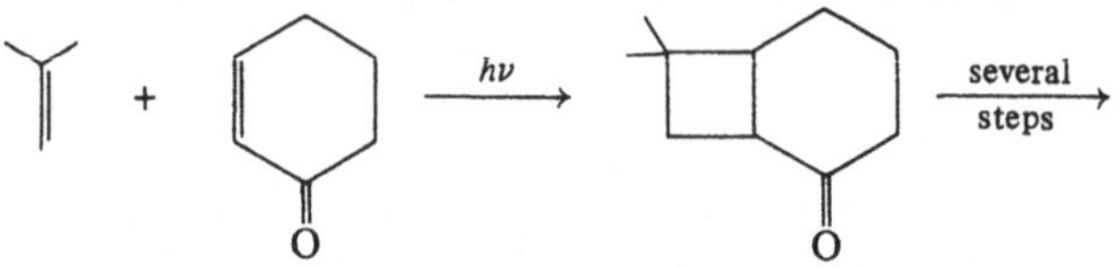

hν — several steps → (**68**) (6.65)

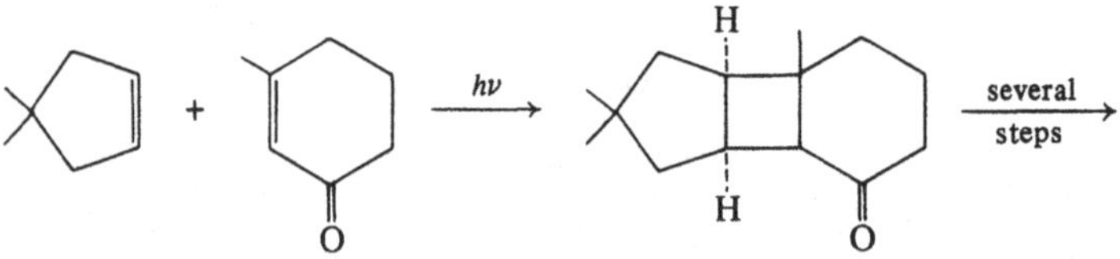

hν — several steps → (**69**) (6.66)

Photodimerisation of enones may also have immense biological significance. One of the causes of ultraviolet damage to DNA is believed to be photodimerisation of thymine units in the DNA chains. Thymine itself undergoes a ready photodimerisation.[91]

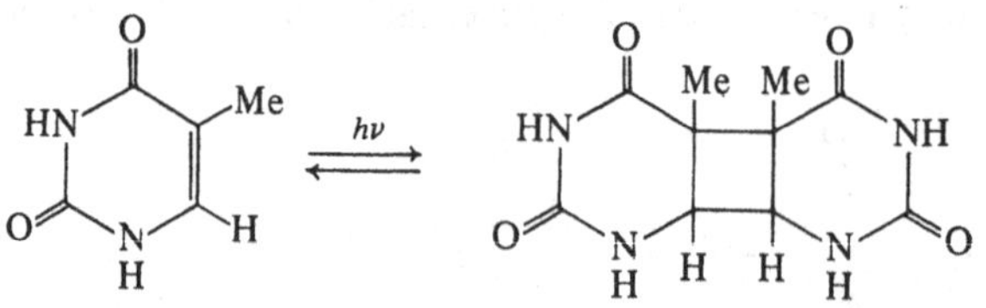

hν (6.67)

Photoaddition of maleic acid derivatives to olefins and to aromatic compounds is a very general route to cyclobutanes.[84] The additions have been extended to acetylenes to give cyclobutenes. Photochemical 2 + 2 additions involving a variety of heterocyclic systems have also found wide application.[84e]

It is apparent from the example shown in (6.65) that the preferred regio-isomer in the enone photoadditions is not consistent with the intermediacy of the most stabilised diradical (**70**) that one would associate with a ground state diradical addition. The excited state is the triplet $n \rightarrow \pi^*$ enone and the important frontier orbital interaction is between the singly occupied enone π^* orbital and the empty alkene π^* orbital. The orbital coefficients are as shown (**71**) and clearly, frontier orbital theory explains the observed orientation.[85]

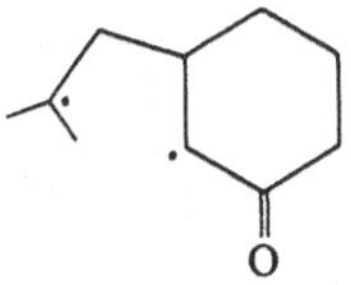

(**70**) 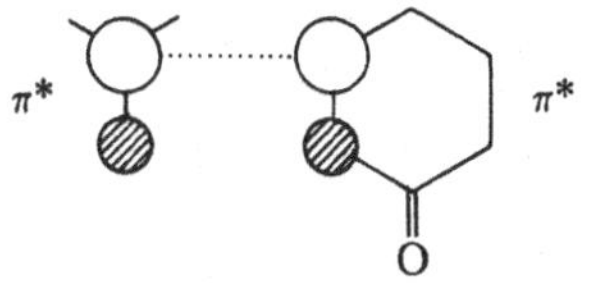 (**71**)

Another very general photocycloaddition is that of aldehydes, ketones and quinones to olefins to give oxetanes (6.68, 6.69).[84] This is known as the Paterno-Buchi reaction, following its discovery by Paterno and the initial investigations into its mechanism by Buchi.

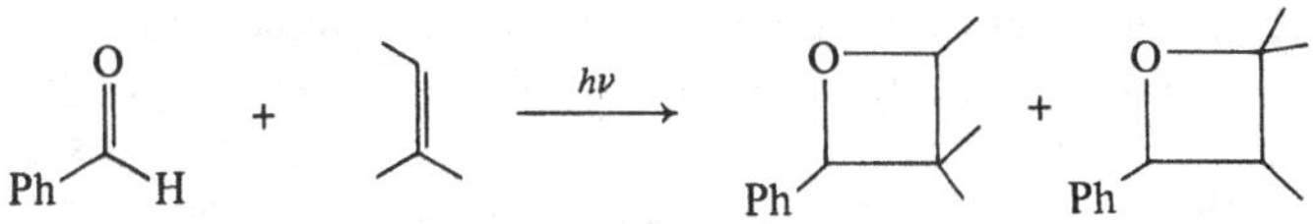

Major product (6.68)

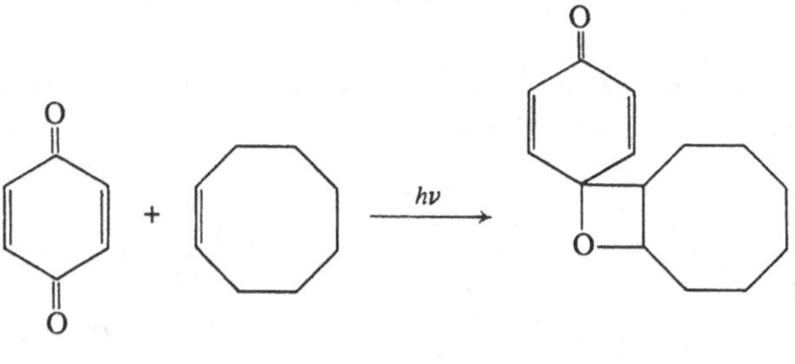

(6.69)

The mechanism depends on the reactants. For aliphatic or aromatic ketones and alkyl alkenes the reaction involves $n \rightarrow \pi^*$ excitation of the carbonyl group; the excited singlet rapidly undergoes intersystem crossing to a triplet which then adds to the olefin. Loss of configuration about the alkene occurs and mixtures of isomeric oxetanes, having all possible orientations, are formed but the oxetane expected from the most stable diradical intermediate predominates. Frontier orbital theory is also consistent with this orientation. The orbitals closest in energy are most likely the singly occupied lone pair orbital of the $n \rightarrow \pi^*$ excited ketone and the occupied HOMO of the alkene, and so the preferred frontier orbital interaction is as shown (**72**).[85] This leads to the more stable diradical, unlike the case for the enone photoadditions discussed above.

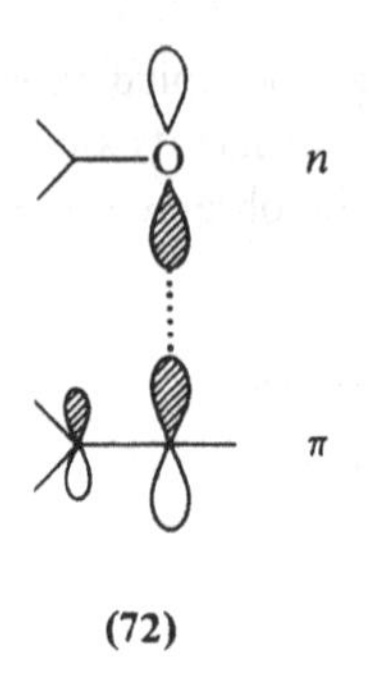

(72)

With aliphatic ketones and alkenes bearing electron-withdrawing groups the singlet $n \rightarrow \pi^*$ excited ketone is involved and alkene configuration is retained. For electron-rich alkenes (enol ethers), either singlet or triplet, a $n \rightarrow \pi^*$ excited ketone may be involved.

The triplet carbonyl can transfer its energy to the olefin if the latter has a triplet state of suitable energy. Side products from triplet-sensitised reactions of the olefins (for example, dimerisation) are therefore often observed.

The reaction is generally efficient for aromatic aldehydes and ketones, and for quinones, since these have lower triplet state energies than most simple alkenes. Aliphatic aldehydes and ketones have higher triplet energies and few examples of such additions are known (those of polyfluoroalkyl carbonyl compounds are an exception). The carbonyl compounds act as triplet sensitisers for dienes and other conjugated olefins which have low triplet energies (see, for example, p. 181).

Photoaddition of carbonyl compounds to acetylenes leads to α,β-unsaturated ketones, presumably via oxetenes (73, 6.70).

$$\text{PhCHO} + \text{RC}{\equiv}\text{CR} \xrightarrow{h\nu} \left[\text{oxetene (73)} \right] \longrightarrow \text{Ph(H)C}{=}\text{C(R)C(=O)R} \qquad (6.70)$$

(73)

6.8. Reactions catalysed by transition metals

The presence of transition-metal compounds has been found to divert reactions from their normal courses in a number of cases. In many of these the product of a formally disallowed reaction is formed under very mild conditions. There are several examples of this type of catalysis in 2 + 2 cycloadditions and fragmentations.[92]

A typical example is the dimerisation of norbornadiene which occurs in the presence of iron, nickel and cobalt compounds (fig. 6.28)[92a]

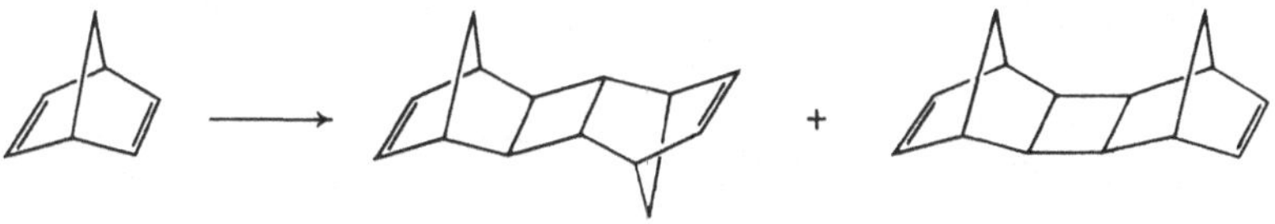

10 : 1

Fig. 6.28

The isomerisation of quadricyclene (**74**) to norbornadiene illustrates a metal-catalysed retro 2 + 2 addition (6.71). In this case the thermal reaction has a half-life of 14 hours at 140 °C, but in the presence of 2 mole % of $[Rh^{I}(norbornadiene)Cl]_2$ the half-life is only 45 minutes at –26 °C.[93]

(6.71)

(**74**)

An interesting type of reaction which formally combines forward and reverse 2 + 2 additions is the dismutation of olefins (fig. 6.29)[92a] The reaction can be brought about at elevated temperatures in the presence of heterogeneous metal-oxide catalysts. It can also take place under very much milder conditions in the presence of homogeneous transition-metal catalysts, when it is often called a metathesis.

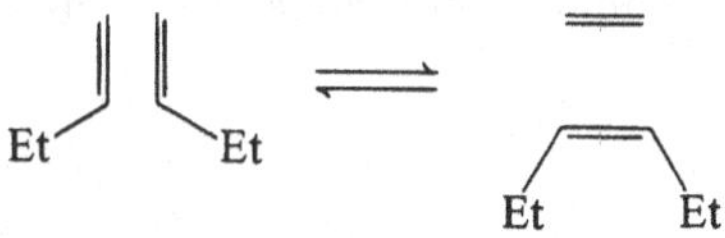

Et Et ⇌ Et Et

Fig. 6.29

Free cyclobutanes are probably not intermediates in the metathesis; metal-carbene complexes have been implicated.[94] An example of metathesis is the conversion of 2-pentene into an equilibrium mixture of 2-pentene, 2-butene and 3-hexene in the presence of tungsten catalysts (fig. 6.30). The equilibrium is established within a few seconds at room temperature.

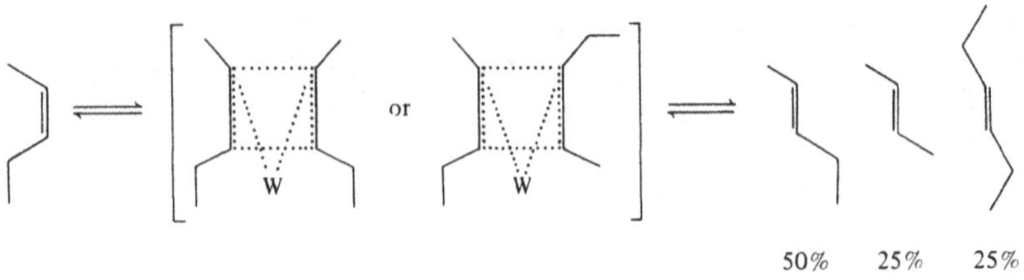

or W W

50% 25% 25%

Fig. 6.30

Metathesis also provides a route to catenanes, which have two carbocyclic rings interlocked. Large cyclic olefins (**75**) can combine by this reaction to give conformationally mobile cyclic dienes (**76**). If disproportionation of these dienes. to regenerate the cyclic monoenes, occurs after twisting of the alkyl chains, the two cyclic monoenes may be interlocked (**77**, 6.72).[95]

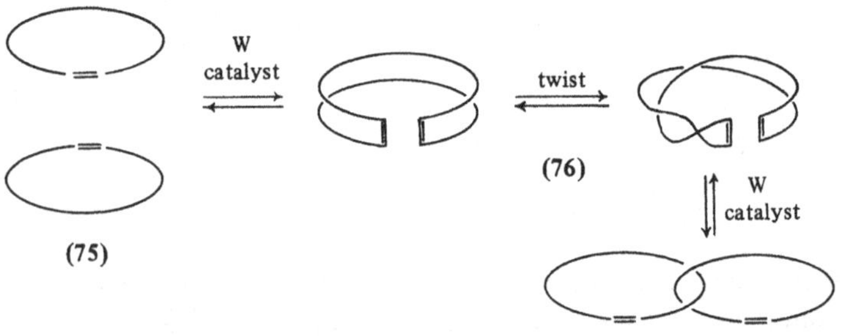

W catalyst twist W catalyst

(**75**) (**76**) (**77**) (6.72)

The origin of the catalytic effect of transition metals was discussed in chapter 3. There is increasing evidence that at least some of these catalysed reactions are stepwise. Thus, for example, the transformation of cubane into tricyclo-octadiene is catalysed by small amounts of certain rhodium(I) compounds of the type $[Rh^1 \text{ diene Cl}]_2$ (6.73).[96]

(6.73)

O Rh CO Cl

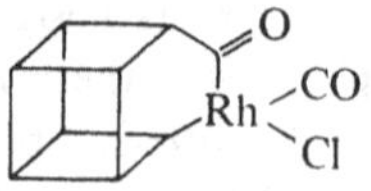

(78)

With stoichiometric amounts of related rhodium complexes $[Rh(CO)_2Cl]_2$, products (78) resulting from oxidative addition of cubane to the rhodium complex can be isolated. It therefore seems likely that the catalysed rearrangement of cubane, and related catalysed reactions, proceed in a stepwise manner in which a key step is oxidative addition (6.74).[96]

$$\xrightarrow{Rh^{I}} \quad Rh^{III} \longrightarrow \qquad (6.74)$$

6.9. Other cycloadditions

In this section, cycloadditions involving more than six electrons are discussed. These are far less common than those involving six or fewer electrons and very little detailed mechanistic work of the type done on the Diels–Alder reaction or 1,3-dipolar addition has yet been carried out. The concerted or stepwise nature of these reactions is therefore largely a matter for speculation. The general success of the Woodward–Hoffmann rules in rationalising the occurrence of concerted or stepwise reactions has been so great, however, that frequently the mere observation of an allowed process is taken to imply a concerted reaction; and similarly, disallowed reactions are assumed to be stepwise. This is too sweeping a generalisation.

Concerted cycloadditions of open-chain polyenes leading to large rings are unlikely from entropy considerations (the configuration and conformation of the polyene will be critical). An alternative concerted mode leading to a smaller ring will normally be more favourable. If the two polyenes interact so as to give an intermediate, this is more likely to collapse to a small ring product than to a large one, again for entropy reasons; just as, for example, the radical addition of olefins to dienes leads to more 2 + 2 than 2 + 4 addition. Cyclic polyenes are more likely to undergo cycloadditions leading to large rings but such cyclic systems as cyclo-octatetraenes, azacyclo-octatetraenes and oxepins exist in equilibrium with bicyclic valence tautomers (chapter 3) and frequently undergo cycloadditions via these tautomers (6.75). Tropones, some tropolones, and 1*H*-azepines do not usually undergo such valence isomerisations and have therefore more often been observed to behave as 6π electron components. On the other hand, frontier orbital effects will tend to favour the formation of large ring products. This is because the terminal HOMO and LUMO coefficients of a polyene are usually the largest and it is these coefficients which determine the periselectivity (see p. 100).

1,3-Dipolar systems with additional conjugation constitute π systems of more than four electrons. Just as 1,3-dipoles parallel dienes in their pericyclic reactions such extended dipolar systems should parallel the isoelectronic polyenes.

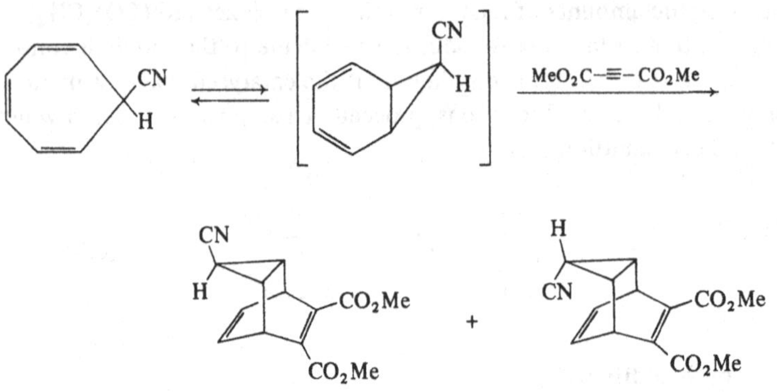

$MeO_2C-\equiv-CO_2Me$

CN H CN H CO_2Me CO_2Me + H CN CO_2Me CO_2Me (6.75)

6 + 2 *Electron cycloadditions*

${}_{\pi}6_s + {}_{\pi}2_s$ Cycloaddition is thermally disallowed and significantly, 6 + 2 additions have only rarely been observed. The addition of nitrosobenzene to cycloheptatriene **(79)** and *N*-ethoxycarbonylazepine **(80)** is presumed to result from stepwise processes which are preferred over concerted 4 + 2 addition because of the highly polarised nitroso group (6.76).[97]

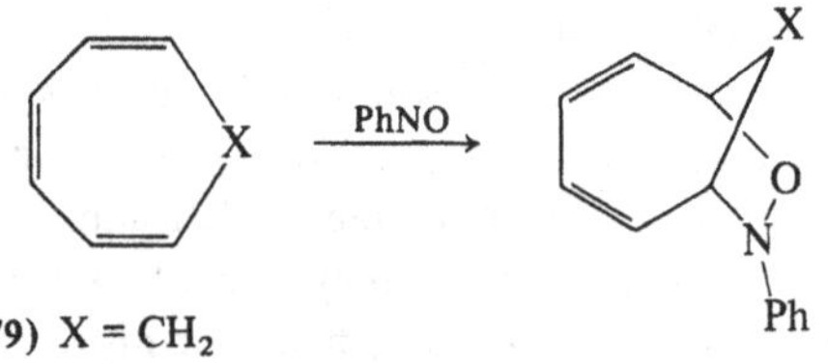

PhNO X O N Ph (6.76)

(79) X = CH_2
(80) X = NCO_2Et

The 6 + 2 adduct **(81)** from chlorosulphonyl isocyanate and cycloheptatriene (6.77) can reasonably be explained as arising from the intermediate **(82)** which may be formed directly. Alternatively it may arise by rearrangement through the same zwitterion, of initially formed adducts **(83)** and **(84)**.[98] These adducts could conceivably result from concerted ${}_{\pi}6_s + {}_{\pi}2_a$ and ${}_{\pi}2_s + {}_{\pi}2_a$ processes which are formally allowed for this cumulene.[99]

+ $ClSO_2N{=}C{=}O$ ⟶ O C N · SO_2Cl

(81) (6.77)

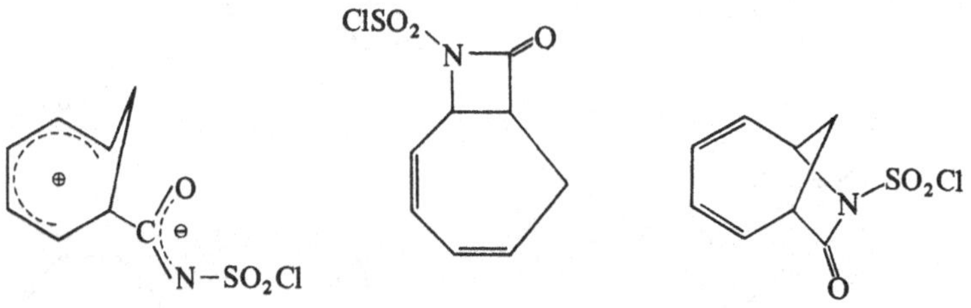

(82) (84) (83)

Addition of the 1,5-dipole (**85**) to acetylenedicarboxylic esters to give the diazepine (**86**) is a $6\pi + 2\pi$ process which most likely proceeds through a zwitterion formed by attack of the side-chain terminal nitrogen on the highly electrophilic acetylene (6.78).[100]

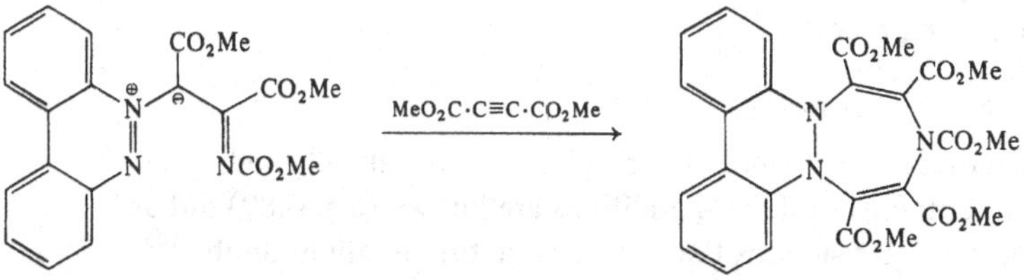

$MeO_2C{\cdot}C{\equiv}C{\cdot}CO_2Me$ (6.78)

(85) (86)

Sulphur dioxide adds readily to *cisoid*-hexatrienes by a linear cheletropic ${}_{\pi}6_a + {}_{\omega}2_s$ process (6.79–6.80). The reverse reaction is a linear cheletropic extrusion of sulphur dioxide with conrotatory twisting of the terminal methylenes.[101] With the cyclic triene (**87**) sulphur dioxide forms only the 1,4-adduct (**88**, 6.81). In this case, antarafacial addition to the triene is geometrically impossible and the alternative non-linear ${}_{\pi}6_s + {}_{\omega}2_a$ process does not compete with the concerted ${}_{\pi}4_s + {}_{\omega}2_s$ linear cheletropic addition to a diene component. The extreme unfavourableness of non-linear cheletropic eight-electron processes is illustrated by the fact that sulphur dioxide is eliminated 60 000 times more slowly (at 180 °C) from (**89**) than from (**88**) (where an allowed retro-${}_{\pi}4_s + {}_{\omega}2_s$ process is possible). The geometry of (**89**) makes conrotatory twisting of the

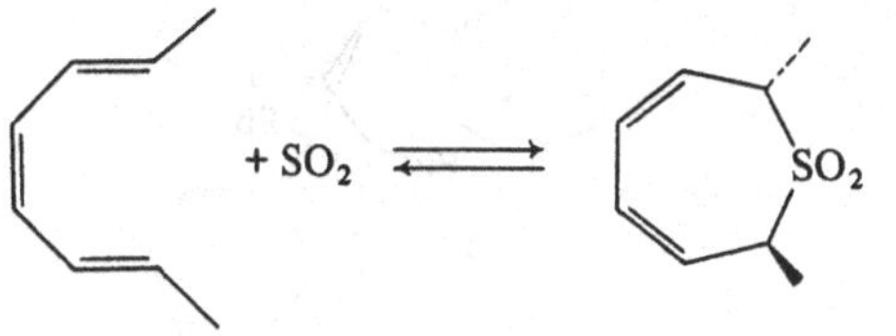

$+ SO_2 \rightleftharpoons$ (6.79)

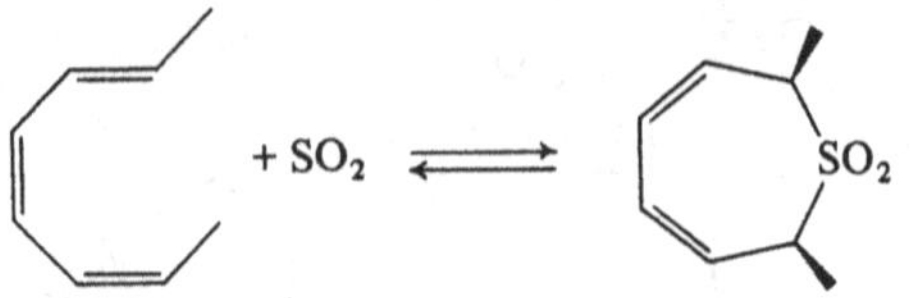

$$+ SO_2 \rightleftharpoons SO_2 \quad (6.80)$$

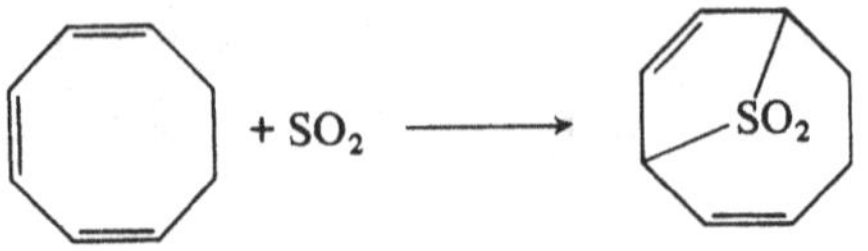

$$+ SO_2 \longrightarrow SO_2 \quad (6.81) \qquad SO_2 \quad (6.81)$$

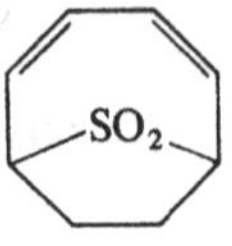

(87) (88) (89)

two methylene groups (retro-$_{\pi}6_a$ + $_{\omega}2_s$) impossible so that, if concerted, the elimination would have to be a non-linear cheletropic process (retro-$_{\pi}6_s$ + $_{\omega}2_a$). This particular extrusion is therefore either stepwise or, if concerted, the non-linear cheletropic extrusion has an energy barrier of the same order of magnitude as a stepwise process.[102]

4π + 4π Cycloadditions

$_{\pi}4_s$ + $_{\pi}4_s$ Cycloaddition should only be photochemically allowed. Several examples of photochemical 4 + 4 additions are known (e.g. 6.82) but as in other photochemical reactions the concerted nature is still in doubt.[103]

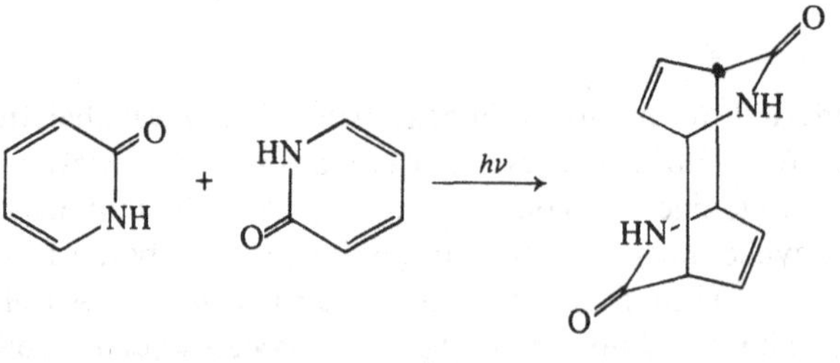

(6.82)

One example of a presumably stepwise thermal 4 + 4 cycloaddition is the dimerisation of diphenylisoindenone **(90)** illustrated in (6.83).[104] The highly reactive isoindenones normally function as dienes and can be trapped by dienophiles.[105]

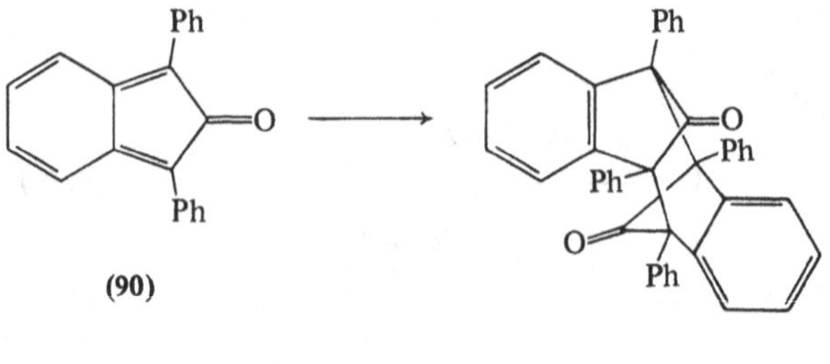

(6.83)

(90)

8 + 2 Electron cycloadditions

${}_{\pi}8_s + {}_{\pi}2_s$ Cycloadditions are thermally allowed but have rarely been observed.[105] The examples in fig. 6.31 may be concerted, but these are also cases where the dipolar intermediates would be particularly stabilised.[106] The example in fig. 6.32 can be considered as an extended $8\pi + 2\pi$ dipolar cycloaddition.[107]

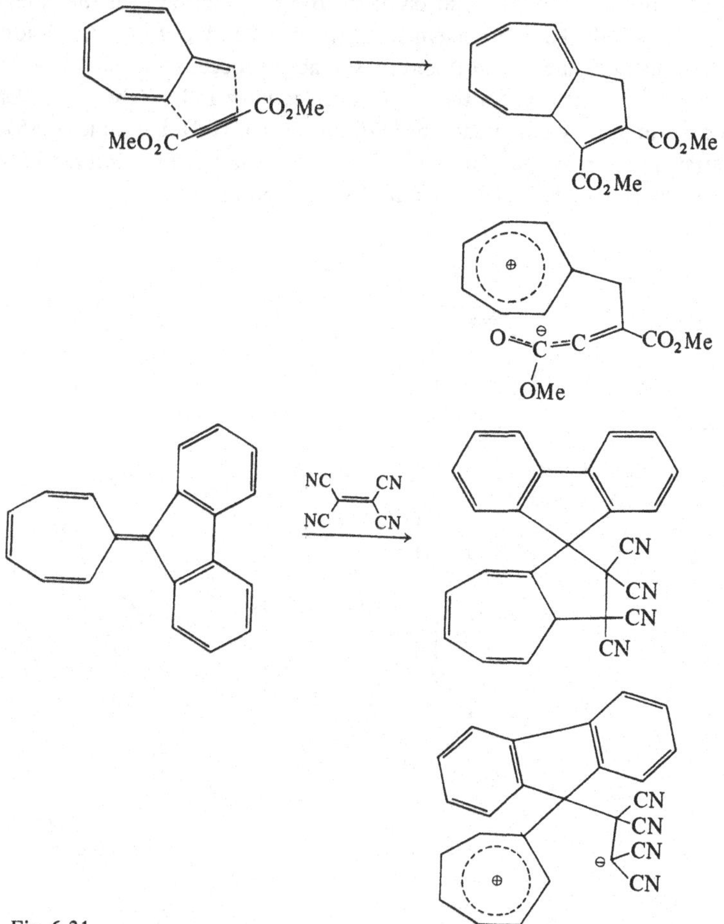

Fig. 6.31

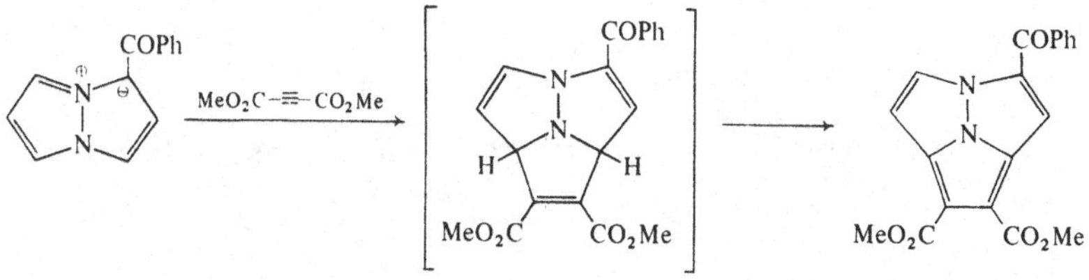

Fig. 6.32

6π + 4π Cycloadditions

Thermal ${}_{\pi}6_s + {}_{\pi}4_s$ cycloadditions (6.84) were discovered after the prediction that they would be allowed. Several examples are now known. Originally only the 6 + 6 dimers (**92**) were isolated from the dimerisation of *N*-substituted *IH*-azepines but reinvestigation following the recognition that this mode of addition was disallowed led to the isolation of the 6 + 4 adducts (**91**) as the kinetically controlled products. Rearrangement of the 6 + 4 adduct to the 6 + 6 adduct involves a suprafacial 1,3-shift and presumably proceeds via a diradical intermediate.[108] The reason for the 6 + 4 periselectivity and for the formation of the *exo*-dimer only has already been discussed (p. 100). Example (6.85) also illustrates the preference for *exo*-addition; secondary orbital interactions (light dashes) disfavour the transition state for *endo*-addition (**93**).

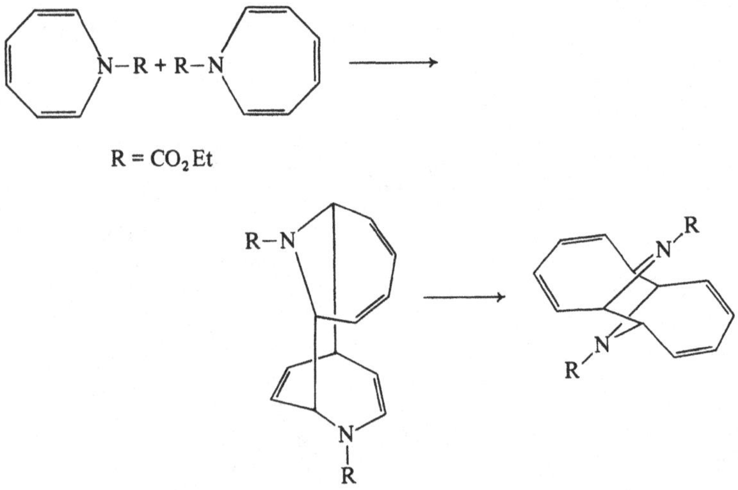

N–R + R–N

$R = CO_2Et$

R–N

N

R

N

R

R

N

6 + 4 Adduct (**91**) 6 + 6 Adduct (**92**) (6.84)

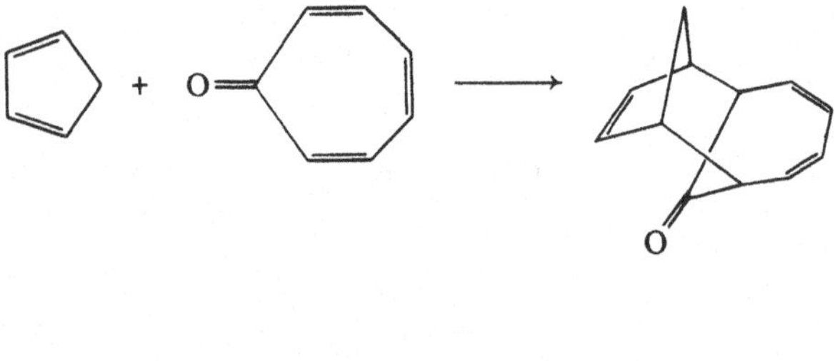

\+ O

O

(6.85)

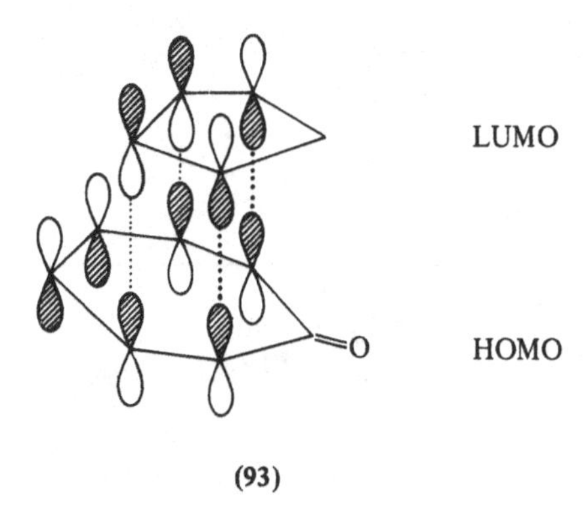

(93)

$6\pi + 4\pi$ Cycloadditions in which the 6π component is a 1,5-dipole have also been observed with the oxido pyridinium betaines **(94)**. With 2π electron dipolarophiles the $4\pi + 2\pi$ 1,3-dipolar cycloadduct is produced.[109]

(94)

Fulvenes are interesting molecules which display varying periselectivity. They illustrate a situation where the largest ring is not necessarily favoured by frontier orbital effects. The orbital energies and coefficients are given in fig. 6.33.[110] With a 4π electron reactant, fulvene can in principle act as a 6π or 2π component. If the 4π component (diene or 1,3-dipole) has a low-energy LUMO the fulvene will react as a 2π system (6.86 and 6.87) because the fulvene HOMO has zero coefficient on C–6.[111]

On the other hand, electron-rich dipoles and dienes,[112] for which interaction involving the fulvene LUMO is dominant, force the fulvene to act as a 6π component since in the LUMO the C–6 coefficient is large (6.88 and 6.89). The situation is quite complicated, however, because interaction between diene HOMO and fulvene NLUMO appear to be important in some cases. Such an interaction is believed to explain the periselectivity of addition of cyclo-

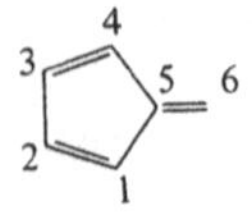

Fulvene

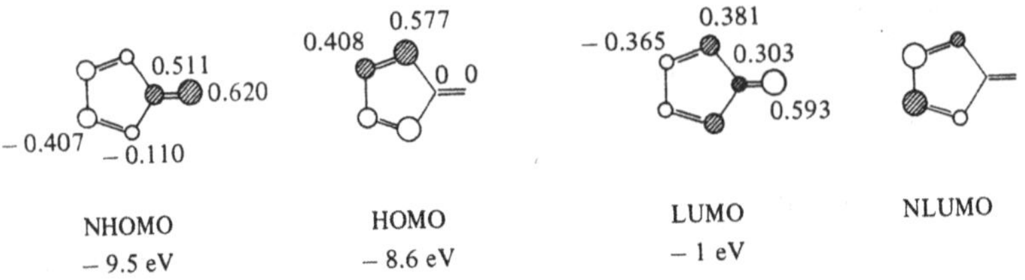

NHOMO
− 9.5 eV

HOMO
− 8.6 eV

LUMO
− 1 eV

NLUMO

Fig. 6.33. Energies and coefficients for orbitals of fulvene.

pentadiene to dimethylfulvene (6.90).[113] This reaction ought to be diene HOMO-fulvene LUMO controlled but appreciable diene HOMO-fulvene NLUMO interaction can lead to addition by virtue of the large coefficients at C-2 and C-3 and zero coefficient at C-6 in this orbital. Powerful electron-releasing groups at C-6 in fulvenes raise the NHOMO but not the HOMO energy so that the former becomes the more important occupied fulvene frontier orbital. For such fulvenes, reaction as a 6π component occurs with electron-deficient dienes.[111]

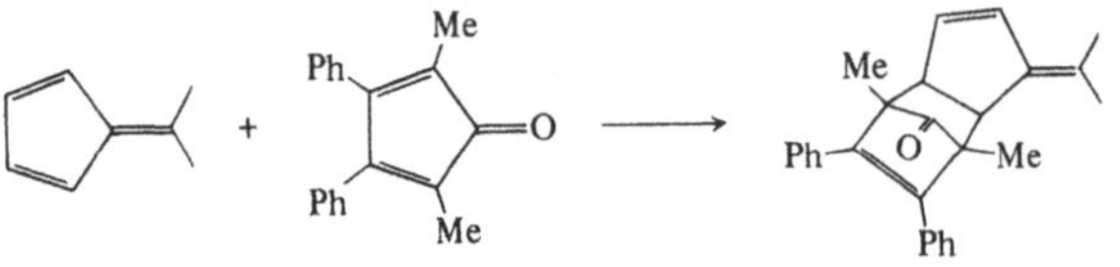

(6.86)

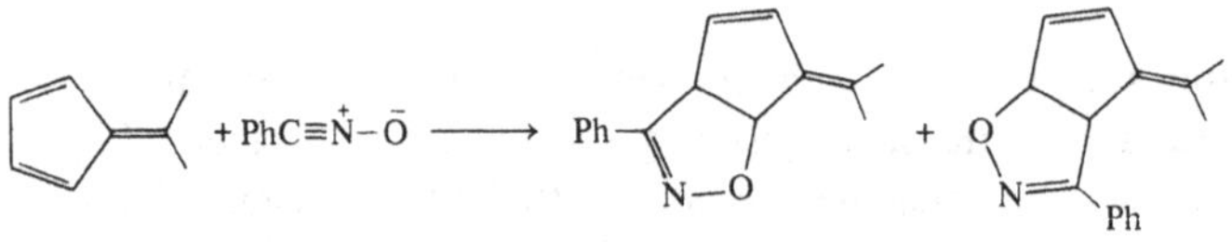

(6.87)

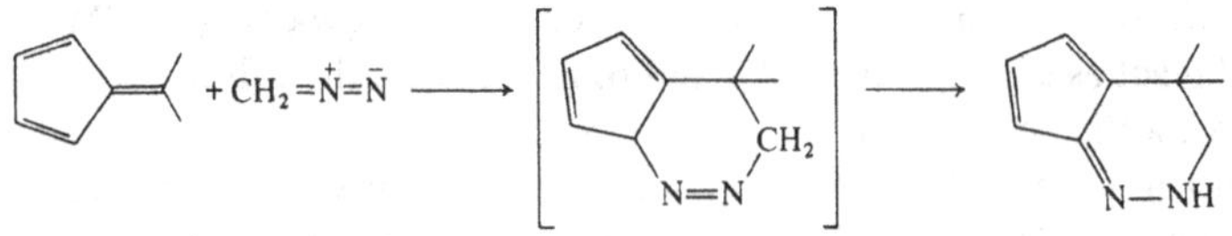

(6.88)

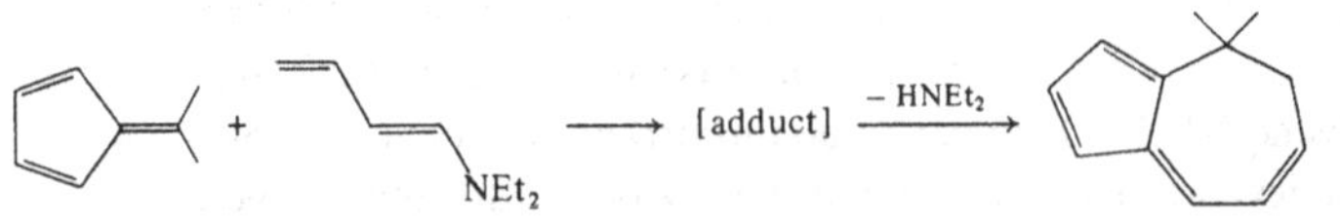

(6.89)

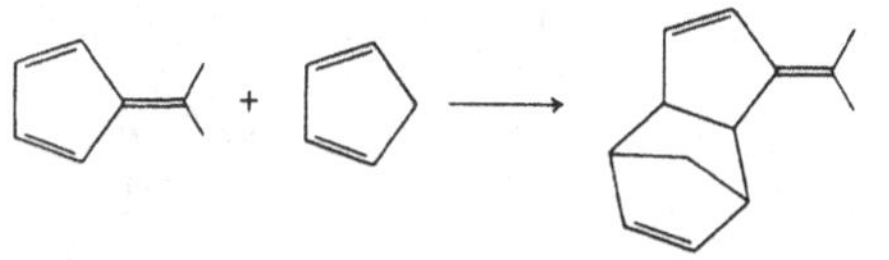

(6.90)

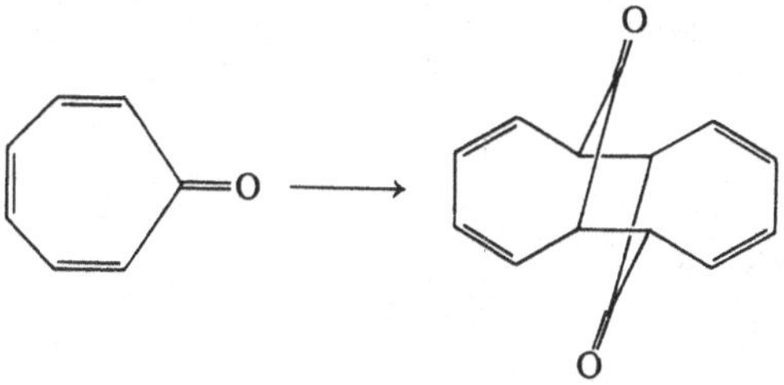

(6.91)

(95)

$6\pi + 6\pi$ Cycloadditions

Tropone gives a 6 + 6 dimer **(95)** when irradiated (6.91). Photochemical $_{\pi}6_s + _{\pi}6_s$ cycloaddition is allowed, but there is no evidence that the dimer **(95)** is formed concertedly.[114]

Cycloadditions with more than 12 electrons

Such cycloadditions become successively rarer. Addition of the sesquifulvene **(96)** to tetracyanoethylene provides an example of an allowed $12\pi + 2\pi$ cycloaddition (6.92).[115] A dipolar analogue for this reaction involves the 2-substituted naphthotriazine **(97)** which gives the stable aromatic product **(98)** with dimethyl acetylenedicarboxylate after spontaneous dehydrogenation of the $12\pi + 2\pi$ adduct (6.93).[100] Frontier orbital theory very nicely rationalises this unexpected observation; the major interaction will involve the naphthotriazine HOMO and the electron-deficient acetylene LUMO and significantly the largest naphthotriazine HOMO coefficients occur at C-6 and C-7 rather than at the 1,3-nitrogens of the *peri* azimine bridge **(99)**.[116]

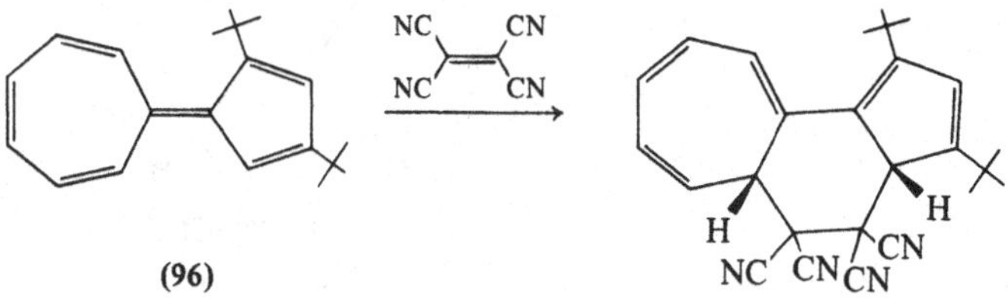

NC CN NC CN

H H NC CN CN CN

(96)

(6.92)

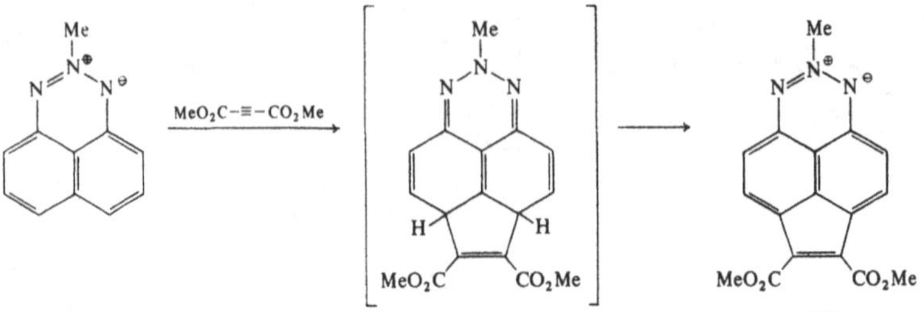

(6.93)

(97) (98)

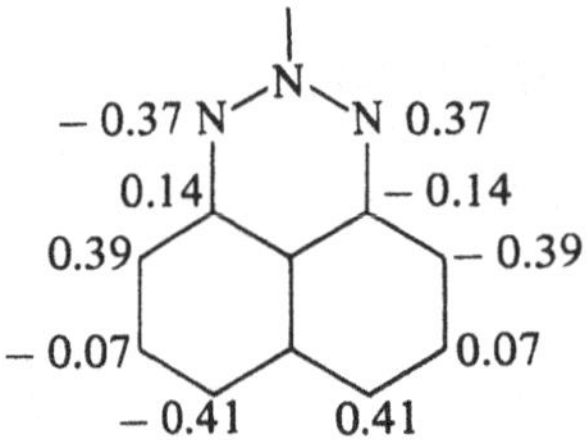

HOMO

(99)

The heptafulvalene (**100**) undergoes $14\pi + 2\pi$ addition with tetracyano-ethylene. Orbital symmetry dictates that if this addition is concerted one component should be antarafacial, and indeed the stereochemistry of the adduct is consistent with a ${}_{\pi}14_{a} + {}_{\pi}2_{s}$ process (6.94).[15]

An attempt to observe $16\pi + 2\pi$ cycloaddition of acetylenedicarboxylic esters to the fulvene (**101**) gave the *cis*- and *trans*-isomers of the spiro-adducts (**102**) by a stepwise process.[117]

Multicomponent cycloadditions

For entropy reasons, multicomponent additions involving more than six electrons are very unlikely unless more than one of the components is situated within the same molecule. The photochemically induced cycloaddition of 2-butyne to dihydrophthalic anhydride can be rationalised as a photochemically allowed ${}_{\pi}2_{s} + {}_{\pi}2_{s} + {}_{\pi}2_{s} + {}_{\pi}2_{s}$ addition, the orthogonal π bonds of the butyne being considered as separate 2π components (6.96). A stepwise mechanism involving an intermediate carbene (**103**) seems equally plausible, however.[118]

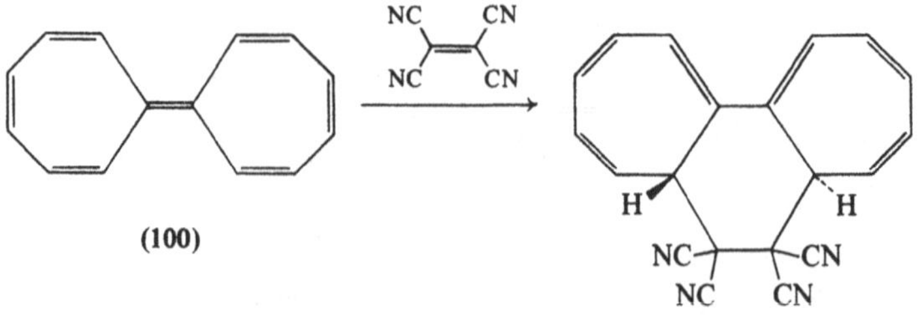

(6.94)

(100)

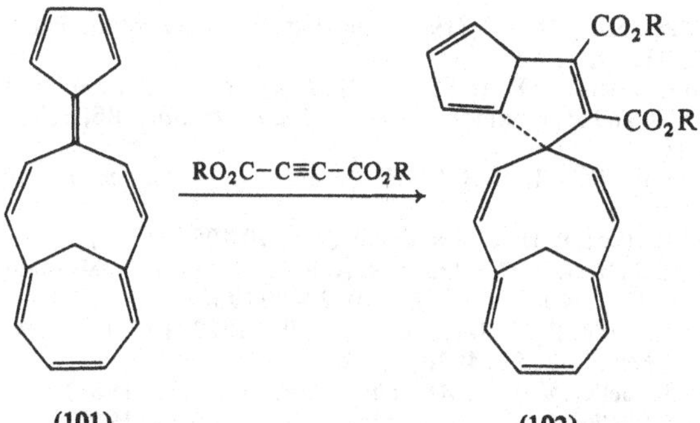

RO$_2$C−C≡C−CO$_2$R

CO$_2$R

CO$_2$R

(6.95)

(101) **(102)**

hν

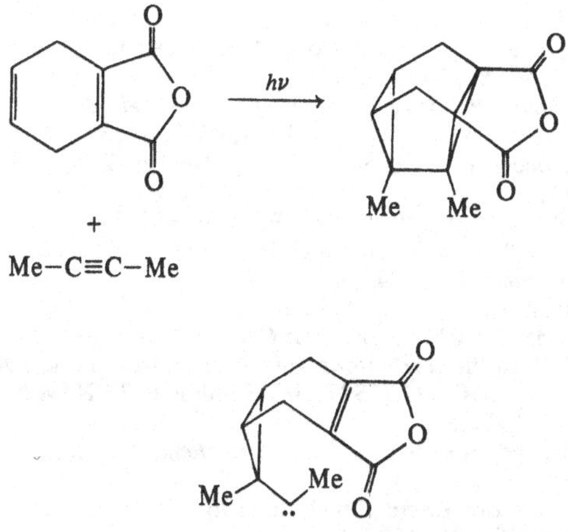

(6.96)

+

Me−C≡C−Me

Me Me

Me Me

(103)

Multicomponent cycloadditions catalysed by metals are also possible. Examples are the nickel-catalysed tetramerisation of acetylene to give cyclooctatetraene (the Reppe synthesis), and catalysed cyclotrimerisation of acetylenes to give benzene derivatives.[119] Mechanisms for these reactions are largely speculative.

References

1. Reviews: (*a*) Huisgen, R., Grashey, R. and Sauer, J., in *The chemistry of alkenes*, ed. S. Patai, p. 741, Interscience, London, 1964; (*b*) Bartlett, P. D., *Q. Rev.*, **24**, 473 (1970); (*c*) Roberts, J. D. and Sharts, C. M., *Org. React.*, **12**, 1 (1962); (*d*) Bartlett, P. D., *Science*, **159**, 833 (1968).
2. Benson, S. W., *Thermochemical kinetics*, Table A11, Wiley, New York, 1968; Quick, L. M., Knecht, D. A. and Back, M. H., *Int. J. chem. Kinet.*, **4**, 61 (1972).
3. Montgomery, L. K., Schueller, K. E. and Bartlett, P. D., *J. Am. chem. Soc.*, **86**, 622 (1964).
4. Bartlett, P. D., Dempster, C. J., Montgomery, L. K., Schueller, K. E. and Wallbillich,

G. E. H., *J. Am. chem. Soc.*, **91**, 405 (1969); Bartlett, P. D. and Wallbillich, G. E. H., *J. Am. chem. Soc.*, **91**, 409 (1969).
5. Bartlett, P. D., Montgomery, L. K. and Seidel, B., *J. Am. chem. Soc.*, **86**, 616 (1964); Bartlett, P. D. and Montgomery, L. K., *J. Am. chem. Soc.*, **86**, 628 (1964); see also reference 1*b*.
6. Bartlett, P. D., Hummel, K., Elliott, S. P. and Minns, R. A., *J. Am. chem. Soc.*, **94**, 2898 (1972).
7. Swenton, J. S. and Bartlett, P. D., *J. Am. chem. Soc.*, **90** 2056 (1968).
8. Bartlett, P. D., Wallbillich, G. E. H., Wingrove, A. S., Swenton, J. S., Montgomery, L. K. and Kramer, B. D., *J. Am. chem. Soc.*, **90**, 2049 (1968).
9. Bartlett, P. D. and Wheland, R., *J. Am. chem. Soc.*, **92**, 3822 (1970).
10. Little, J. C., *J. Am. chem. Soc.*, **87**, 4020 (1965).
11. Bartlett, P. D. and Schueller, K. E., *J. Am. chem. Soc.*, **90**, 6077 (1968).
12. Bartlett, P. D. and Schueller, K. E., *J. Am. chem. Soc.*, **90**, 6071 (1968).
13. Grant, D., McKervey, M. A., Rooney, J. J., Samman, N. G. and Step, G., *Chem. Communs*, 1972, 1186; Lenoir, D., *Tetrahedron Lett.*, 1972, 4049.
14. Ziegler, K., Sauer, H., Bruns, L., Froitzheim-Kuhlhorn, H. and Schneider, J., *Justus Liebigs Annln Chem.*, **589**, 122 (1954); Cope, A. C., Howell, C. F. and Knowles, A., *J. Am. chem. Soc.*, **84**, 3190 (1962).
15. Woodward, R. B. and Hoffmann, R., *Angew. Chem., Int. Edn Engl.*, **8**, 781 (1969).
16. Kraft, K. and Koltzenburg, G., *Tetrahedron Lett.*, 1967, 4357, 4723.
17. Padwa, A., Koehn, W., Masaracchia, J., Osborn, C. L. and Trecker, D. J., *J. Am. chem. Soc.*, **93**, 3633 (1971).
18. Wheland, R. and Bartlett, P. D., *J. Am. chem. Soc.*, **95**, 4003 (1973).
19. Bartlett, P. D., Cohen, G. M., Elliot, S. P., Hummel, K., Minns, R. A., Sharts, C. M. and Fukunaga, J. Y., *J. Am. chem. Soc.*, **94**, 2899 (1972).
20. Epiotis, N. D., *J. Am. chem. Soc.*, **94**, 1935 (1972).
21. Reviews: (*a*) Gompper, R., *Angew. Chem., Int. Edn Engl.*, **8**, 312 (1969); (*b*) Muller, L. L. and Hamer, J., *1,2-Cycloaddition reactions*, Interscience, New York, 1967; (*c*) Huisgen, R., *Acc. chem. Res.*, **10**, 117 (1977); (*d*) Reinhoudt, D. N., *Adv. heterocyclic Chem.*, **21**, 253 (1977).
22. Williams, J. K., Wiley, D. W. and McKusick, B. C., *J. Am. chem. Soc.*, **84**, 2210 (1962).
23. Kuehne, M. E. and Foley, L., *J. org. Chem.*, **30**, 4280 (1965).
24. Huisgen, R., *Acc. chem. Res.*, **10**, 199 (1977).
25. Brannock, K. C., Bell, A., Burpitt, R. D. and Kelly, C. A., *J. org. Chem.*, **26**, 625 (1961); Brannock, K. C., Bell, A., Burpitt, R. D. and Kelly, C. A., *J. org. Chem.*, **29**, 801 (1964).
26. Huisgen, R., in *Topics in heterocyclic chemistry*, ed. R. N. Castle, p. 223, Inter-Science, New York, 1969.
27. Proskow, S., Simmons, H. E. and Cairns, T. L., *J. Am. chem. Soc.*, **88**, 5254 (1966).
28. For further discussion of the nature of intermediates in 2 + 2 cycloaddition see references 21*c*,*d* and Epiotis, N. D., Yates, R. L., Carlberg, D. and Bernardi, F., *J. Am. chem. Soc.*, **98**, 453 (1976).
29. Epiotis, N. D., *Angew. Chem., Int. Edn Engl.*, **13**, 751 (1974), and references therein.
30. For further discussion see reference 29 and Epiotis, N. D., *J. Am. chem. Soc.*, **94**, 1924 (1972).
31. Houk, K. N. and Munchausen, L. L., *J. Am. chem. Soc.*, **98**, 937 (1976).
32. Hoffmann, R. W. and Häuser, H., *Angew. Chem., Int. Edn Engl.*, **3**, 380 (1964).
33. Koerner von Gustorf, E., White, D. V., Leitich, J. and Henneberg, D., *Tetrahedron Lett.*, 1969, 3113; Koerner von Gustorf, E., White, D. V., Kim, B., Hess, D. and Leitich, J., *J. org. Chem.*, **35**, 1155 (1970).
34. Koerner von Gustorf, E., White, D. V. and Leitich, J., *Tetrahedron Lett.*, 1969, 3109.
35. Brannock, K. C., Burpitt, R. D., Davis, H. E., Pridgen, H. S. and Thweatt, J. G., *J. org. Chem.*, **29**, 2579 (1964).
36. Acheson, R. M., *Adv. heterocyclic Chem.*, **1**, 125 (1963).

37. For a comprehensive review of heterocumulene cycoladditions, see Ulrich, H., *Cycloaddition reaction of heterocumulenes*, Academic Press, New York, 1968. For reviews of allene and other cumulene cycloadditions, see Fischer, A., in *The chemistry of alkenes*, ed. S. Patai, p. 1025, Interscience, London, 1964, and Taylor, D. R., *Chem. Rev.*, **67**, 317 (1967).
38. Review: Holder, R. W., *J. chem. Ed.*, **53**, 81 (1976).
39. Montaigne, R. and Ghosez, L., *Angew. Chem., Int. Edn Engl.*, **7**, 221 (1968).
40. Huisgen, R. and Otto, P., *Tetrahedron Lett.*, 1968, 4491.
41. Huisgen, R., Feiler, L. A. and Otto, P., *Chem. Ber.*, **102**, 3405, 3444 (1969); Huisgen, R., Feiler, L. A. and Binsch, G., *Chem. Ber.*, **102**, 3460 (1969); Huisgen, R. and Feiler, L. A., *Chem. Ber.*, **102**, 3391, 3428 (1969); Huisgen, R. and Otto, P., *Chem. Ber.*, **102**, 3475 (1969).
42. Baldwin, J. E. and Kapecki, J. A., *J. Am. chem. Soc.*, **92**, 4868, 4874 (1970).
43. Reference 42 gives a slightly different picture of the secondary interaction which is claimed to explain the observed isotope effects. See also Wagner, H. U. and Gompper, R., *Tetrahedron Lett.*, 1970, 2819, for an alternative two-stage dipolar mechanism for keten cycloaddition.
44. (*a*) Houk, K. N., Strozier, R. W. and Hall, J. A., *Tetrahedron Lett.*, 1974, 897; (*b*) Sustmann, R., Ansmann, A. and Vahrenholt, F., *J. Am. chem. Soc.*, **94**, 8099 (1972); (*c*) Inagaki, S., Minato, T., Yamabe, S., Fujimoto, H. and Fukui, K., *Tetrahedron*, **30**, 2165 (1974).
45. Huisgen, R. and Otto, P., *J. Am. chem. Soc.*, **90**, 5342 (1968).
46. Otto, P., Feiler, L. A. and Huisgen, R., *Angew. Chem., Int. Edn Engl.*, **7**, 862 (1968).
47. Huisgen, R. and Otto, P., *J. Am. chem. Soc.*, **91**, 5922 (1969).
48. Huisgen, R., Davies, B. A. and Morikawa, M., *Angew. Chem., Int. Edn Engl.*, **7**, 862 (1968).
49. Kagan, H. B. and Luche, J. C., *Tetrahedron Lett.*, 1968, 3093.
50. Gomes, A. and Joullié, M. M., *Chem. Communs*, 1967, 935.
51. Brady, W. T. and Dorsey, E. D., *Chem. Communs*, 1968, 1638.
52. Cripps, H. N., Williams, J. K. and Sharkey, W. H., *J. Am. chem. Soc.*, **81**, 2723 (1959).
53. Kiefer, E. F. and Okamura, M. Y., *J. Am. chem. Soc.*, **90**, 4187 (1968), and references therein.
54. Baldwin, J. E. and Roy, U. V., *Chem. Communs*, 1969, 1225.
55. Moore, W. R., Bach, R. D. and Ozretich, T. M., *J. Am. chem. Soc.*, **91**, 5918 (1969).
56. Dolbier, W. R. and Dai, S.-H., *J. Am. chem. Soc.*, **92**, 1774 (1970), and references therein.
57. Levek, T. J. and Kiefer, E. F., *J. Am. chem. Soc.*, **98**, 1875 (1976); Beetz, T. and Kellogg, R. M., *J. Am. chem. Soc.*, **95**, 7925 (1974).
58. For a review of acid-catalysed cyclodimerisations via vinyl cation intermediates, see Griesbaum, K., *Angew Chem., Int. Edn Engl.*, **8**, 933 (1969); see also Griesbaum, K. and Seiter, W., *J. org. Chem.*, **41**, 937 (1976).
59. Griesbaum, K., Naegele, W. and Wanless, G. G., *J. Am. chem. Soc.*, **87**, 3151 (1965).
60. Criegee, R. and Moschel, A., *Chem. Ber.*, **92**, 2181 (1959).
61. Reviews: Frey, H. M. and Walsh, R., *Chem. Rev.*, **69**, 103 (1969); Frey, H. M. and Walsh, R., *Adv. phys. org. Chem.*, **4**, 170 (1966).
62. Gerberich, H. R. and Walters, W. D., *J. Am. chem. Soc.*, **83**, 3935, 4884 (1961).
63. Cocks, A. T., Frey, H. M. and Stevens, I. D. R., *Chem. Communs*, 1969, 458.
64. Baldwin, J. E. and Ford, P. W., *J. Am. chem. Soc.*, **91**, 7192 (1969).
65. Rieber, N., Alberts, J., Lipsky, J. A. and Lemal, D. M., *J. Am. chem. Soc.*, **91**, 5668 (1969).
66. Closs, G. L. and Pfeffer, P. E., *J. Am. chem. Soc.*, **90**, 2452 (1968).
67. Noyce, D. S. and Banitt, E. H., *J. org. Chem.*, **31**, 4043 (1966); Sultanbawa, M. U. S., *Tetrahedron Lett.*, 1968, 4569.
68. Paquette, L. A., Wyvratt, M. J. and Allen, G. R., *J. Am. chem. Soc.*, **92**, 1763 (1970).

69. White, E. H., Wiecko, J. and Wei, C. C., *J. Am. chem. Soc.*, **92**, 2167 (1970); Güsten, H. and Ullman, E. F., *Chem. Communs*, 1970, 28.
70. McCapra, F., *Chem. Communs*, 1968, 155; McCapra. F., *Prog. org. Chem.*, **8**, 231 (1973).
71. Review: Stark, B. P. and Duke, A. J., *Extrusion reactions*, Pergamon Press, Oxford, 1967.
72. Reviews: (*a*) Kirmse, W., *Carbene chemistry*, 2nd edn, Academic Press, New York, 1971; (*b*) Lwowski, W., *Nitrenes*, Interscience, New York, 1970; (*c*) Gilchrist, T. L. and Rees, C. W., *Carbenes, nitrenes and arynes*, Nelson, London, 1969; (*d*) Bethell, D., *Adv. phys. org. Chem.*, **7**, 153 (1969); (*e*) Marchand, A. P , in *The chemistry of double-bonded functional groups*, ed. S. Patai, p. 533, Wiley, London, 1977.
73. Hoffmann, R., *J. Am. chem. Soc.*, **90**, 1475 (1968).
74. Skell, P. S. and Woodworth, R. C., *J. Am. chem. Soc.*, **78**, 4496 (1956).
75. Neureiter, N. P., *J. Am. chem. Soc.*, **88**, 558 (1966); Paquette, L. A., *Acc. chem. Res.*, **1**, 209 (1968).
76. Bordwell, F. G., Williams, J. M., Hoyt, E. B. and Jarvis, B. B., *J. Am. chem. Soc.*, **90**, 429 (1968); Dittmer, D. C., Levy, G. C. and Kuhlmann, G. E., *J. Am. chem. Soc.*, **91**, 2097 (1969); Tatsumi, K., Yoshioka, Y., Yamaguchi, K. and Fueno, T., *Tetrahedron*, **32**, 1705 (1976).
77. Hartzell, G. E. and Paige, J. N., *J. org. Chem.*, **32**, 459 (1967).
78. Clark, R. D. and Helmkamp, G. K., *J. org. Chem.*, **29**, 1316 (1964).
79. Freeman, J. P. and Graham, W. H., *J. Am. chem. Soc.*, **89**, 1761 (1967).
80. Carpino, L. A. and Kirkley, R. K., *J. Am. chem. Soc.*, **92**, 1784 (1970).
81. Heine, H. W., Myers, J. D. and Peltzer, E. T., *Angew. Chem. Int. Edn Engl.*, **9**, 374 (1970).
82. Zimmerman, H. E. and Sousa, L. R., *J. Am. chem. Soc.*, **94**, 834 (1972), and references therein.
83. Felix, D., Müller, R. K., Horn, U., Joos, R., Schreiber, J. and Eschenmoser, A., *Helv. chim. Acta*, **55**, 1276 (1972).
84. Reviews: (*a*) Chapman, O. L. and Lenz, G., in *Organic photochemistry*, vol. 1, p. 283, Arnold, London, 1967; (*b*) Warrener, R. N. and Bremner, J. B., *Rev. pure appl. Chem.*, **16**, 117 (1966); (*c*) Dilling, W. L., *Chem. Rev.*, **66**, 373 (1966); (*d*) Lamola, A. A. and Turro, N. J., *Technique of organic chemistry*, vol. 14, Interscience, New York, 1969; see also Vollmer, J. L. and Servis, K. L., *J. chem. Ed.*, **47**, 491 (1970); (*e*) Barltrop, J. A. and Coyle, J. D., *Excited states in organic chemistry*, Wiley, London, 1975.
85. Fleming, I., *Frontier orbitals and organic chemical reactions*, Wiley, London, 1976.
86. Schmidt, G. M. J., *Pure appl. Chem.*, **27**, 647 (1971).
87. Arnold, D. R. and Abraitys, V. Y., *Chem. Communs*, 1967, 1053.
88. Yamazaki, H. and Cvetanović, R. J., *J. Am. chem. Soc.*, **91**, 520 (1969); Yamazaki, H., Cvetanović, R. J. and Irwin, R. S., *J. Am. chem. Soc.*, **98**, 2198 (1976).
89. Eaton, P. E., *Acc. chem. Res.*, **1**, 50 (1968).
90. Corey, E. J., Mitra, R. B. and Uda, H., *J. Am. chem. Soc.*, **86**, 485 (1964); Corey, E. J. and Nozoe, S., *J. Am. chem. Soc.*, **86**, 1652 (1964).
91. Deering, R. A. and Setlow, R. B., *Biochim. biophys. Acta*, **68**, 526 (1963).
92. (*a*) Mango, F. D., *Adv. Catalysis*, **20**, 291 (1969); (*b*) Schrauzer, G. N., *Adv. Catalysis*, **18**, 373 (1968); (*c*) Pettit, R., Sugahara, H., Wristers, J. and Merk, W., *Disc. Faraday Soc.*, **47**, 71 (1969).
93. Hogeveen, H. and Volger, H. C., *J. Am. chem. Soc.*, 89, 2486 (1967).
94. For a discussion see Calderon, N., Ofstead, E. A. and Judy, W. A., *Angew. Chem., Int. Edn Engl.*, **15**, 401 (1976).
95. Wolovsky, R., *J. Am. chem. Soc.*, **92**, 2132 (1970); Ben-Efraim, D. A., Batich, C. and Wasserman, E., *J. Am. chem. Soc.*, **92**, 2133 (1970).
96. Cassar, L., Eaton, P. E. and Halpern, J., *J. Am. chem. Soc.*, **92**, 3515 (1970).
97. Burns, P. and Waters, W. A., *J. chem. Soc., (C)*, 1969, 27; Murphy, W. S. and McCarthy, J. R., *Chem. Communs*, 1968, 1155.

98. Paquette, L. A., Kirschner, S. and Malpass, J. R., *J. Am. chem. Soc.*, **91**, 3970 (1969).
99. See Moriconi, E. J., Hummel, C. F. and Kelly, J. F., *Tetrahedron Lett.*, 1969, 5325, and references therein, for possible examples of concerted additions of this cumulene.
100. Gait, S. F., Rance, M. J., Rees, C. W., Stephenson, R. W. and Storr, R. C., *J. chem. Soc. Perkin I*, 1975, 556.
101. Mock, W. L., *J. Am. chem. Soc.*, **89**, 1281 (1967); Mock, W. L., *J. Am. chem. Soc.*, **91**, 5682 (1969).
102. Mock, W. L., *J. Am. chem. Soc.*, **92**, 3807 (1970).
103. For the specific example shown see Paquette, L. A. and Slomp, G., *J. Am. chem. Soc.*, **85**, 765 (1963).
104. Holland, J. M. and Jones, D. W., *Chem. Communs*, 1969, 587.
105. Blatt, K. and Hoffmann, R. W., *Angew. Chem., Int. Edn Engl.*, **8**, 606 (1969).
106. Doering, W. von E. and Wiley, D. W., *Tetrahedron*, **11**, 183 (1960); Prinzbach, H., Seip, D., Knothe, L. and Faisst, W., *Justus Liebigs Annln Chem.*, **698**, 34 (1966).
107. Boekelheide, V. and Fedoruk, N. A., *Proc. natn. Acad. Sci. USA*, **55**, 1385 (1966); Boekelheide, V. and Fedoruk, N. A., *Chem. Abstr.*, **65**, 13683 (1966).
108. Paul, I. C., Johnson, S. M., Barrett, J. H. and Paquette, L. A., *Chem. Communs*, 1969, 6; Paquette, L. A., Barrett, J. H. and Kuhla, D. E., *J. Am. chem. Soc.*, **91**, 3616 (1969).
109. Dennis, N., Katritzky, A. R. and Takeuchi, Y., *Angew. Chem., Int. Edn Engl.*, **15**, 1 (1976).
110. Houk, K. N., George, J. K. and Duke, R. E., *Tetrahedron*, **30**, 523 (1974).
111. Houk, K. N. and Luskus, L. J., *J. org. Chem.*, **38**, 3836 (1973); Caramella, P., Frattini, P. and Grünanger, P., *Tetrahedron Lett.*, 1971, 3817; Asao, T., Machiguchi, T., Kitamura, T. and Kitahara, Y., *Chem. Communs*, 1970, 89; Harmon, R. E., Barta, W. D., Gupta, S. K. and Slomp, G., *J. chem. Soc. (C)*, 1971, 3645.
112. Houk, K. N. and Luskus, L. J., *Tetrahedron Lett.*, 1970, 4029; Dunn, L. C., Chang, Y-M. and Houk, K. N., *J. Am. chem. Soc.*, **98**, 7095 (1976).
113. Padden-Row, M. N., Gell, K. and Warrener, R. N., *Tetrahedron Lett.*, 1975, 1975.
114. Mukai, T., Tezuka, T. and Akasaki, Y., *J. Am. chem. Soc.*, **88**, 5025 (1966).
115. Prinzbach, H. and Knöfel, H., *Angew. Chem., Int. Edn Engl.*, 8, 881, (1969).
116. Mok, K. L. and Nye, M. J., *J. chem. Soc. Perkin I*, 1975, 1810.
117. Prinzbach, H. Knothe, L. and Dieffenbacher, A., *Tetrahedron Lett.*, 1969, 2093.
118. Huisgen, R., *Angew. Chem., Int. Edn Engl.*, **7**, 321 (1968), and references therein.
119. Review: Reppe, W., Kutepow, N. and Magin, A., *Angew. Chem., Int. Edn Engl.*, 8, 727 (1969).

7 Sigmatropic rearrangements and related reactions

Uncatalysed thermal rearrangements occurring intramolecularly and involving a six-membered cyclic transition state are very common and include an enormous variety of structural types. Some of these, such as the Cope and Claisen rearrangements and 1,5-hydrogen shifts in conjugated dienes, have been known for many years; others have only recently been recognised as being of this type. The development of the theory of concerted reactions has led to the realisation that they represent only one of many possible classes of similar rearrangements, some thermally and some photochemically induced, which could occur in anions and cations as well as in neutral molecules. The unifying features of all these reactions are that they are concerted, uncatalysed and involve a bond migration through a cyclic transition state in which an atom or a group is simultaneously joined to both termini of a π electron system. Woodward and Hoffmann have given the name *sigmatropic rearrangements* to such reactions, the adjective 'sigmatropic' indicating movement of a sigma bond.

7.1. Nomenclature

There is a formal system of nomenclature for sigmatropic rearrangements which is widely used. Consider the Cope rearrangement of 1,5-hexadiene (fig. 7.1): in this reaction a σ bond is broken and one is made, and two π bonds migrate. The termini of the σ bond have moved to carbon atoms 3 and 3′, according to the numbering shown, with the original termini at carbons 1 and 1′. This change is then defined as a sigmatropic reaction of order [3, 3], both termini having moved to the *third* carbon atom along the π system. In general, a change of order $[i, j]$ involves a movement of a σ bond to a new position where its termini are i-1 and j-1 atoms removed from the original position.

A few other examples will illustrate the system. The feasibility of the reactions will be discussed later; the examples are given to show the use of the numbering system. In each case the numbering starts at the original termini of the σ bond.

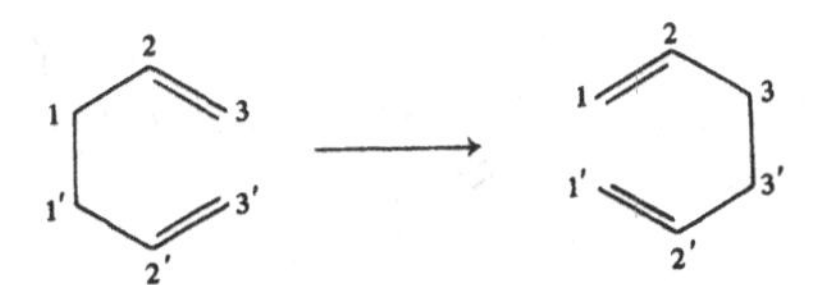

Fig. 7.1

The ylide rearrangement (*a*) is a sigmatropic change of order [3, 2], since the new σ bond is two and one atoms away from the original position.

(*a*)

The migration of hydrogen through an allyl system (*b*) is a rearrangement of order [1, 3] because the univalent hydrogen has moved from C-1 to C-3 ($j = 3$); the other end of the σ bond is still attached to hydrogen and so $i = 1$.

(*b*)

Similarly, any migration in which an atom or a group moves unchanged through a π system will be a rearrangement of order $[1, j]$.

The norcaradiene rearrangement (*c*) is a [1, 5] shift, the alkyl group at C-1

(*c*)

migrating to C-5. The numbering must go through the π system, even though an alternative numbering through the adjacent tetrahedral carbon would give a smaller figure.

A simple rule for determining the order $[i, j]$ is as follows: count the number of atoms in each of the fragments formally produced by breaking the migrating σ bond. This gives i and j directly.

7.2. Hydrogen migrations[1]

Selection rules

Consider the $[1, j]$ shift of a hydrogen atom between the ends of a polyene.

$$\begin{matrix} R^1 \\ R^2 \end{matrix} \!\!>\! \overset{H}{\overset{|}{C}}(CH{=}CH)_nCH{=}C\!<\!\!\begin{matrix} R^3 \\ R^4 \end{matrix} \longrightarrow \begin{matrix} R^1 \\ R^2 \end{matrix}\!\!>\!C{=}CH(CH{=}CH)_n\overset{H}{\overset{|}{C}}\!<\!\!\begin{matrix} R^3 \\ R^4 \end{matrix}$$

In a cyclic transition state, the bonding hydrogen orbital must overlap simultaneously with orbitals on both the terminal carbon atoms. These carbon orbitals must also overlap with the other p orbitals of the polyene chain. The orbital on C-1 eventually becomes a p orbital of the polyene while that on C-*j* becomes the sp^3 bonding orbital of the new C—H bond.

There are two stereochemically distinct ways in which the overlap can take place in the transition state. The spherically symmetrical hydrogen orbital can overlap with the p lobes on the same side of the π system (*suprafacial* overlap) or on opposite sides (*antarafacial* overlap). The geometry of the two transition states is different (fig. 7.2); the suprafacial migration is characterised by a plane of symmetry, and the antarafacial migration by a twofold axis.

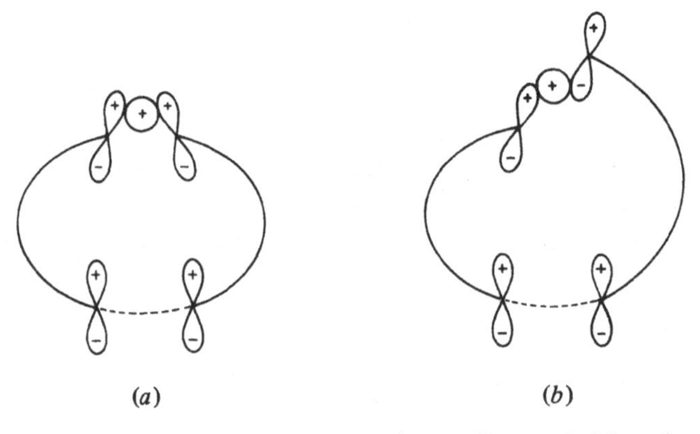

Fig. 7.2. (*a*) Suprafacial overlap, Hückel type; (*b*) antarafacial overlap, Möbius type.

Selection rules for such migrations can most easily be derived by applying the 'aromatic transition state' concept. The signs of the lobes in fig. 7.2 are put in according to the method described in chapter 2; that is, with the maximum number of adjacent overlapping lobes having the same sign. The transition state can then be classed as being of the Hückel type (no sign inversions in the cycle) or of the Möbius type (one sign inversion). The advantage of the method for these systems is that it does not require a knowledge of the symmetries of the HOMO and LUMO of the reactants. Selection rules for hydrogen migrations can then be deduced.

The Hückel-type transition state is favoured when 2, 6, . . . , $(4n + 2)$

electrons participate, and the Möbius type when 4, 8, . . . , 4*n* electrons participate. Thus *it is the number of electrons, not the number of atoms, which*

$$\overset{H\oplus}{R_2CCR_2'} \longrightarrow \overset{\oplus H}{R_2CCR_2'}$$

$$\overset{H\ominus}{R_2CCR_2'} \longrightarrow \overset{\ominus H}{R_2CCR_2'}$$

Fig. 7.3. Transition state for suprafacial [1,2] shift.

determines the selection rules. This is particularly important to bear in mind when considering migrations in charged polyenes. For example, a [1, 2] hydrogen shift in a carbenium ion is favoured as a suprafacial process, since the two electrons (those of the σ bond) participate; but a [1, 2] hydrogen shift in a carbanion is not favoured as a suprafacial process, since four electrons (two from the anion and two from the σ bond) participate (fig. 7.3). A general formulation of the selection rules for thermal sigmatropic shifts of hydrogen, based on the number of participating electrons, is given in table 7.1.

Table 7.1. Selection rules for sigmatropic hydrogen migrations

Number of electrons	Polyene: Neutral	Cation	Anion	Thermally allowed migration
2	–	[1, 2]	–	Suprafacial
4	[1, 3]	[1, 4]	[1, 2]	Antarafacial
6	[1, 5]	[1, 6]	[1, 4]	Suprafacial
$4n$	[1, $(4n-1)$]	[1, $4n$]	[1, $(4n-2)$]	Antarafacial
$4n+2$	[1, $(4n+1)$]	[1, $(4n+2)$]	[1, $4n$]	Suprafacial

The selection rules can also be derived from frontier orbital theory. In chapter 2, sigmatropic hydrogen migrations were discussed in terms of interacting C—H bonds and π systems. A comparison of figs. 2.10 and 2.11 shows that the interaction depends upon the symmetry of the HOMO and LUMO of the polyene. For transition states in which a total of $4n+2$ electrons participates, the interactions are as shown in fig. 7.4 and for $4n$ systems they are as shown in fig. 7.5. The first clearly leads to suprafacial overlap, and the second to antarafacial overlap.

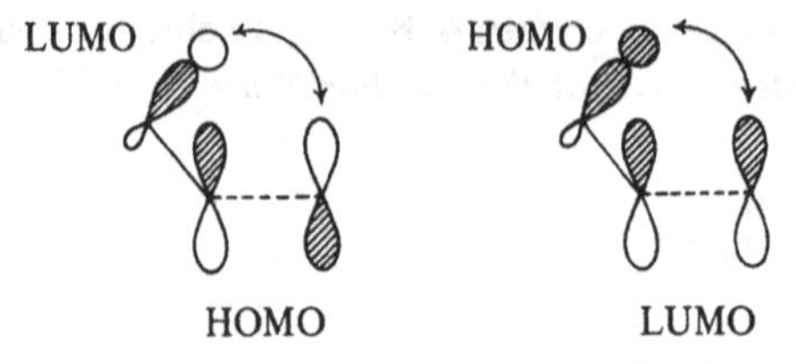

Fig. 7.4. Frontier orbital interactions for sigmatropic hydrogen shifts involving $4n + 2$ electrons.

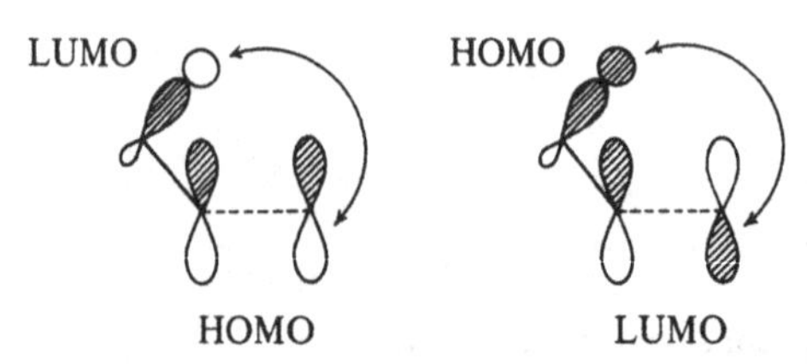

Fig. 7.5. Frontier orbital interactions for sigmatropic hydrogen shifts involving $4n$ electrons.

Another approach is to construct an interaction diagram for the orbitals involved in the transition state.[2] For a suprafacial shift of hydrogen in an allyl system the relevant orbitals are the hydrogen 1s orbital and the π orbitals of the allyl group; the interaction diagram is shown in fig. 7.6.

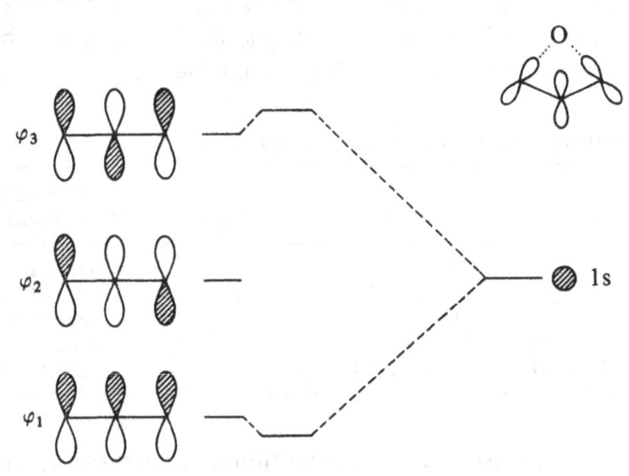

Fig. 7.6. Orbital interaction diagram for suprafacial [1,3] hydrogen migration in an allyl system.

Interaction of the s orbital will be most effective with π orbitals closest in energy and of the right phase relationship at both termini. In this case only ψ_1 and ψ_3 are of the right phase to overlap, and both are well separated in energy from the s orbital, so there will be no effective pericyclic bonding in the transition state.

An advantage of this approach is that it allows the possibility of introducing substituent effects: substituents which alter the energies of the allyl π orbitals relative to the s orbital will have the effect of increasing the interaction either with ψ_1 or with ψ_3. This point is developed later in connection with sigmatropic alkyl shifts.

Examples

The selection rules predict that suprafacial [1, 5] hydrogen shifts in neutral polyenes are thermally allowed, whereas [1, 3] and [1, 7] shifts must go by an antarafacial process. This implies that thermal [1, 3] shifts of hydrogen are unlikely to be observed, since the transition state for antarafacial migration would be very strained and difficult to attain. On the other hand, [1, 5] shifts should be facile. The experimental observations fit in with this pattern; whereas concerted uncatalysed [1, 3] hydrogen shifts have not been established, [1, 5] shifts in dienes (fig. 7.7) are well known, and there is evidence to support the view that they are concerted reactions.[1]

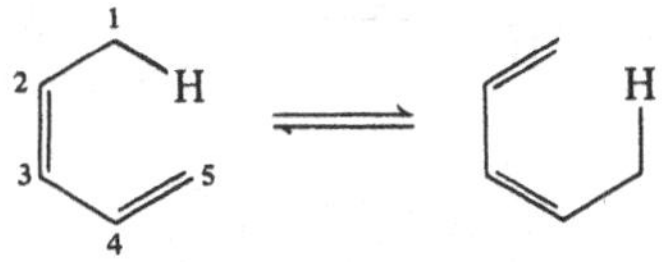

Fig. 7.7. [1,5] hydrogen shift.

In acyclic dienes, the [1, 5] shifts have activation energies of about 134 kJ mol^{-1} and they show large deuterium isotope effects (k_H/k_D is about 5 at 200 °C in *cis*-1,3-pentadiene). The stereospecific suprafacial nature of the migration has been demonstrated with the diene (**1**).[3] As equation 7.1 shows, the optically active starting material gave the two isomers expected from a suprafacial [1, 5] shift, but gave neither of the isomers that would result from an antarafacial migration.

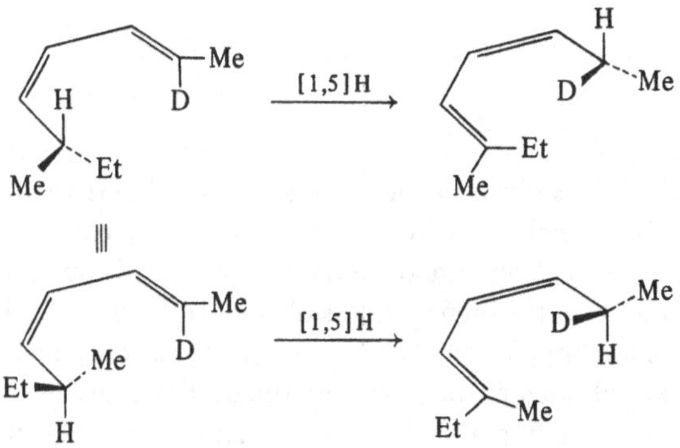

(7.1)

(**1**)

In both cyclic and acyclic dienes which can achieve the necessary geometry the [1, 5] shift is commonly observed, the activation energy being lowest for transition states involving minimal distortion. This is particularly so in cyclopentadienes and indenes, where a [1, 5] shift simply involves the movement of hydrogen to an adjacent carbon atom. As a result, activation energies for hydrogen migration in cyclopentadienes are considerably lower (by about 40 kJ mol^{-1}) than in open-chain systems. The preference for [1, 5] over [1, 3] shifts in such systems is demonstrated by the thermal rearrangement of 1-deuterioindene (**2**). When this was heated at 200 °C, the deuterium became 'scrambled' over all three non-benzenoid carbons (7.2). The presence of 2-deuterioindene in the product mixture indicates that the system prefers to rearrange by successive [1, 5] shifts rather than by [1, 3] shifts, even though [1, 5] shifts involve the intermediacy of isoindenes.[4]

H D 1 2 3 4 5 [1,5]D H D [1,5]H D

(7.2)

(**2**)

In cyclohexa-1,3-dienes the [1, 5] hydrogen migration must be to a non-adjacent carbon atom, so that the activation energies are comparable with those for acyclic dienes. Thus, the activation energy for rearrangement of 5-methylcyclohexa-1,3-diene (**3**) is 143 kJ mol^{-1}. The greater flexibility of seven-

Me [1,5]H Me

(7.3)

(**3**)

membered rings lowers the activation energy slightly: for 2-methylcycloheptadiene[1] the value is 124 kJ mol^{-1}. As all such thermal reactions are equilibria, the composition of the equilibrium mixture obviously depends upon the nature of the substituents in the dienes. A less obvious influence of substituents is that they can considerably affect the activation of the rearrangement, particularly if they are situated at the migration origin. For example, activation energies for [1, 5] hydrogen migration in the cycloheptatrienes (**4**) and (**5**) are 135 kJ mol^{-1} and 108 kJ mol^{-1}, respectively.[1]

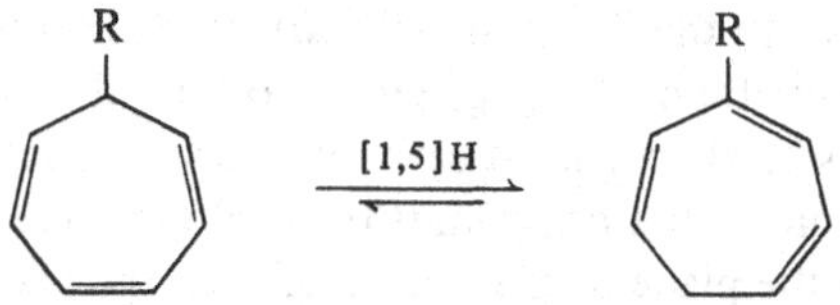

R R

[1,5]H

(7.4)

(4) R = Me

(5) R = OMe

The simplest analogy for the reaction outside all-carbon systems appears to be the rearrangement of the unsaturated enol (7) formed by tautomerisation of the β,γ-unsaturated carbonyl compound (6). The overall result of the rearrangement is that the double bond is brought into conjugation with the carbonyl group (7.5). Such a rearrangement might be expected to be thermally induced, or acid catalysed, the acid promoting the enolisation of the carbonyl compound. Thus, the [1, 5] shift in the enol is a possible mechanism, though not the only one, for the movement of a double bond into conjugation with a carbonyl group.

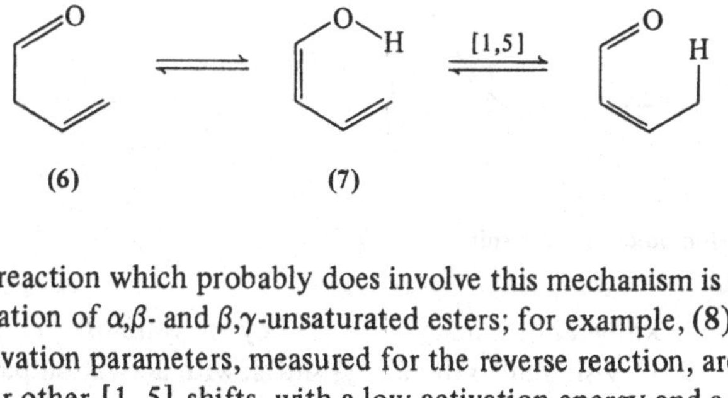

O O H [1,5] O H

(7.5)

(6) (7)

One reaction which probably does involve this mechanism is the thermal equilibration of α,β- and β,γ-unsaturated esters; for example, (8) ⇌ (9), (7.6).[5] The activation parameters, measured for the reverse reaction, are similar to those for other [1, 5] shifts, with a low activation energy and a negative entropy of activation.

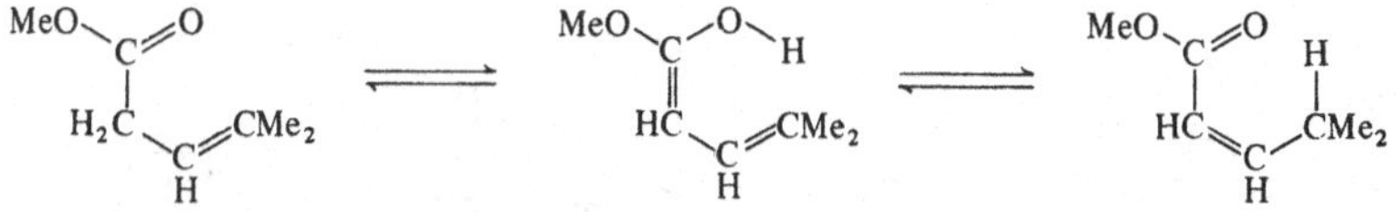

MeO C O H₂C C CMe₂ H — MeO C O H HC C CMe₂ H — MeO C O H HC C CMe₂ H

(8) (9) (7.6)

It is a general feature of pericyclic reactions that a cyclopropane ring can often participate in place of a double bond. Indeed, the olefin-like character of cyclopropane derivatives is well known; for example, in their ability to transmit conjugation. Various pictures of the bonding in cyclopropane have been constructed to explain this behaviour. In the Walsh model **(10)**, each carbon has an sp^2 orbital directed towards the centre of the ring, and a p orbital the lobes of which overlap with the p orbitals on the adjacent carbons. These p lobes can overlap with other p lobes in a conjugated π system if the cyclopropane ring is suitably orientated; that is, with the plane of the ring parallel to the plane in which the rest of the π system lies.

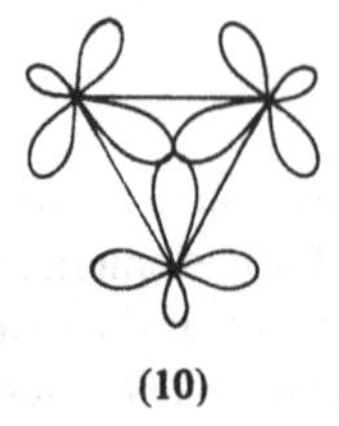

(10)

Thus, a well-documented extension of the [1, 5] shift occurs in systems where one of the double bonds is replaced by a cyclopropane ring. Such reactions have been called *homodienyl* [1, 5] hydrogen shifts (fig. 7.8).[1] The transition state for such rearrangements will be very similar to that for dienyl shifts, and the selection rules will similarly apply.

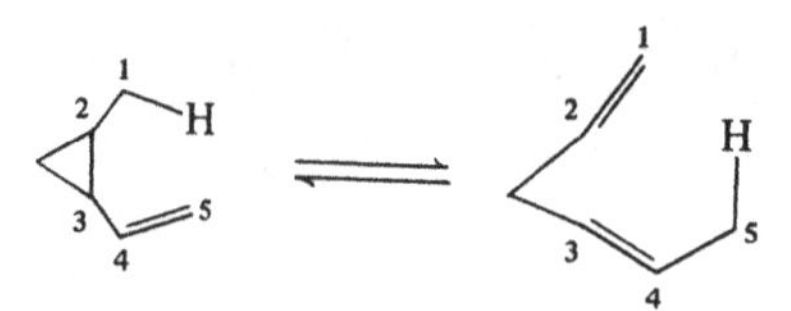

Fig. 7.8. Homodienyl [1,5] shift.

Such reactions are known both in acyclic and in cyclic systems, and appear to be mechanistically very similar to the dienyl shifts, with activation parameters of the same order. An example of the rearrangement in a cyclic system is the thermal equilibration of bicyclo[6,1,0]non-2-ene **(11)** and *cis,cis*-1,4-cyclononadiene at 150–170 °C (7.7).

H H (7.7)

(11)

The geometry of the transition state requires that the migrating hydrogen and the methylene group of the cyclopropane are *anti*, with the plane of the ring parallel to that of the π bond. This constraint makes the transition state increasingly difficult to achieve as the ring size is reduced. In rings of smaller than seven atoms the activation energy is too great for the reaction to be a significant one.

The oxygen analogue of the reaction, the rearrangement of a cyclopropyl ketone, is also known (7.8).[6] This reaction provides a possible route for the isomerisation of γ,δ-unsaturated carbonyl compounds (7.9). In aromatic substrates the reaction is called the abnormal Claisen rearrangement; it is described with the Claisen rearrangement in §7.4.

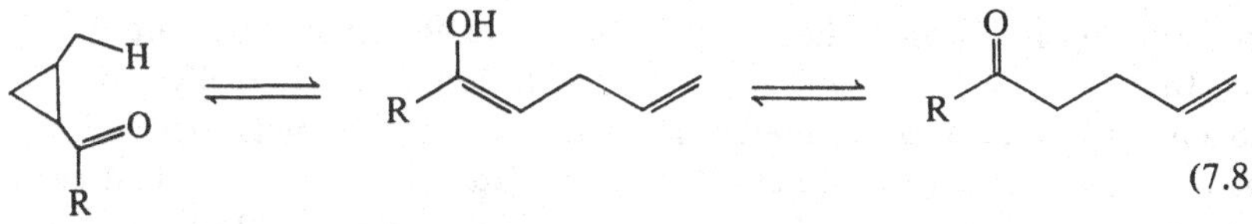

(7.8)

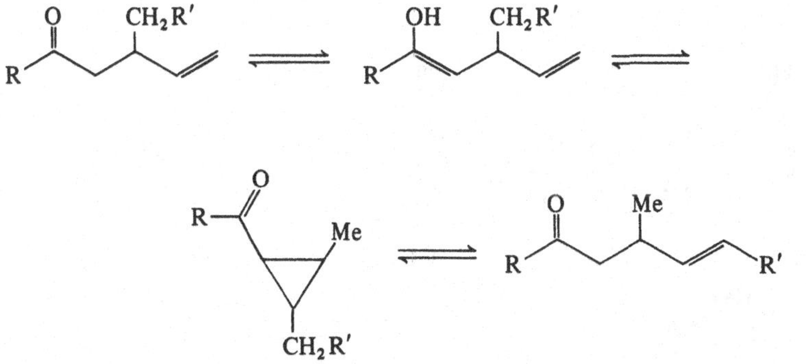

(7.9)

There is no hard and fast dividing line between [1, 5] shifts in which one of the participating bonds is a π bond or a 'π-type' bond, as in the systems just described ($n = 0$ or 1 in structure **12**), and those in which it is a σ bond ($n > 1$ in **12**; or structure **13**). Mechanistically, all such reactions are closely related. The analogous reaction of σ systems such as (**13**) is the retro-ene reaction (§7.8).

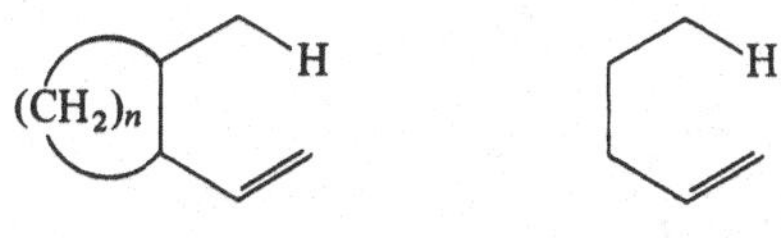

(**12**) (**13**)

Antarafacial [1, 7] shifts are also thermally allowed, and the geometry of the transition state is not inaccessible (fig. 7.9).

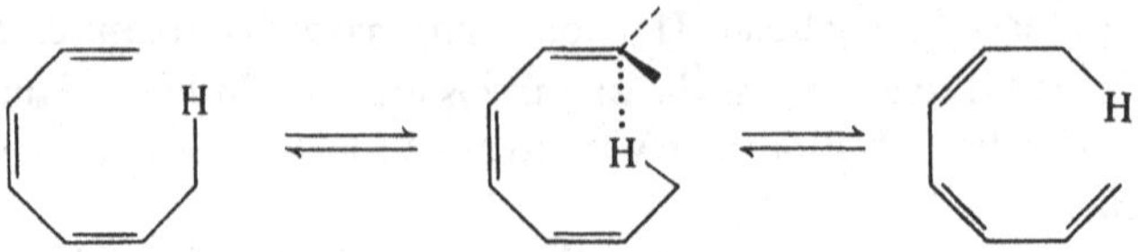

Fig. 7.9. Antarafacial [1,7] shift.

A few reactions are known which probably do involve [1, 7] shifts of this type; one is the thermal rearrangement of calciferol to precalciferol (chapter 3, equation 3.29). In acyclic systems where there is competition between [1, 5] suprafacial and [1, 7] antarafacial shifts, and where both transition state geometries are readily attainable, the [1, 7] shift occurs preferentially and with a much lower activation energy. The entropy of activation is large and negative, indicating a highly ordered transition state. Thus, the [1, 7] hydrogen shift takes place in the triene **(14)** with an activation energy of 63 kJ mol^{-1}, and an activation entropy of -105 J mol^{-1} K^{-1}.

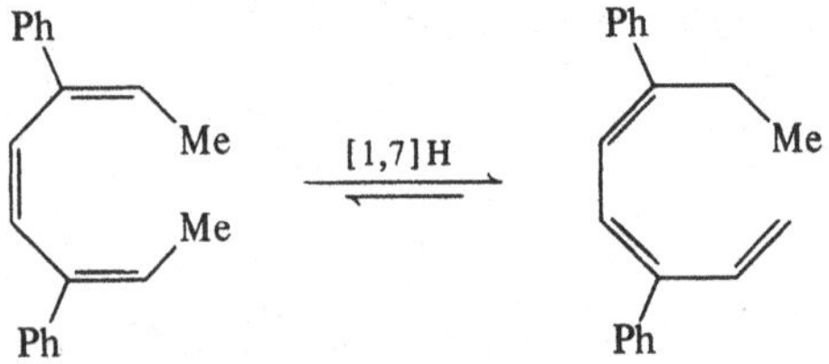

(14) $\xrightarrow{[1,7]H}$ (7.10)

In charged systems, suprafacial [1, 2] hydrogen shifts in cations are the best known. A few examples are known of hydrogen shifts in 1,3-dipoles: one is the rearrangement of the ylide **(15)** which takes place slowly at room temperature. The reaction appears to be intramolecular because no deuterium is incorporated from a deuteriated solvent, and the rate is relatively insensitive to solvent polarity. This therefore appears to be a [1, 4] hydrogen shift, which is allowed as a suprafacial process.[7]

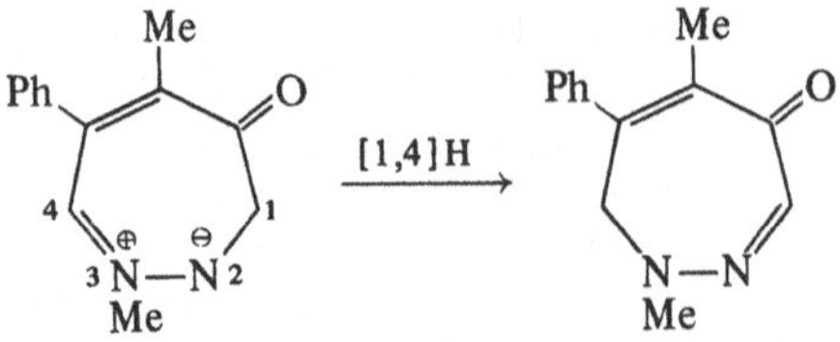

(15) $\xrightarrow{[1,4]H}$ (7.11)

7.3. Migrations of atoms or groups other than hydrogen

Sigmatropic migrations are not confined to hydrogen atoms. Groups of nearly every type have been found to migrate; rearrangements of aryl and alkyl groups are quite common, and there is an increasing number of examples of migration of vinyl, acyl and alkoxycarbonyl groups. Organometallic functions such as trialkylsilyl groups also migrate readily.

In attempting to understand and predict the course of such reactions we are faced with several problems in addition to those we have dealt with for hydrogen migrations. When the migration involves an alkyl group, there is the question of whether its configuration is retained or inverted in the product. There are also the problems of how readily different groups migrate, why they should migrate at different rates, and whether the relative rates are likely to be the same in all cases. Answers to at least some of these questions are beginning to emerge, as the experimental results described later in the section will show.

Selection rules

In general, four types of transition states can be envisaged for sigmatropic migrations of the order $[1, j]$, two in which the migrating group moves suprafacially and two in which it moves antarafacially. These are illustrated in fig. 7.10, and the selection rules, based on their classification as aromatic or antiaromatic transition states, are given in table 7.2.

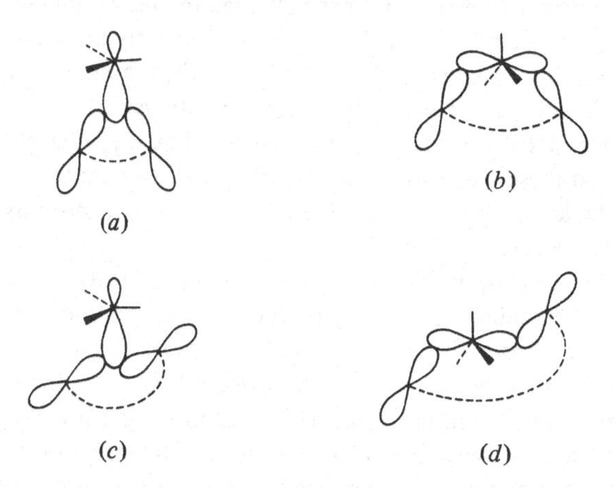

Fig. 7.10. Transition states for sigmatropic migration of alkyl groups. (*a*) Suprafacial migration with retention; (*b*) suprafacial migration with inversion; (*c*) antarafacial migration with retention; (*d*) antarafacial migration with inversion.

The predictions of the orbital symmetry rules are straightforward when applied to migrations in which the transition states involve relatively little strain. The difficult cases to predict are those where the allowed transition

Table 7.2. Selection rules for sigmatropic group migrations of order [1, j]

Number of electrons	Neutral	Polyene: Cation	Anion	Thermally allowed migration
2	–	[1, 2]	–	Suprafacial, with retention
4	[1, 3]	[1, 4]	[1, 2]	Antarafacial, with retention; Suprafacial, with inversion
6	[1, 5]	[1, 6]	[1, 4]	Suprafacial, with retention; Antarafacial, with inversion
$4n$	$[1, 4n - 1)]$	$[1, 4n]$	$[1, (4n - 2)]$	Antarafacial, with retention; Suprafacial, with inversion
$4n + 2$	$[1, (4n + 1)]$	$[1, (4n + 2)]$	$[1, 4n]$	Suprafacial, with retention; Antarafacial, with inversion

states are sterically difficult to achieve. This is particularly true for the four-electron transition states, in which both of the allowed modes of migration are unattractive from a steric point of view. The problem is exactly analogous to that discussed earlier for 2 + 2 cycloadditions (p. 173): the reactions could go concertedly by an allowed but sterically unfavourable pathway or by a forbidden but sterically more attractive pathway, or they could be stepwise. We can get further insight into these reactions by considering the orbital interactions in the transition state, and as an example, [1, 3] shifts in neutral systems will be considered in more detail.

Figs. 7.11 and 7.12 show orbital interaction diagrams for [1, 3] suprafacial alkyl shifts; the first is for migration with retention of configuration in the migrating group and the second for migration with inversion.

In fig. 7.11 the orbital of the migrating group interacts with ψ_1 and ψ_3 of the allyl group framework. The interactions will result in a slight lowering of the energy of ψ_1 and a slight raising of the energy of ψ_3; the effect on the energy of the migrating group will be negligible because the two interactions will cancel out. Thus the transition state for the reaction which is formally a forbidden one on the basis of the Woodward-Hoffmann rules actually shows a very slight net stabilising interaction because ψ_1 will be filled and ψ_3 will be vacant.[8]

When the migrating group migrates with inversion, as shown in fig. 7.12, there is a strong interaction between ψ_2 of the allyl group and the migrating

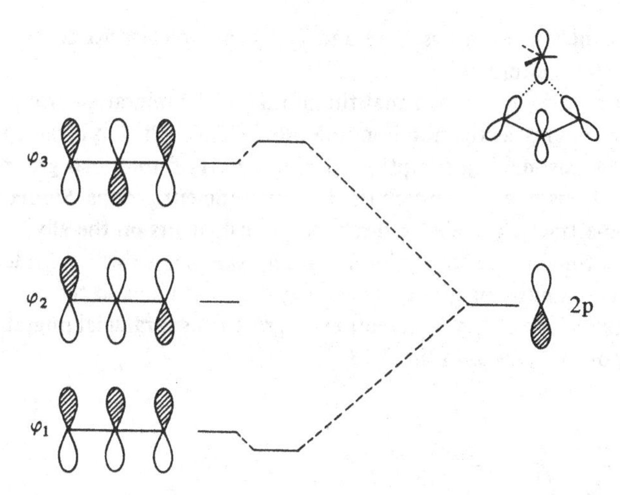

Fig. 7.11. Orbital interaction diagram for [1,3] suprafacial migration with retention.

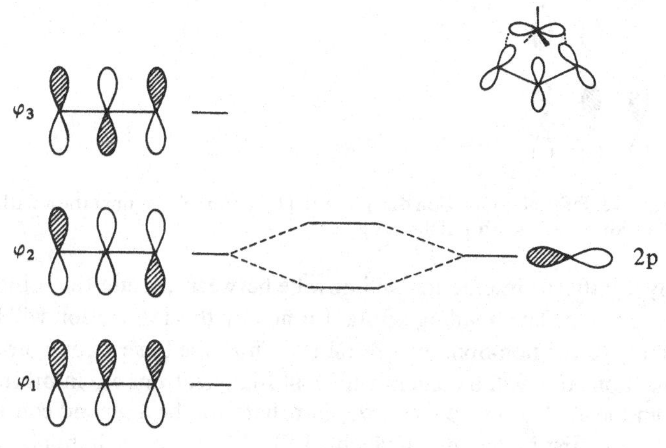

Fig. 7.12. Orbital interaction diagram for [1,3] suprafacial migration with inversion.

group orbital. The four electrons involved can occupy bonding orbitals, and the transition state is strongly stabilised (much more so than in the previous case). This reaction is allowed by the Woodward-Hoffmann rules. Thus the energies of the various possible modes of [1,3] alkyl migration appear to lie in the order: migration with inversion (allowed) < migration with retention (forbidden) < stepwise. Of course this takes no account of the strain involved in the transition states of such migrations; steric interactions are likely to be important in such tight transition states, and the ordering in particular systems may well be changed for this reason. Antarafacial [1,3] shifts (which can be analysed in the same way as in figs. 7.11 and 7.12) are even less attractive because of such steric interactions.

In this analysis we have assumed that the orbital of the migrating group is at about the same energy as the non-bonding orbital ψ_2 of the allyl framework. This is probably a reasonable assumption for migrations of non-polar groups, but not for migrations of electron-rich or electron-deficient groups. Similarly, strongly electron-attracting or electron-releasing substituents on the allyl framework affect the relative energy levels. As an example, we will consider a system in which the migrating group is relatively electron rich and the allyl system is electron deficient. The interaction diagram for suprafacial migration with retention now appears as in fig. 7.13.

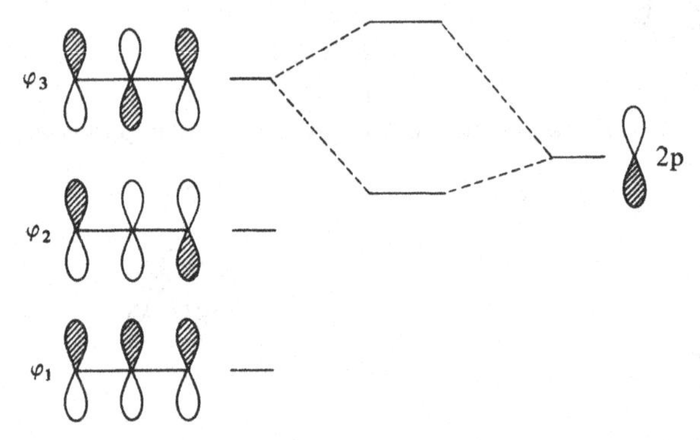

Fig. 7.13. Orbital interaction diagram for [1,3] suprafacial migration with retention by an electron-rich group.

The only significant interaction will now be between ψ_3 and the orbital on the migrating group. The bonding orbital formed by this interaction will be close in energy to the non-bonding orbital ψ_2. Thus, the lowest configuration for the transition state will be one in which all four electrons are in orbitals derived from the allyl group (ψ_1 and ψ_2) but there will be a second configuration, close to the first in energy, in which all four electrons are in bonding orbitals (derived from ψ_1 and from the interaction of the migrating group orbital with ψ_3). As the wave function contains components of both configurations, the result will be an increase in the bonding character of the transition state compared with that of a non-polar transition state; that is, the transition state takes on more of the character expected for an 'allowed' reaction. How important this is in determining the stereoselectivity of the migration will obviously depend on the exact ordering of the energy levels. Changing the polarity of the transition state in this way for a suprafacial migration with inversion has very little effect; this can be seen by constructing the interaction diagram as a modification of fig. 7.12. In the same way, we can deduce the likely consequences of having a transition state in which the migrating group is electron deficient and the allyl framework electron rich. In such a case the transition state for migration with retention will involve a major interaction between ψ_1 of the allyl group and the migrating group orbital, and for migration with inversion, the interaction with ψ_2 will be less than in fig. 7.12, resulting in less bonding character. Again, migration with retention is likely to be more favourable than in the non-polar transition state.

Examples: neutral species

[1,3] *Shifts.* These are not common, but mechanistic studies of the reaction have produced some elegant experimental work. One example of a suprafacial shift with inversion of configuration, due to Roth and Friedrich, is the thermal rearrangement of the bicyclohexane **(16)**. The substituted methylene bridge migrates in such a way that the methyl group is almost exclusively *exo* in the product.[9]

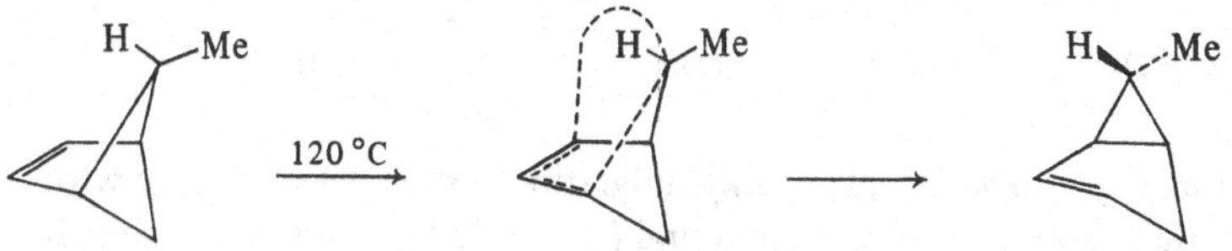

H Me — 120 °C → H Me → H Me (7.12)

(16)

Another striking example of suprafacial migration with inversion, due to Berson and his colleagues, is rearrangement of the bicyclo-octenone **(17)**.[8] This gives stereoselectively the ketone **(18)** which is the less stable of the two possible products of [1,3] rearrangement (this is because of interaction between the methyl group and a hydrogen atom in a bridging methylene group). The migration with inversion may be particularly favoured here by overlap with the carbonyl π orbital in the transition state **(19)** and this is reflected in the activation energy of 156 kJ mol^{-1} which is lower than for other [1,3] alkyl shifts.

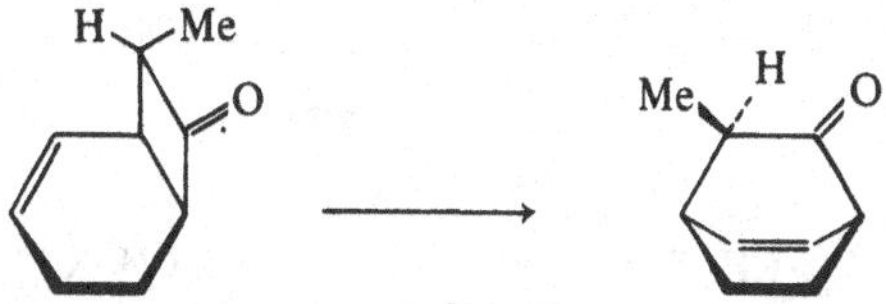

H Me O → Me H O (7.13)

(17) **(18)**

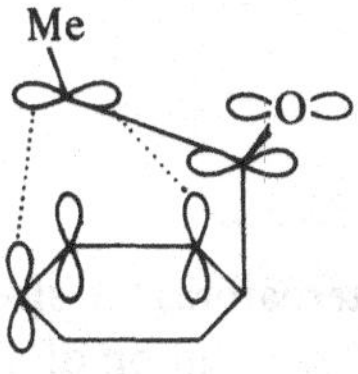

Me O

(19)

The result is quite different in related systems containing substituents which hinder the migration with inversion; the bicyclo-octene **(20)**, for example, gives the products of migration with inversion **(21)** and with retention **(22)** in a ratio of 2.4 : 1, the activation energy being 195 kJ mol^{-1}. In more hindered systems the migration with retention is even more favoured. These results can be interpreted as indicating that the forbidden transition state becomes energetically competitive, or, perhaps less likely, that the mechanism changes to a diradical one.

Me H X H → Me X + Me X (7.14)

(20) (X = $OSiMe_3$) **(21)** **(22)**

Another example in which [1,3] alkyl migration occurs preferentially with retention is the rearrangement of compound **(23)** to **(24)**.[10] This is a system in which the cyano groups might sufficiently lower the energy levels of the allyl framework to make the migration with retention the favoured pathway, although of course it is also a system in which radical intermediates would be particularly well stabilised. Radical intermediates have been detected by CIDNP in 1,3-allylic shifts **(25)** to **(26)**, X being a substituent bearing a lone pair which helps to stabilise the radical.[11]

Me H Ph Me NC CN —250 °C→ Me H Me Ph CN CN (7.15)

(23) **(24)**

CH_2X ⇌ $\cdot CH_2X$ ⇌ CH_2X (7.16)

(25) **(26)**

Examples of 1,3-migration from heteroatoms are also known. A [1,3] sigmatropic shift from oxygen to carbon, with retention of configuration at the migrating group, is illustrated in equation 7.17.[12] 1,3-Migrations from nitrogen to oxygen in imidates (the *Chapman* rearrangement, equation 7.18) are definitely stepwise processes. When an aryl group migrates, a zwitterionic intermediate is probably involved; the less common 1,3-migration of alkyl groups is an intermolecular process. (Note, however, that migration of an *allyl* group allows the reaction to attain the characteristics of a [3,3] shift; p. 280.)

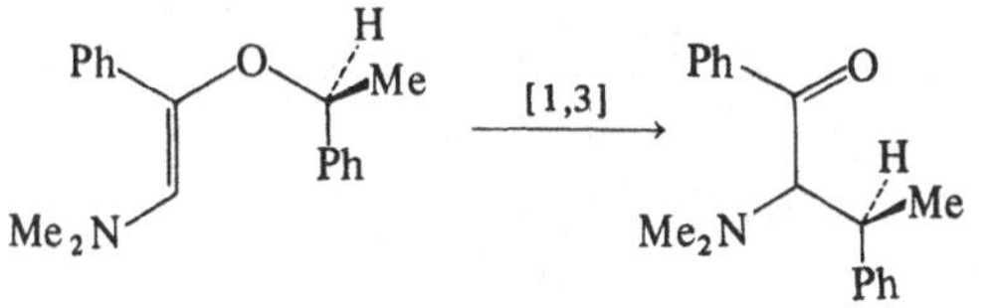

(7.17)

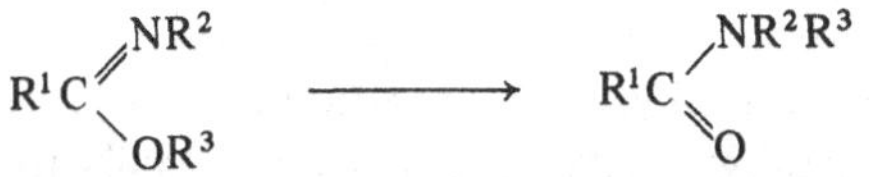

(7.18)

The above examples of 1,3-shifts clearly illustrate that we cannot simply equate 'symmetry-allowed' with 'most favourable' in trying to predict the course of a reaction. When the symmetry-allowed pathway is different from the sterically most favourable pathway, the choice depends critically upon the type of system involved and the nature of the substituents.

[1,5] *Shifts.*[1] In the case of [1,5] shifts the symmetry-allowed pathway involves migration with retention, and the required transition state is quite accessible to most systems. Evidence that migration of an alkyl group does go with retention is provided by the rearrangement of the spirodienes shown in equations 7.19 and 7.20. The stereospecific [1,5] alkyl migrations are followed by rapid [1,5] hydrogen shifts, to give the products having the stereochemistry shown.[13]

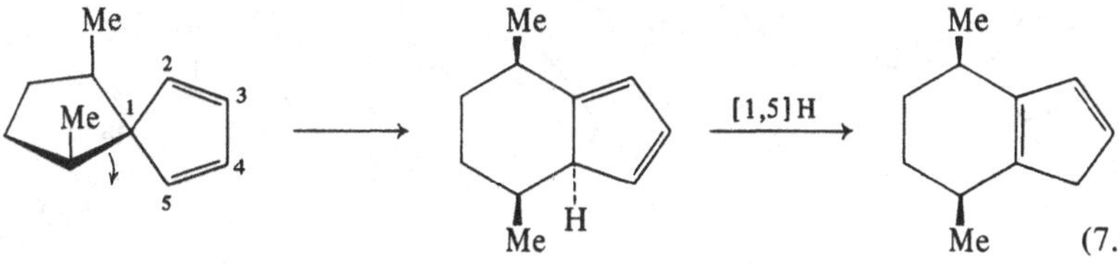

(7.19)

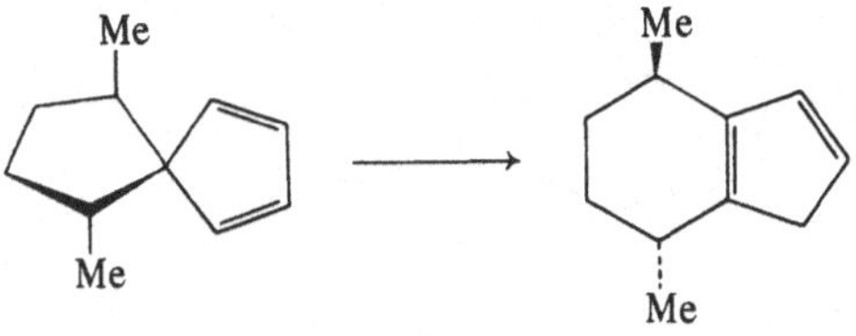

(7.20)

This is undoubtedly the pathway followed by most [1,5] alkyl shifts. There is at least one case, however, where the opposite stereochemistry is observed: the esters (27) and (28) are interconverted at 180 °C in a reaction which apparently involves *inversion* at the migrating group.[14] The reason is that the cyclopropane is epimerising faster than it is undergoing sigmatropic rearrangement; the rearrangement does in fact go with retention of configuration.[14]

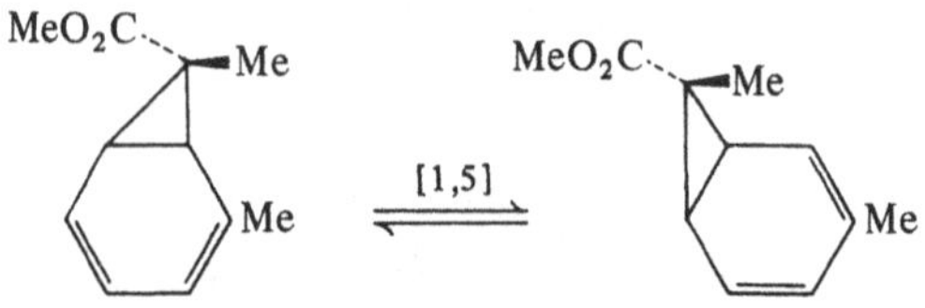

MeO_2C Me Me [1,5] MeO_2C Me Me (7.21)

(27) (28)

It is a common feature that [1,5] alkyl shifts are accompanied by migration of hydrogen or other groups. Hydrogen migrates much more readily than does an alkyl group, the difference in activation energies being about 84 kJ mol^{-1} in comparable systems. There are now sufficient experimental data to allow some comparison of the relative ease of migration of a wide range of groups. Groups which migrate particularly readily include those in which the carbon atom of the migrating group is sp^2 hybridised; that is, acyl, imidoyl, aryl and vinyl groups. D. W. Jones and his colleagues have made a careful study of [1,5] shifts in indenes (equation 7.22) and have suggested that the electron-accepting ability of the unsaturated migrating group (that is, lower π^* energy) parallels its tendency to migrate.[15] This is explained by a secondary interaction (29) between the HOMO of the isoindene framework and the LUMO (π^*) of the migrating group as the transition state is approached.

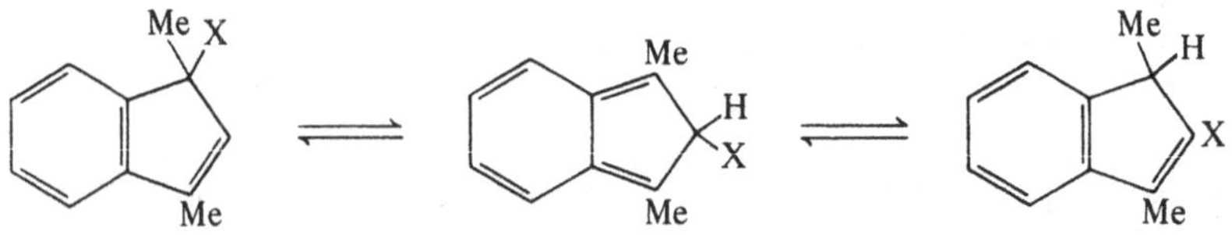

Me X Me Me H X Me Me H X Me

(7.22)

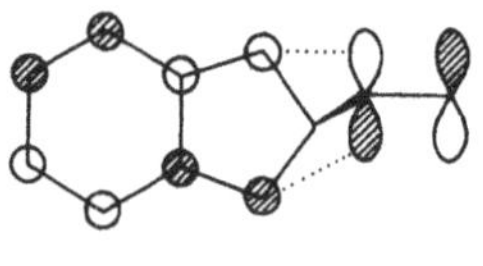

(29)

In the indene system the order in which different groups migrate is CHO > COMe > H > CH=CH_2 > CO_2 Me > CN > Me, and it is likely that a similar order is followed in other ring systems. Other groups which undergo [1,5] shifts very readily include Me_3Sn, Me_3Si and NO_2.[1, 16]

[1,5] Shifts occur most readily in five-membered rings, both carbocyclic and heterocyclic, since these rearrangements require movement of the group to an adjacent atom in the ring. [1,5] Migrations in cyclohexadienes require about 40-60 kJ mol^{-1} more activation energy than in cyclopentadienes, and this makes migrations of alkyl and similar groups difficult in six-membered rings. Activation energies are influenced by other factors, such as the nature of the substituents on the dienyl skeleton, and whether aromatic character is created or destroyed by the migration. Several examples of [1,5] shifts which involve the isomerisation of a heteroaromatic system to a non-aromatic system are now known; three such examples are shown in equations 7.23-7.25.[17]

N Me [1,5] crotyl 500 °C → N H Me [1,5] H → N H Me (7.23)

N—N Ph Ph N Ph [1,5] Ph 800 °C → N=N Ph Ph N Ph − N_2 → Ph NH Ph

(7.24)

N Ph N Me [1,5] Me 900 °C → N Ph Me N [1,5] Ph → N Me N Ph (7.25)

Another factor which arises in some [1,5] shifts is a choice between two different types of migration. This is illustrated by the reactions of 3*H*-pyrazoles, which can be prepared by the cycloaddition of diazo compounds to activated acetylenes. The substituents at the 3-position can undergo [1,5] migration either to the adjacent carbon or to nitrogen (fig. 7.14). It seems that the more readily the group migrates, the more likely it is to move to nitrogen: thus, acetyl groups move to nitrogen, alkoxycarbonyl groups give both types of product, and alkyl or aryl groups move exclusively to carbon (the *van Alphen* rearrangement).[18] It is notable that the development of aromatic character in the shift to nitrogen is not the determining factor in every case.

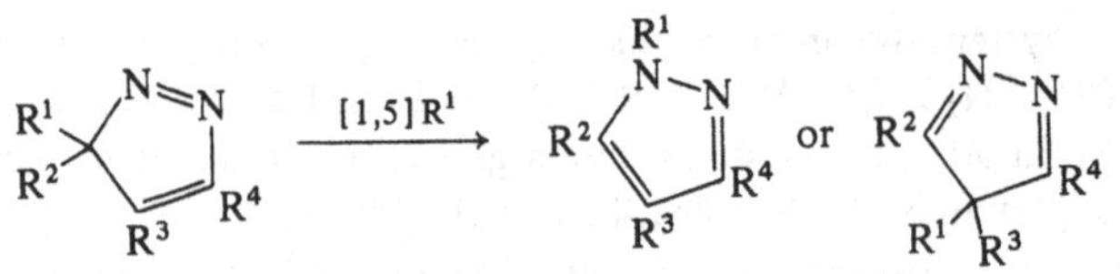

Fig. 7.14. [1,5] shifts in 3*H*-pyrazoles.

[1,7] *and higher order shifts.*[1] Examples of shifts of higher order than [1,5] are relatively rare. [1,7] Shifts of alkyl groups are symmetry-allowed as suprafacial processes in which the configuration of the migrating group is inverted. This inversion has been shown to occur in the thermal equilibration of the bicyclononatrienes in equation 7.26[19] (but note that, as with the re-arrangement shown in equation 7.21, orbital symmetry control can be masked by other factors in reactions of this type).

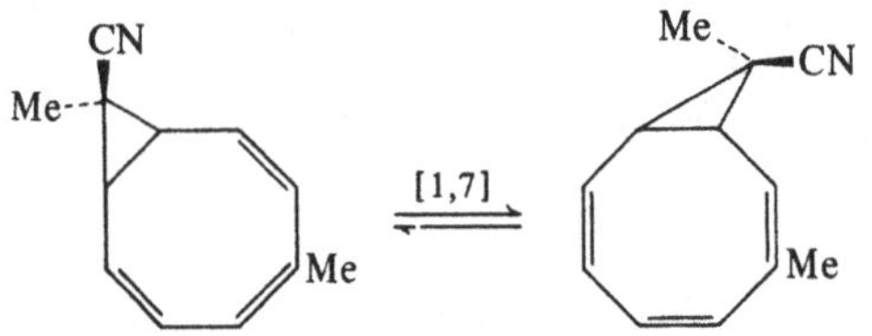

(7.26)

Shifts of higher order are normally precluded by the impossibility of attaining a cyclic transition state, but systems can be designed in which these constraints no longer apply. The [1,9] methyl shift shown in equation 7.27 has been suggested as a step in a more complex rearrangement sequence: here the transition state for suprafacial migration is easy to attain because the methyl group moves to an adjacent atom.[20]

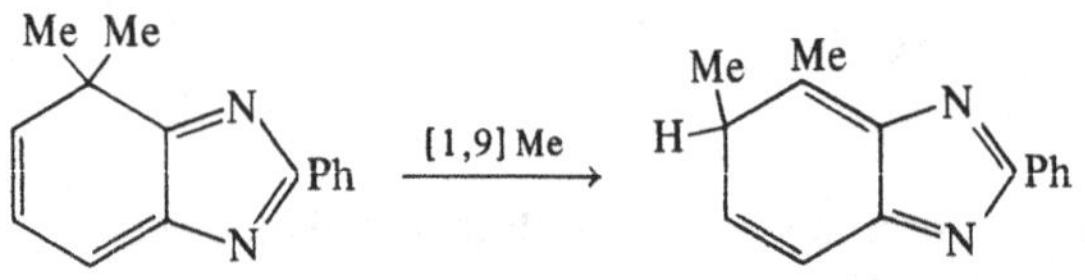

(7.27)

Charged species

Alkyl or aryl shifts in carbenium ions and other electron-deficient species represent a large and important group of molecular rearrangements. In this group are Wagner–Meerwein rearrangements of terpenes, the dienone-phenol rearrangement, the pinacol-pinacolone rearrangement, and expansion and contraction of small rings via carbenium ion intermediates. The electron-deficient centre need not be a carbenium ion; the requirement is only that it has a vacant p orbital. Thus, [1,2] shifts in singlet carbenes, nitrenes and nitrenium ions are of the same general type (fig. 7.15).

The simple theory predicts that [1,2] shifts to electron-deficient centres should proceed suprafacially and with retention of configuration, through a two-electron Hückel-type transition state. Molecular orbital calculations have been carried out which support this prediction. It is an oversimplification to represent all these reactions as going through free intermediates of the types shown in fig. 7.15, which then rearrange; the anion associated with the intermediate may play a major part in determining the product distribution, and in some cases the [1,2] shift and the loss of the anion are probably concerted. The bridged species may even be an intermediate, especially if it involves a migrating aryl group. Despite these variations, the simple qualitative picture is useful. In these reactions the migrating group moves with retention of configuration, as predicted. This has been shown for a carbenium ion rearrangement[21] and for several other [1,2] shifts.

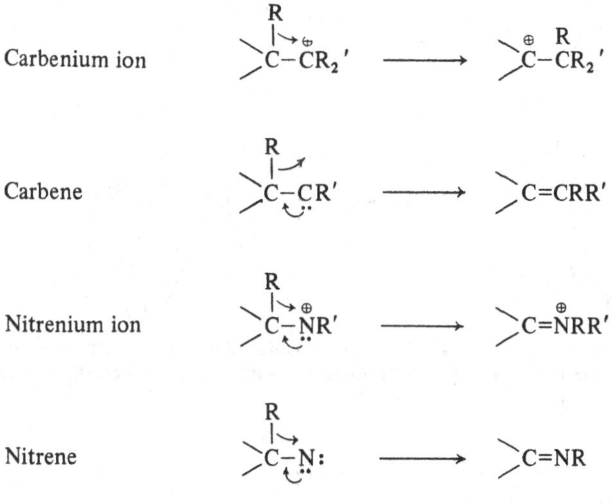

Carbenium ion $\quad >C(R)-\overset{\oplus}{C}R_2' \longrightarrow >\overset{\oplus}{C}-C(R)R_2'$

Carbene $\quad >C(R)-\ddot{C}R' \longrightarrow >C{=}CRR'$

Nitrenium ion $\quad >C(R)-\overset{\oplus}{\ddot{N}}R' \longrightarrow >C{=}\overset{\oplus}{N}RR'$

Nitrene $\quad >C(R)-\ddot{N}: \longrightarrow >C{=}NR$

Fig. 7.15. [1,2] shifts to electron-deficient centres.

The theory is also useful in that it predicts that concerted [1,2] shifts to electron-rich centres should be unfavourable. This is reflected in the rarity of 1,2-shifts of hydrogen or alkyl groups in radicals and anions; when such rearrangements are observed in anions, they appear to be stepwise (§7.5).[24] 1,2-Migrations of aryl groups, which are fairly common, may also be stepwise, although it seems unlikely that discrete bridged intermediates such as the radical (**30**) are involved.

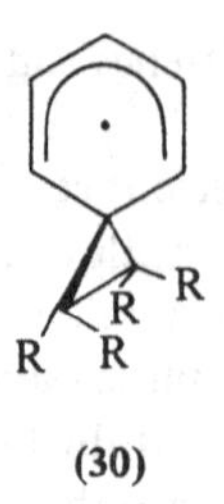

(30)

[1,4] Shifts in cationic polyenes are allowed either as antarafacial processes where the configuration of the migrating group is retained, or as suprafacial processes with inversion at the migrating centre. These possibilities are illustrated in fig. 7.16. Migrations of this type occur in systems where the migrating group is constrained to move in a suprafacial manner, and the predicted inversion of configuration has been observed. One example is the rearrangement of the bicyclohexenyl cation **(31)**.[22]

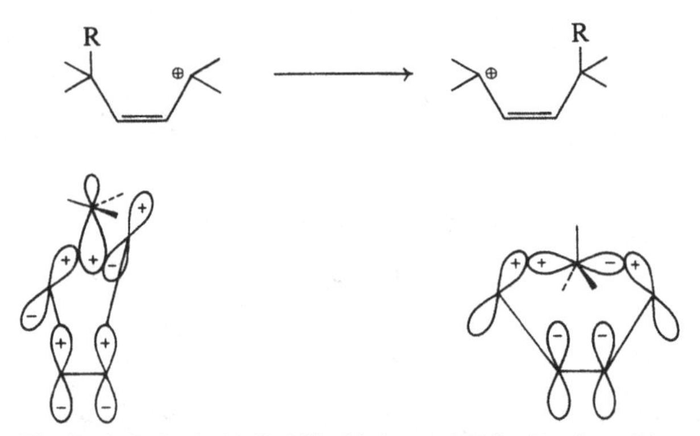

Fig. 7.16. Cationic [1,4] shift. (*a*) Antarafacial migration with retention of configuration; (*b*) suprafacial migration with inversion of configuration. Both are Möbius systems.

(31)

The selection rules also predict that suprafacial [1,6] shifts should be allowed in cationic π systems, with retention of configuration at the migrating centre; but it is difficult to conceive of a case where [1,2] shifts would not supervene.

One example of a rearrangement which can be regarded either as a [1,2] shift or as a [1,6] shift is the migration of the methyl group in the labelled benzenonium ion (7.28). The label is 'scrambled' over the benzene ring.[23]

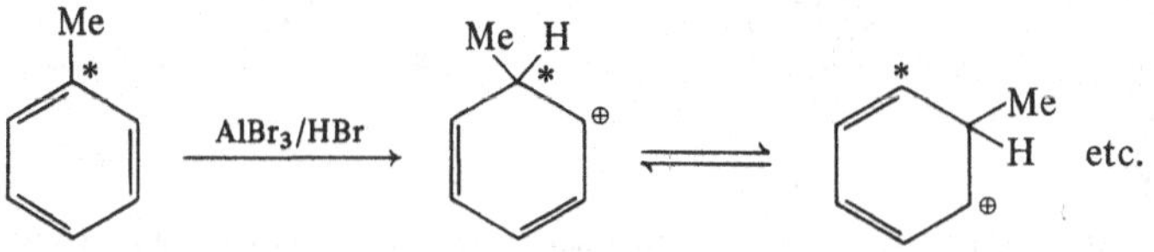

$AlBr_3/HBr$ (7.28)

Suprafacial [1,4] shifts in anions are also symmetry-allowed with retention of configuration at the migrating group. Alkyl shifts in 2-alkoxypyridine-1-oxides (**32**) are reactions of this type: the reactions are first order, with a high negative entropy of activation, and the alkyl group migrates with retention.[24]

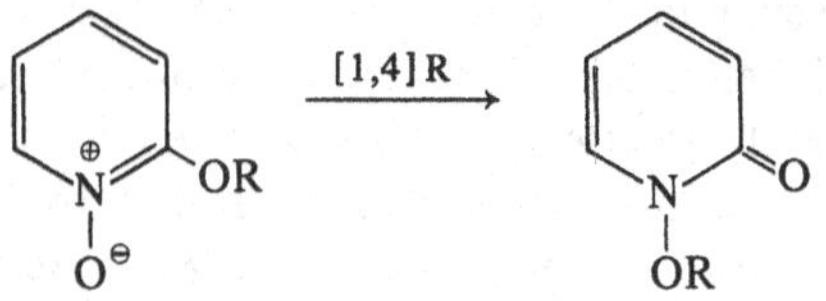

[1,4] R (7.29)

(**32**)

7.4. [3,3] Sigmatropic changes. The Cope and Claisen rearrangements

Selection rules

[3,3] Sigmatropic changes are an important group of thermal rearrangements which involve a six-membered cyclic transition state. This transition state (**33**) can be considered as two interacting allyl systems.

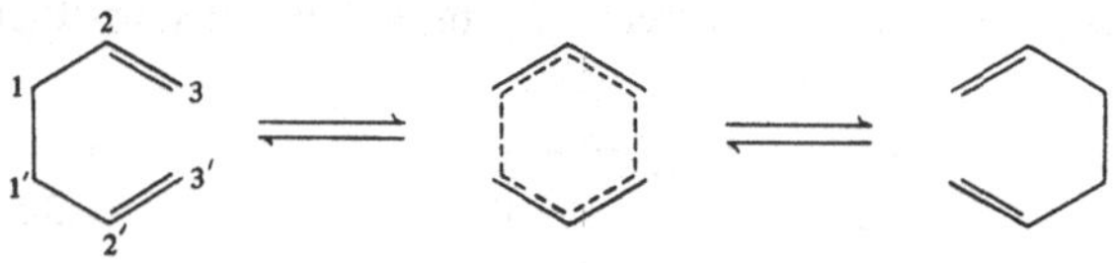

(**33**)

Various geometries are possible for the transition state, which can be classified according to whether each of the allyl systems interacts with lobes of the other system on the same side (suprafacially) or on opposite sides (antarafacially). Three transition states are shown in fig. 7.17; all are classed as Hückel systems on the basis of the 'aromatic transition state' approach, and all three are therefore thermally allowed.

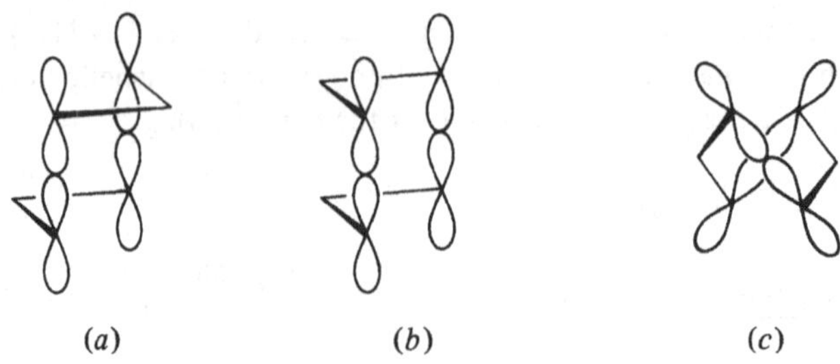

Fig. 7.17. Allowed transition states for thermal [3,3] shifts. (*a*) Chair (suprafacial, suprafacial); (*b*) boat (suprafacial, suprafacial); (*c*) twist (antarafacial, antarafacial).

Of these three types of transition states, the antarafacial, antarafacial one is much less likely to be found than the others, because it involves twisting of the allyl systems. The chair and boat forms of the suprafacial, suprafacial transition state are both relatively strain free. Of the two, the chair form might be expected to be more favoured because the six p lobes lie in a quasi-planar arrangement. The interaction of the central p lobes of the two allyl systems in the boat form does appear to have a slight destabilising effect: this has been shown by Woodward and Hoffmann with the aid of correlation diagrams, and by Dewar, by means of molecular orbital calculations.[31] Both the chair and the boat forms are found in practice, as the examples will show.

The Cope rearrangement[25]

1,5-Dienes isomerise on heating temperatures up to about 300 °C. The reaction is normally reversible and gives an equilibrium mixture of starting material and product. The temperature needed to bring about the reaction depends on the substituents; a conjugating substituent R (acyl, phenyl, etc.) lowers the energy of the transition state and the rearrangement (7.30) goes at 165–185 °C.

The reaction shows characteristics typical of a concerted process. It has a large negative entropy of activation, is relatively insensitive to substituent and

R R (7.30)

solvent effects, and is highly stereoselective. In acyclic 1,5-dienes, the evidence is that the transition state prefers a chair to a boat conformation. Doering and Roth showed this with *meso*-3,4-dimethyl-1,5-hexadiene (**34**) which rearranged almost exclusively (99.7 %) to *cis,trans*-2,4-octadiene at 225 °C (7.31 and 7.32).[26] This stereochemistry is consistent only with a chair conformation for the transition state: a boat would give the *cis,cis*- (7.33) or *trans,trans*-octadiene

(7.34). Similarly, optical activity is retained in the rearrangement since asymmetry is induced at a new centre. The optically active hexadiene (**35**) rearranged

225 °C

(7.31)

(**34**)

(7.32)

cis, trans

but

(7.33)

cis, cis

(7.34)

trans, trans

on heating to give a mixture of two other hexadienes, (**36**) and (**37**), which were formed in the ratio 87 : 13. Both compounds (**36**) and (**37**) were of about 90 % optical purity, showing that the rearrangement was highly stereoselective (7.35, 7.36).[27] The stereochemistry and absolute configurations of the products are consistent with the chair conformations shown for the reaction and place an upper limit of about 3 % on the contribution of boat conformations. The difference in energy of the two types of transition state in the Cope rearrangement is probably about 25 kJ mol^{-1}.

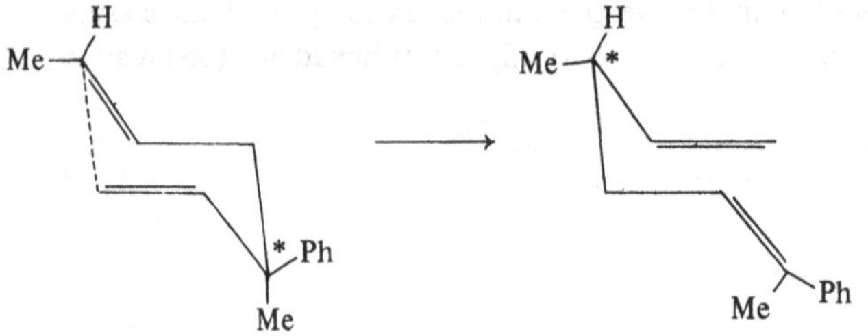

(R) **(35)** → (S) **(36)** 87 (7.35)

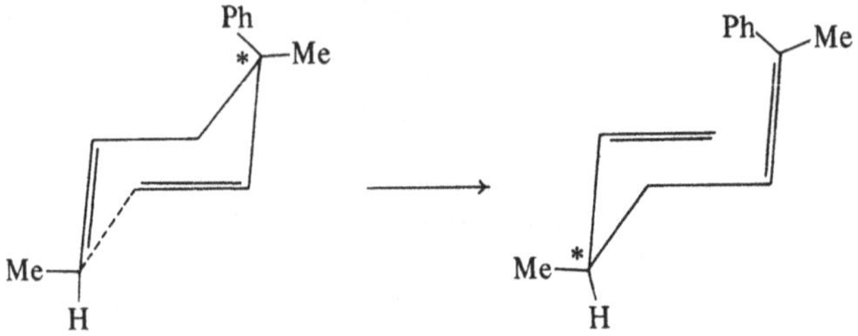

→ (R) **(37)** 13 (7.36)

In some cyclic systems the chair transition state is sterically impossible to attain, and the Cope reaction still goes but by a boat transition state. The Cope rearrangement of *cis*-1,2-divinylcyclopropane **(38)** and *cis*-1,2-divinylcyclobutane **(39)** must involve boat transition states, but both go extremely readily because of the relief of strain in the small rings (7.37, 7.38). The rearrangement of divinylcyclopropane **(38)** can be regarded as an electrocyclic ring closure in which the cyclopropane ring takes the place of a double bond; however, reactions of this sort have normally been classed as Cope rearrangements.[28]

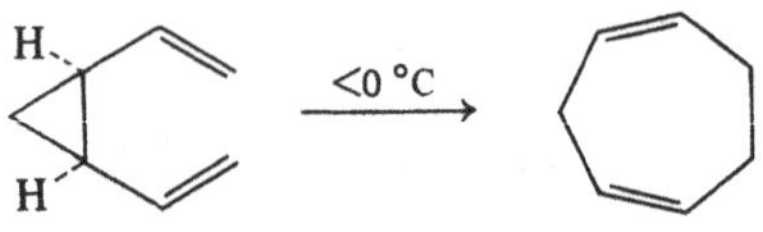

(38) $\xrightarrow{<0\ ^\circ C}$ (7.37)

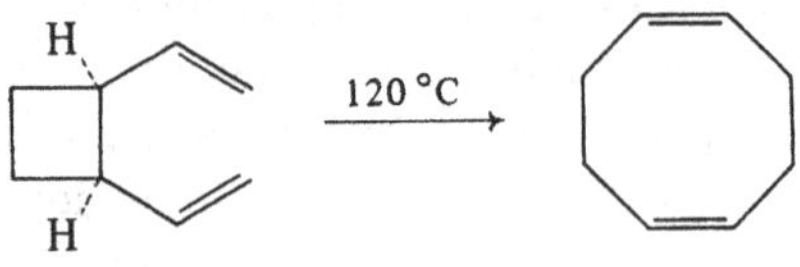

(39) $\xrightarrow{120\ ^\circ C}$ (7.38)

There are several systems related to **(38)** which undergo Cope rearrangement very readily, and in which the products have the same structure as the starting materials. Compounds which undergo such 'degenerate' rearrangements include the structures **(40)** in which a bridging group maintains the *cisoid* form (fig. 7.18)

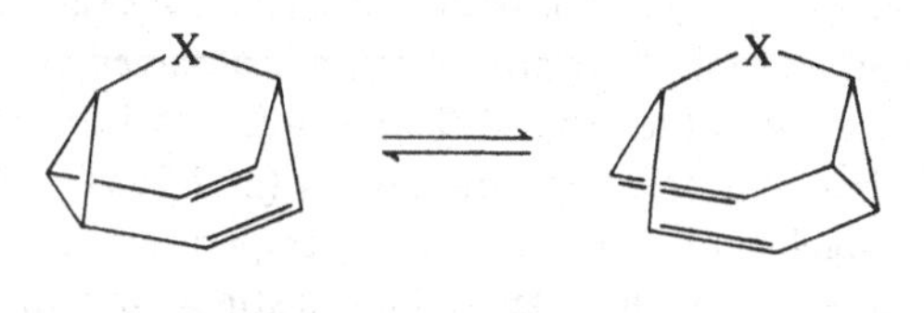

(40)

Fig. 7.18. Degenerate Cope rearrangements for *cis*-divinylcyclopropane derivatives (40).

X	*Trivial name*	*(kJ mol⁻¹)* $\Delta G^{\ddagger}$ *for rearrangement*[29]
CH=CH	Bullvalene	53.5
CH_2CH_2	Dihydrobullvalene	39.7
CH_2	Barbaralane	32.6
CO	Barbaralone	40.1
Direct bond	Semibullvalene	23.0

Dihydrobullvalene, semibullvalene, barbaralane and barbaralone each have two identical valence tautomers. For bullvalene, however, there are not two, but 1 209 600 possible arrangements; the cyclopropane can be at any three adjacent carbons (7.39).

etc.

(7.39)

Azabullvalenes **(41)** have potentially the same number of interconvertible forms as bullvalene, but now not all of these are equivalent in energy. The rearrangements have been found to show a strong preference for structures in which the nitrogen is at the junction of a double bond rather than at a bridgehead, so that the number of attainable arrangements is reduced to 28 (7.40).

R N R N R N etc.

(7.40)

(41)

These rearrangements bear a close relationship to the electrocyclic interconversion of norcaradienes and cycloheptatrienes and similar systems, and as in these cases, most of the information about them has come from a study of the effect of temperature change on the n.m.r. spectra. The principle of the method has been described earlier (chapter 3). With bullvalene, for example, an n.m.r. spectrum corresponding to the 'frozen' structure is obtained at -85 °C; there are two sharp signals in the ratio 3 : 2, the larger (at 5.65δ) corresponding to the six olefinic hydrogens and the smaller (at 2.58δ) to the three cyclopropyl and the bridgehead hydrogens. As the temperature is increased these signals gradually become broader and less distinct, and are then replaced by a new maximum which appears as a single sharp peak (4.22δ) at 100 °C and above. Above 100 °C therefore, the interconversion of the bullvalene tautomers occurs too rapidly for the individual forms to be distinguished, and an average structure, in which all the hydrogens are equivalent, is recorded. Similar averaging processes have been observed in the n.m.r. spectra of the other bridged species (**40**).

From the data quoted below fig. 7.18 it is clear that the $\Delta G^{\ddagger}$ values for rearrangements of this type are very low. There exists the intriguing possibility that the introduction of substituents might lower the value still further, to the point where $\Delta G^{\ddagger}$ becomes negative and the 'non-classical' structure becomes the favoured one. Hoffmann and Stohrer have calculated that this should be so for the semibullvalene derivatives (**42**).[30]

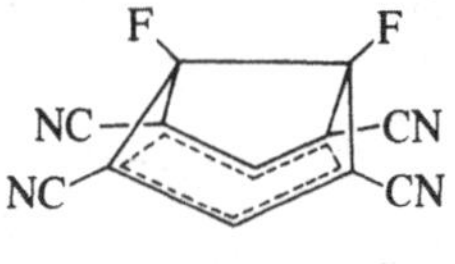

F F
NC CN
NC CN

(**42**)

The course of most Cope rearrangements can thus be described as a concerted [3,3] shift, with the chair transition state being slightly more favourable than the boat. There are, however, several instances where the cyclohexadiene system is so modified by substituents as to make alternative stepwise mechanisms feasible. These alternatives are of two types: (*a*) dissociation into a pair of allylic radicals or into a cation–anion pair, followed by recombination, and (*b*) formation of the new σ bond before the first has started to cleave, leading to a diradical or zwitterionic intermediate. These possibilities are illustrated in fig. 7.19.

A diradical intermediate would be most likely to be observed in the rearrangement of dienes such as 2,5-diphenyl-1,5-hexadiene, where the phenyl groups could stabilise the radical centres. Although the transition state (**43**) for this reaction has the character of a diradical, there is no evidence that it is a discrete intermediate on the reaction pathway.[31]

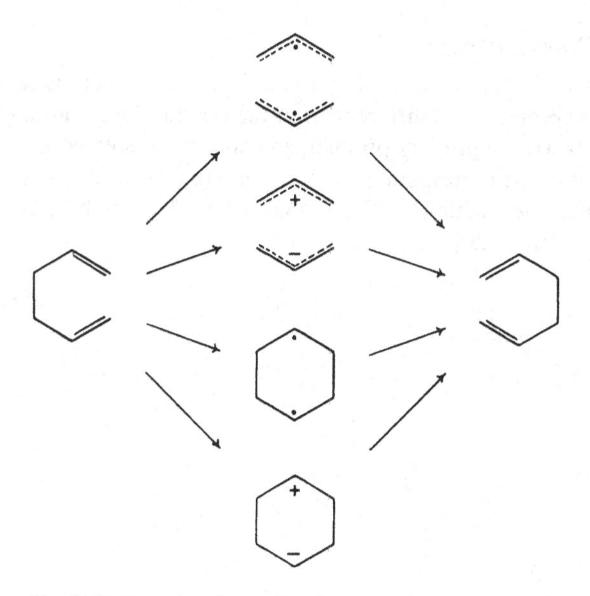

Fig. 7.19. Stepwise alternatives for the Cope rearrangement.

Ph Ph Ph

Ph Ph Ph

(43) (7.41)

The substituents in the diene **(44)**, which undergoes the Cope rearrangement in benzene at 80 °C, are suitably disposed to stabilise a zwitterionic intermediate **(45)**.[32] The intermediate can be trapped as its adduct **(46)** with benzaldehyde.

N R CO_2Me (44)

N ⊕ ⊖ R CO_2Me (45)

R = D

N D CO_2Me

PhCHO R = H

N O Ph CO_2Me (46) (7.42)

The oxy-Cope rearrangement[33]

If the 1,5-diene has a 3-hydroxy substituent, **(47)**, the rearrangement is called the oxy-Cope rearrangement. It is different from the normal Cope rearrangement in two respects: (i) the primary product, an enol, is not isolated since it rapidly tautomerises to the corresponding ketone; and (ii) the [3,3] shift competes with a retro-ene reaction (§7.8), so that the products of both reactions are normally isolated (fig. 7.20).

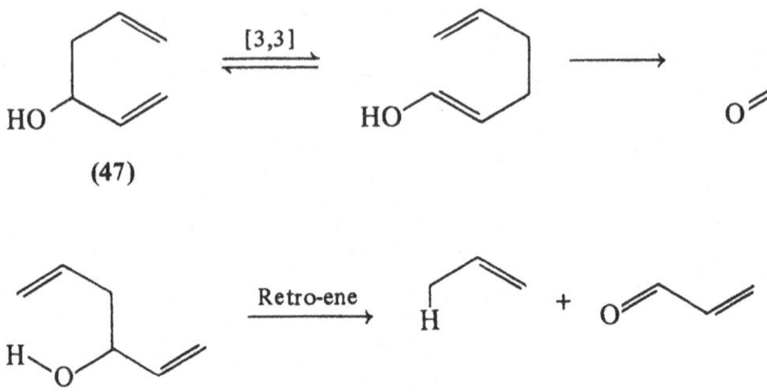

HO [3,3] HO O

(47)

H O Retro-ene H + O

Fig. 7.20. The oxy-Cope rearrangement.

The Cope rearrangement of alkoxide anions (equation 7.43) shows an enormous rate enhancement (10^{10}–10^{17} times) compared with the normal oxy-Cope rearrangement.[34]

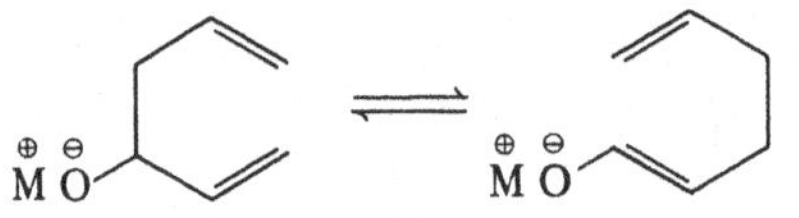

$\overset{\oplus}{M}\,\overset{\ominus}{O}$ ⇌ $\overset{\oplus}{M}\,\overset{\ominus}{O}$ (7.43)

The Claisen rearrangement[25]

The Claisen rearrangement is basically a thermal [3,3] sigmatropic rearrangement of an allyl vinyl ether (fig. 7.21). With aliphatic ethers it closely resembles the Cope rearrangement.

R O ⇌ R O

Fig. 7.21. The aliphatic Claisen rearrangement.

The rearrangement is better known with allyl aryl ethers, however, (fig. 7.22). In these systems the reaction is more complex because the product of rearrangement, an *ortho*-dienone, almost invariably reacts further.

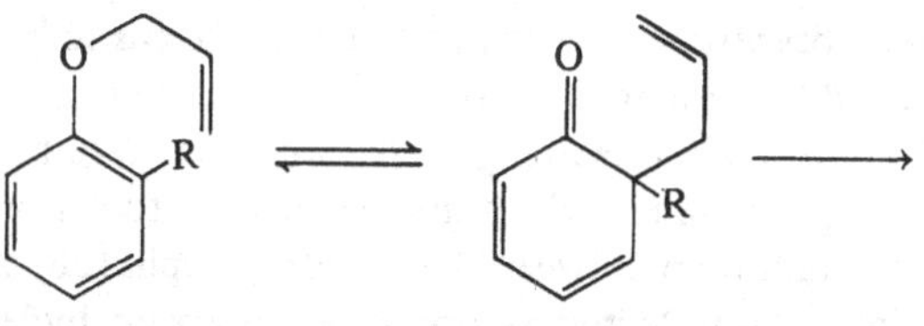

Further reactions

Fig. 7.22

Both in the alkyl and in the aryl series, the reactions show characteristics typical of concerted processes. They are first order and have large negative entropies of activation consistent with cyclic transition states. The allylic group is inverted in the rearrangement, and optical activity is retained if the starting ether is optically active. Polar solvents can increase the rate appreciably–the rearrangement has been observed to be about 100 times as fast in a polar solvent as in a non-polar solvent – and this may indicate some charge separation in the transition state. Similarly, there is a noticeable, but small, substituent effect on the rate: electron-releasing groups in the *para* position increase the rate, and electron-withdrawing groups decrease it. The overall rate difference is only about 20 or 30, however.

The Claisen rearrangement goes through a transition state in the chair conformation. *trans,trans*-Crotyl propenyl ether **(48)** gave more than 97 % of the *threo*-aldehyde **(49)** in the Claisen rearrangement, indicating a preference for the chair transition state (7.44). Reaction in the boat conformation leads to the *erythro*-aldehyde (7.45).

Although the Claisen reaction is formally an equilibrium, the disparity in energies of the starting material and product in aliphatic systems is usually such

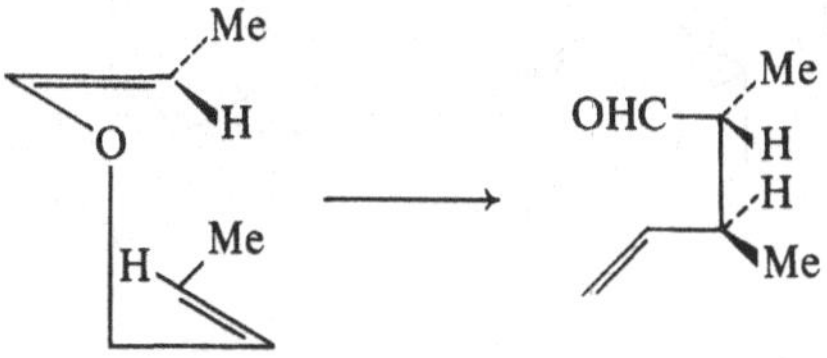

(7.44)

(48) **(49)** *threo*

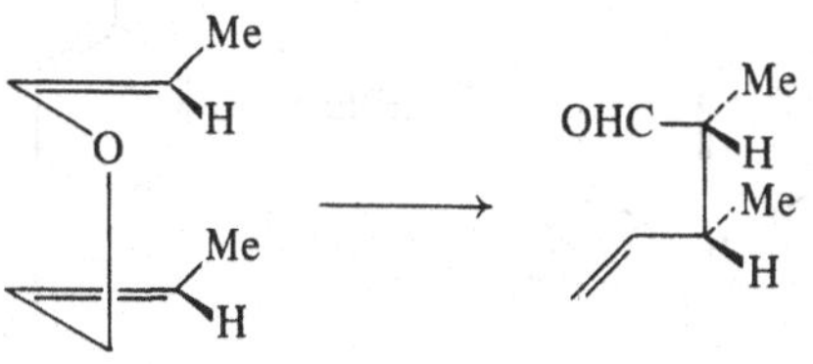

(7.45)

erythro

that the forward reaction goes virtually to completion. This seems to be due mainly to the energy gained in forming the carbonyl group. Even in cases where the reverse reaction (the retro-Claisen rearrangement) would seem to be especially favourable, there is an equilibrium in which the carbonyl tautomer predominates. For example, the equilibrium between 2,5-dihydro-oxepin **(50)** and its Claisen rearrangement product, *cis*-2-vinylcyclopropane carboxaldehyde **(51)**, lies to the right (7.46), even though the reaction involves the formation of the cyclopropane ring; at room temperature the equilibrium ratio of **(50)** to **(51)** is 1 : 19.[35]

(7.46)

(50) **(51)**

With allyl aryl ethers the formal equilibrium is between the allyl ether and the *ortho*-dienone (fig. 7.23). Here the equilibrium favours the ether, because it is aromatic and the dienone is not. In most cases, however, the dienone is removed as it is formed by other reactions. These are of two main types, illustrated for allyl phenyl ether. One is enolisation to give the *o*-allylphenol; the other, Cope rearrangement to give the *para*-dienone which then enolises.

Thus the products normally isolated from aromatic Claisen rearrangements are *o*- and *p*-allylphenols, the allyl group having undergone one inversion to give the *ortho*-isomer and two to give the *para*-isomer.

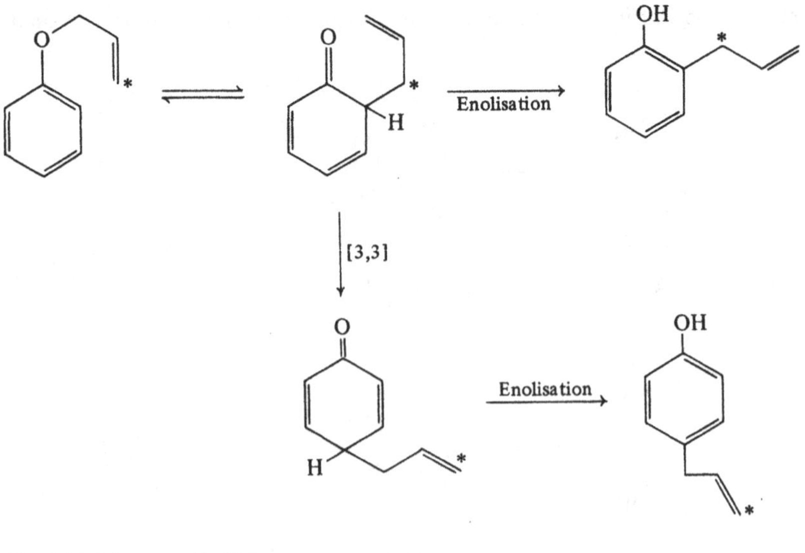

Fig. 7.23. The aromatic Claisen rearrangement.

In unsubstituted phenyl ethers, the enolisation is normally faster than the Cope reaction, so that the product is predominantly or entirely the *ortho*-isomer. On the other hand, when both *ortho*-positions are substituted, enolisation of the *ortho*-dienone is prevented and it undergoes the Cope rearrangement instead. If the *para*-position is substituted as well, none of the dienones can enolise and an equilibrium is set up amongst the three dienones and the ether, with the ether predominating.

This general picture may be complicated in individual cases, especially by steric interactions which inhibit the enolisation of the *ortho*-dienone and therefore favour the *para*-isomer. In such cases, enolisation may become the slow step in the formation of the *ortho*-isomer, and so be subject to isotope and solvent effects. For example, the ether **(52)** rearranges to give both the *ortho*- and *para*-allylphenols. The ratio of products is solvent dependent, solvents of lower polarity favouring the *para*-isomer. The proportion of *para*-isomer is also increased by replacing the aromatic ring protons by deuterium. A possible explanation in this case for the slow enolisation is that in the conformation of the *ortho*-dienone necessary for enolisation, there is a steric interaction between the methyl group on the side chain and the neighbouring ring methyl.

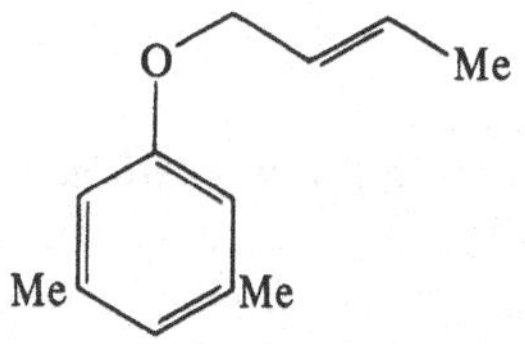

O Me

Me Me

(52)

Two other types of reaction that have been observed in aromatic Claisen rearrangements, and are therefore closely linked to them, are not [3,3] sigmatropic shifts at all. One is the *ortho-ortho*-Claisen rearrangement; the other, the so-called 'abnormal' Claisen rearrangement.[36] These will be described briefly.

ortho-ortho-*Claisen rearrangement*

The Cope rearrangement of the *ortho*- to the *para*-dienone is common in Claisen reactions. Interconversion of two *ortho*-dienones has also been observed, though it is much less common. An example of the reaction is the formation of the two labelled *ortho*-dienones shown from the labelled mesityl allyl ether (**53**, 7.47). The two *ortho*-dienones are thermally interconvertible (7.48).

If the rearrangement involved a concerted migration through the π system of the ring, it would be a [3,5] shift, and would be thermally allowed only as a suprafacial, antarafacial process. The geometry of the transition state would be difficult or impossible to attain. It is therefore probably a stepwise reaction. There is some evidence that this is so. The first step is probably an intramolecular

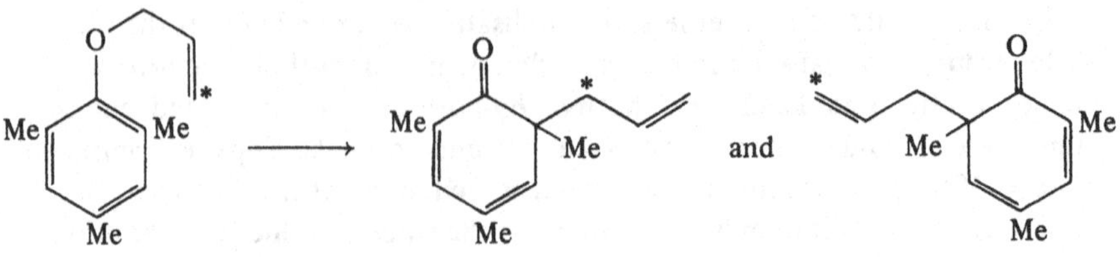

(53)

(7.47)

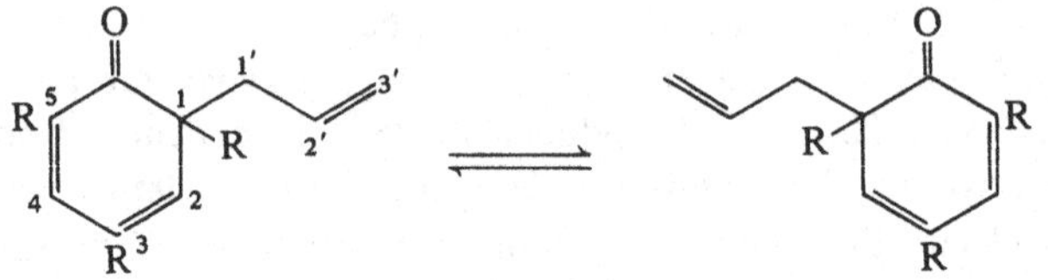

(7.48)

Diels–Alder addition of the allyl π bond to the diene, the adduct then undergoing a stepwise fragmentation to the observed product.

Abnormal Claisen rearrangement[36]

Some allyl phenyl ethers with an alkyl substituent on the end carbon of the allyl group rearrange to give the normal *ortho*-Claisen product plus another isomeric *o*-allylphenol. The second *o*-allylphenol, the 'abnormal' Claisen product, is formed by rearrangement of the normal product; this has been experimentally established (7.49). The abnormal product can be derived from the normal *o*-allylphenol by two [1,5] homodienyl shifts (§ 7.2). These are illustrated for the system (**54**, 7.50).

e.g.

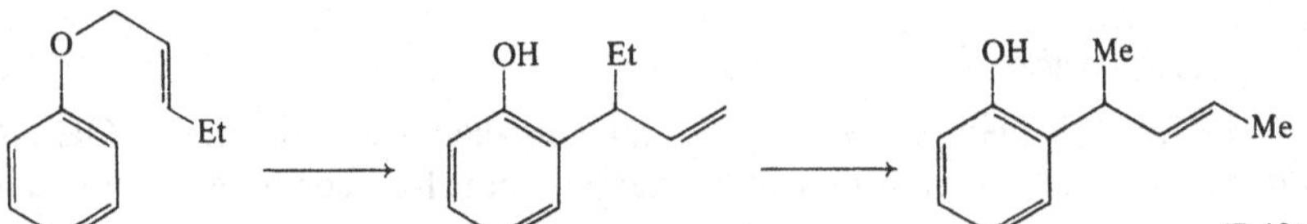

(**54**)

(7.49)

Equilibria of the type (**54**) ⇌ (**55**) may be set up in most Claisen rearrangements, but they cannot lead to the formation of a new product unless there is another alkyl group (here the ethyl group) on the side chain which is able to participate.

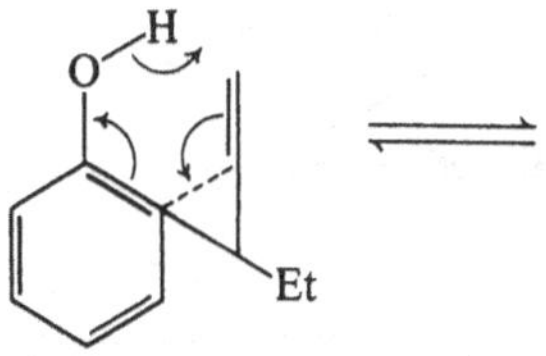

(54)

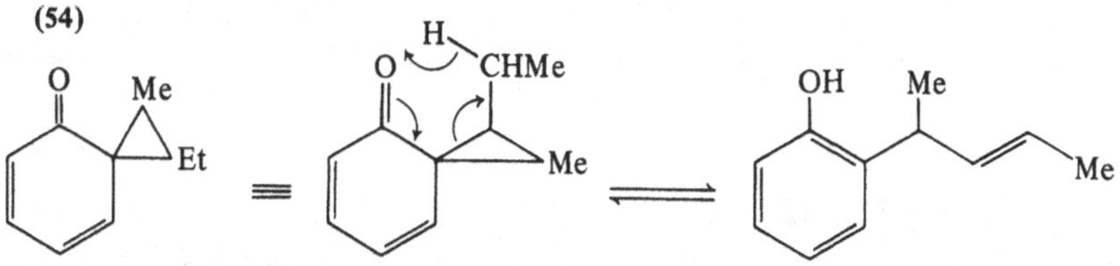

(55)

(7.50)

A reaction closely related to the abnormal Claisen reaction occurs when phenyl propargyl ether (56) is heated (7.51). The product of normal Claisen rearrangement, *o*-allenylphenol, can further rearrange by a [1,5] hydrogen shift and an electrocyclic ring closure to give a chromene (57) which is the observed product.

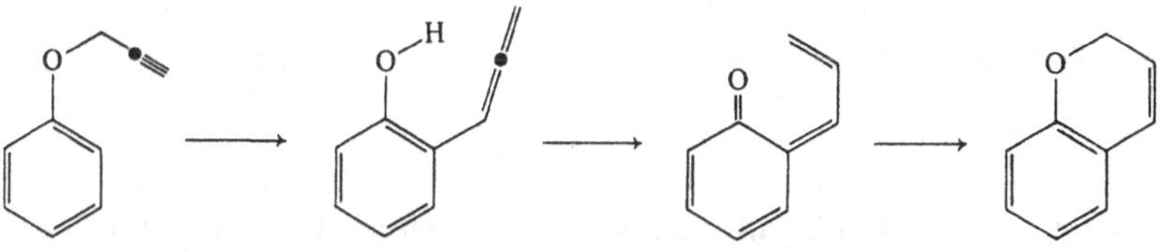

(56) (57)

(7.51)

Modified Claisen rearrangements[25]

The great importance of the Claisen rearrangement, as a method of constructing unsaturated carbon chains, has led to the development of a variety of synthetically useful modifications. Many of these involve the generation of an allylic ether *in situ,* followed by rearrangement. Three examples of such reactions are shown (7.52-7.54). Some Claisen rearrangements can also be appreciably catalysed by the addition of mercury and silver salts, and other Lewis acids.[37]

$Me_2CHCOOCH_2CMe{=}CH_2$ $\xrightarrow{NaH}$ (7.52)

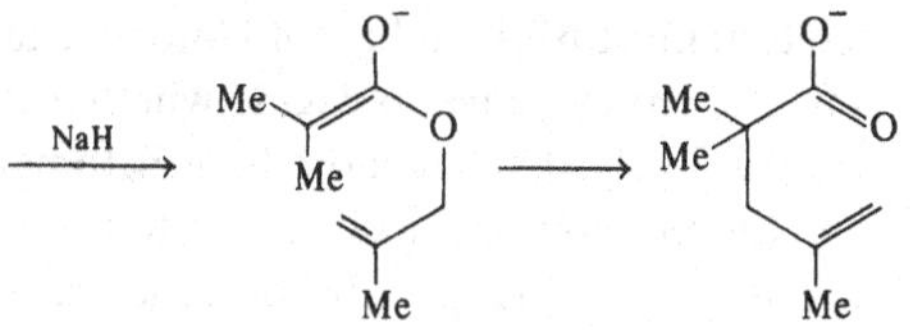

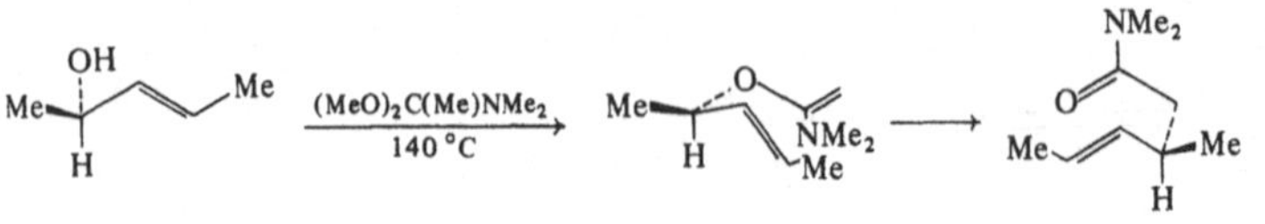

(7.53)

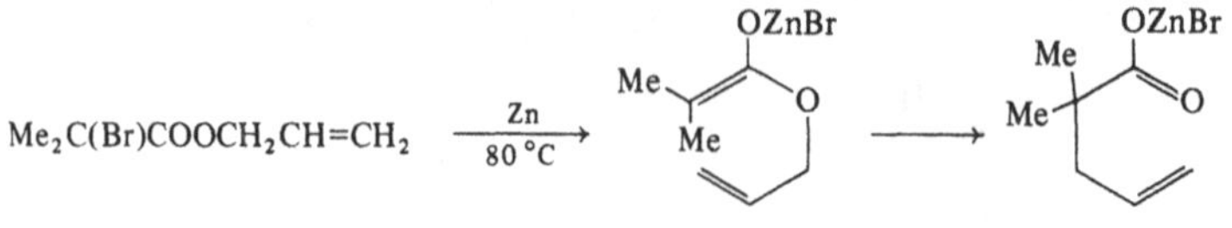

$Me_2C(Br)COOCH_2CH{=}CH_2 \xrightarrow[80\,°C]{Zn}$ (7.54)

Other [3,3] *shifts*

Replacement of one or more of the carbon atoms of a 1,5-hexadiene by heteroatoms does not appear to alter the symmetry requirements for the [3,3] shift. Thus it is possible that concerted thermal rearrangements can take place in any systems of the general form **(58)**, where one or more of the atoms *a* to *f* are heteroatoms. The effect of introducing heteroatoms may be to produce charge separation in the transition state, but still to preserve the overall characteristics of the Cope reaction, as has already been shown with the Claisen rearrangement.

(58)

A useful way of indicating the presence of heteroatoms in the skeleton is to number their position in the 3,3′ system and to indicate it by the appropriate prefix. Thus, a rearrangement involving the system **(59)** is a 2,2′-diaza-[3.3] shift, and one involving the system **(60)** is a 1-oxa-3-aza-[3,3] shift.[38]

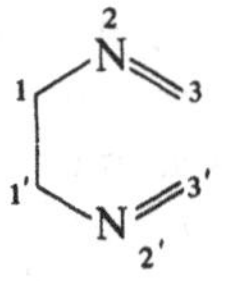

(59) 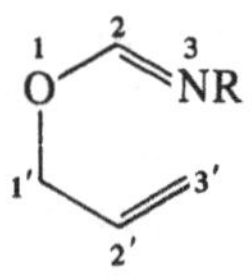 **(60)**

There is an enormous number of possible combinations of heteroatoms in the six-atom skeleton; most of them have no known examples and for many others there is insufficient evidence to decide whether the rearrangements are concerted or not. It is likely that the more heteroatoms there are in the system, the smaller will be the preference for the concerted suprafacial, suprafacial [3,3] shift mechanism over other possibilities, such as dissociation into radicals and recombination. The following examples show rearrangements where there is at least the formal possibility of a [3,3] shift; they illustrate the wide applicability and the synthetic potential of the reactions.

1. The general reaction shown in fig. 7.24 has been called the *amino-Claisen* rearrangement. The reaction has an activation energy about 25 kJ mol^{-1} higher

RN RN

Fig. 7.24

than the Claisen rearrangement and therefore is not so generally observed, because other reactions may compete at the temperature required for rearrangement. For example, *N*-allylaniline gives mainly aniline and propene when it is heated to 275 °C. One example of the reaction, where allylic inversion occurs, is shown in equation 7.55.[39]

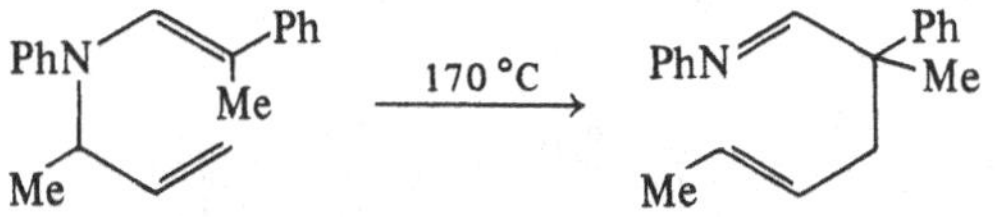

PhN Ph Me Me —170 °C→ PhN Ph Me Me (7.55)

2. The rearrangement shown in fig. 7.25 is the *thio-Claisen* rearrangement.[40] The reaction is much less well known than the Claisen reaction but in aliphatic systems it may have an even lower activation energy. Interpretation of the

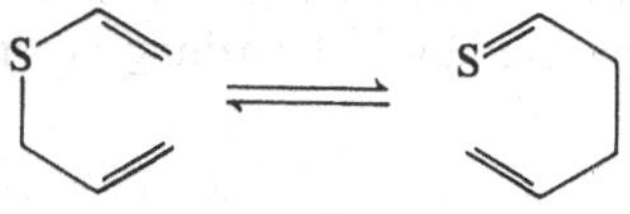

S S

Fig. 7.25

mechanistic data is complicated by the fact that several other reactions compete with the [3,3] shift. One important competing reaction is a 1,3-thioallylic rearrangement of the allyl aryl sulphides, for which a mechanism involving a dipolar intermediate has been suggested; the reaction can be intermolecular, giving crossover products with a mixture of allyl aryl sulphides (7.56).

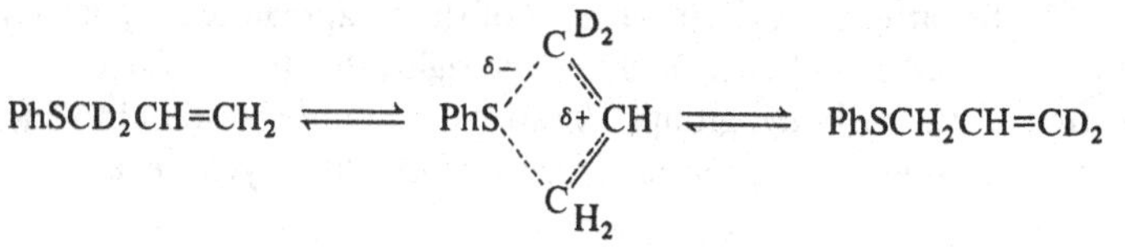

$$PhSCD_2CH{=}CH_2 \rightleftharpoons PhS^{\delta-} \cdots (CD_2 \cdots {}^{\delta+}CH \cdots CH_2) \rightleftharpoons PhSCH_2CH{=}CD_2 \qquad (7.56)$$

Another complication is that the primary products of the thio-Claisen rearrangement are unstable in the reaction conditions and undergo further reaction before they can be isolated. *o*-Allylthiophenol, for example, is unstable even at room temperature. It can only be detected indirectly in the thio-Claisen rearrangement of allyl phenyl sulphide, and cyclisation products are isolated (7.57).

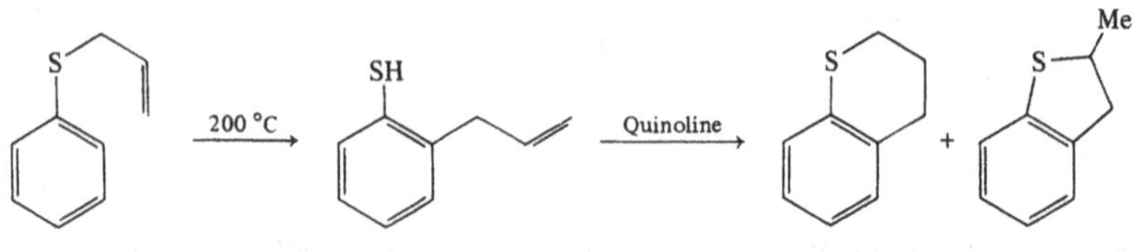

(7.57)

Like the Claisen rearrangement, this rearrangement also takes place with propargyl groups. The sulphide (**61**), for example, rearranges above 80 °C to the allene (**62**) which cyclises in pyridine (7.58).

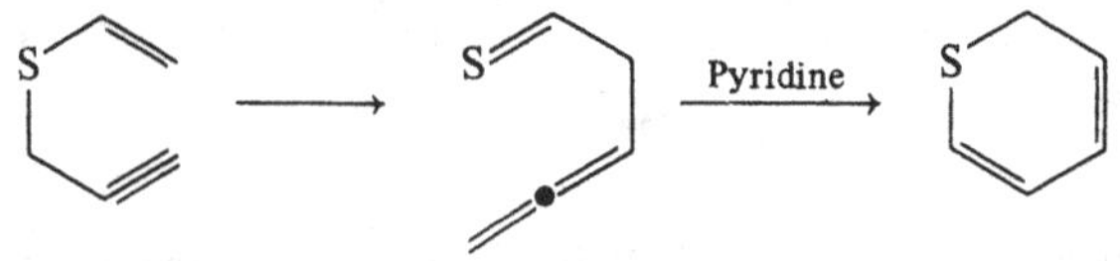

(7.58)

3. The allylic ester rearrangements (fig. 7.26) are examples of a system in which the transition state for the [3,3] shift has considerable charge separation. The concerted nature of the process is supported by ^{18}O labelling experiments.[41]

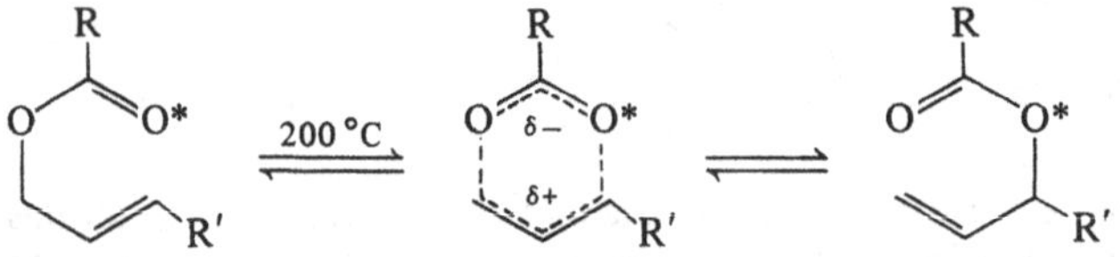

Fig. 7.26

4. Two rearrangements where a concerted [3,3] mechanism is suggested by allylic inversion are those of allyl thionocarbonates (**63**, 7.59)[42] and of imino-esters (**64**, 7.60).[43] The latter is an allylic version of the Chapman rearrangement. In these systems the equilibrium lies strongly to the right, the driving force being the formation of the carbonyl groups. A rare example of a system in which the reverse process is favoured is the rearrangement of the diazacyclohexene (**65**).[44]

(63) $\xrightarrow{20\ °C}$ (7.59)

(64) $\xrightarrow{220\ °C}$ (7.60)

(65) $\xrightarrow{78\ °C}$ (7.61)

5. The Fischer indole synthesis, which is one of the best general methods for preparing indoles, involves the acid-catalysed cyclisation of phenylhydrazones.[45] The role of the acid catalyst is probably to assist the tautomerisation of the hydrazone to the enehydrazine **(66)**, but a catalyst is not essential in all cases (7.62). Several phenylhydrazones give indoles simply on heating in a high-boiling solvent. The key step, in which the new C—C bond is formed, may be a [3,3] rearrangement of **(66)**. Positive evidence for the concerted rearrangement is not strong, but an alternative free radical mechanism has been shown to be unlikely.

$\xrightleftharpoons{245\ °C}$ (66) $\xrightleftharpoons{[3,3]}$ $\xrightarrow{\text{several steps}}$ (83%) (7.62)

7.5. [3,2] Sigmatropic changes. Ylide rearrangements

An important group of six-electron sigmatropic rearrangements involves the participation of five atoms, rather than six, in the transition state. It is notable that the unifying feature of such reactions, based on the Woodward–Hoffmann rules, was recognised only in the late 1960s, and that this mechanistic rationalisation has provided the impetus for rapid advances in the synthetic uses of such reactions. This is particularly true for allylic sulphonium ylide rearrangements, which are now recognised as an important method of forming C—C bonds.[46]

The selection rules are illustrated first for an all-carbon system, and examples of the synthetically more important systems containing heteroatoms are then given.

Consider, as an example, the hypothetical degenerate rearrangement illustrated in fig. 7.27. This is a [3,2] shift involving six electrons. The transition state **(67)** for a suprafacial, suprafacial migration is of the Hückel type, and since six electrons participate, the reaction should be thermally allowed via this transition state.[24]

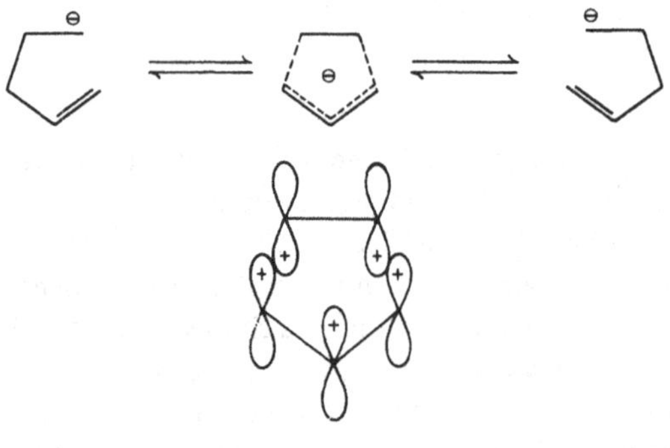

(67)

Fig. 7.27

An example in an all-carbon system is the spontaneous rearrangement of the carbanion **(68**, 7.63).[47]

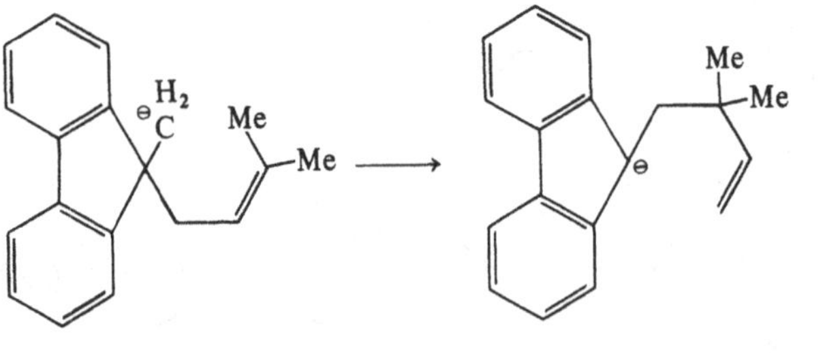

(7.63)

(68)

[3,2] Shifts involving heterocyclic transition states can be envisaged, represented by the general scheme shown in fig. 7.28 where one or more of the atoms *a* to *e* are heteroatoms. Atom *a* need not be an anion; it can be a heteroatom with a lone pair of electrons available to participate in the reaction.

$\overset{\ominus}{a}$ b c d e ⇌ b a c d e (⊖) ⇌ $\overset{\ominus}{b}$ a c d e

Fig. 7.28

Some examples involving heterocyclic transition states are shown below (equations 7,64-7.71). In some of these, the concerted nature of the rearrangement is supported by the observed inversion of the allyl group. In the others, the [3,2] shift seems the most likely mechanism for rearrangement, although alternatives may not have been ruled out.

1. Allyl benzyl ether anions (the *Wittig* rearrangement):[48, 49]

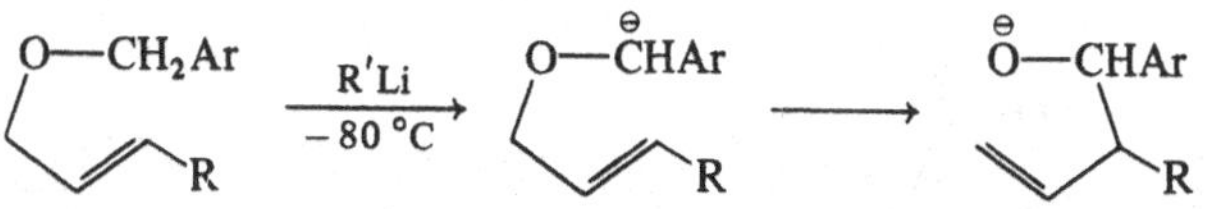

$$\text{O—CH}_2\text{Ar (R)} \xrightarrow[-80\,°\text{C}]{\text{R}'\text{Li}} \text{O—}\overset{\ominus}{\text{C}}\text{HAr (R)} \longrightarrow \overset{\ominus}{\text{O}}\text{—CHAr (R)} \tag{7.64}$$

2. Quaternary ammonium ylides (the *Sommelet-Hauser* rearrangement):[48, 50]

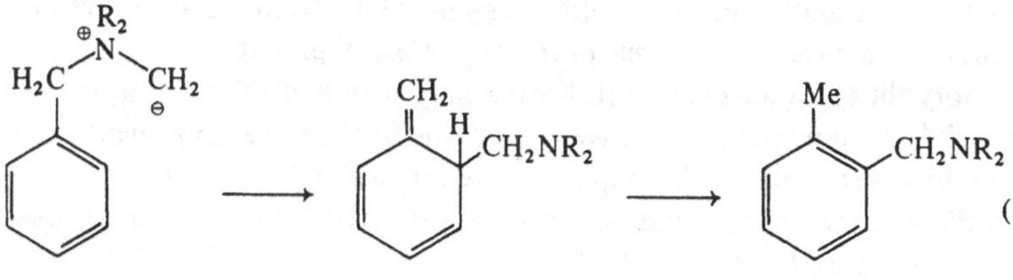

$$\text{H}_2\text{C—}\overset{\oplus}{\text{N}}\text{R}_2\text{—}\overset{\ominus}{\text{C}}\text{H}_2 \longrightarrow \text{CH}_2,\ \text{H},\ \text{CH}_2\text{NR}_2 \longrightarrow \text{Me},\ \text{CH}_2\text{NR}_2 \tag{7.65}$$

$$\overset{\ominus}{\text{N}}\text{—}\overset{\oplus}{\text{S}}\text{R—SCH}_2\text{X} \rightleftharpoons \text{HN—}\overset{\oplus}{\text{S}}\text{R—}\overset{\ominus}{\text{C}}\text{HX} \longrightarrow \text{NH}_2,\ \text{CH(X)SR} \tag{7.66}$$

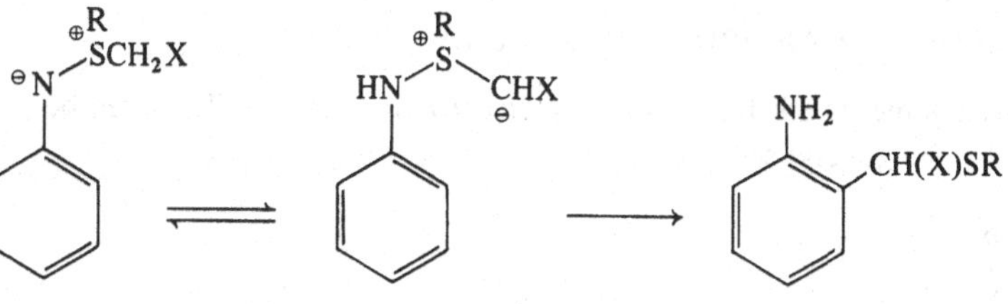

3. Sulphonium ylides:[46]

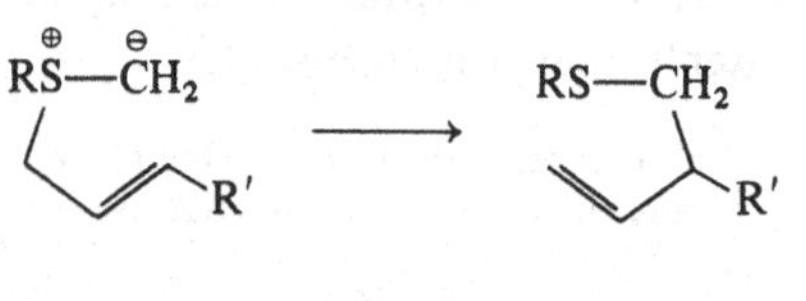

$$\text{R}\overset{\oplus}{\text{S}}\text{—}\overset{\ominus}{\text{C}}\text{H}_2\ (\text{R}') \longrightarrow \text{RS—CH}_2\ (\text{R}') \tag{7.67}$$

4. Allylic quaternary ammonium oxides (the *Meisenheimer* rearrangement):[51]

$$R_2\overset{\oplus}{N}-\overset{\ominus}{O} \xrightarrow{80\text{–}150\,°C} R_2N-O \qquad (7.68)$$

5. Allylic amido-ammonium salts:[52]

$$R_2\overset{\oplus}{N}-\overset{\ominus}{N}COAr \xrightarrow{120\text{–}140\,°C} R_2N-NCOAr \qquad (7.69)$$

6. Allylic phosphinites:[53]

$$O-\ddot{P}Ph_2 \longrightarrow \overset{\ominus}{O}-\overset{\oplus}{P}Ph_2 \qquad (7.70)$$

7. Allylic sulphenates:[54]

$$O-\ddot{S}Ar \longrightarrow \overset{\ominus}{O}-\overset{\oplus}{S}Ar \qquad (7.71)$$

Some of these rearrangements have been known for many years, while others have only recently been discovered. The possible variations in the system are enormous, and have by no means been completely explored.

The theory thus accounts very well for rearrangements of this type in which an allyl group migrates with inversion. Some of these rearrangements also occur, however, if the allyl group is replaced by a simple alkyl group. For example, the Wittig rearrangement occurs not only with allyl benzyl ethers, but also with alkyl and aryl benzyl ethers.†

$$Ar\overset{\ominus}{C}HOR \longrightarrow ArCHRO^{\ominus} \longrightarrow ArCHROH$$

Such rearrangements are 1,2-shifts to an electron-rich centre. They can be represented by the general scheme:

$$\overset{\ominus}{a}-b \longrightarrow a-b^{\ominus}$$

(with c attached to b in each case)

Ylide and amine oxide rearrangements of this type also occur; in ammonium or sulphonium ylides, the reaction is known as the *Stevens* rearrangement[50] and with amine oxides, it is the Meisenheimer rearrangement.†

† It is unfortunate that for historical reasons, both the concerted (six-electron) and stepwise (four-electron) rearrangements bear the same names.

$$\overset{\ominus}{a}-b^{\oplus} \longrightarrow a-b$$
(with c attached to b in each)

Meisenheimer: $a = O, b = NR_2$ Stevens: $a = CR_2, b = NR_2$ or SR

Even when the migrating group is an allyl group, a minor product of the reaction is often found in which allylic inversion has not taken place, and which therefore cannot be formed by a concerted [3,2] shift. These competing routes are illustrated for the Wittig rearrangement of benzyl dimethylallyl ether in fig. 7.29.[55]†

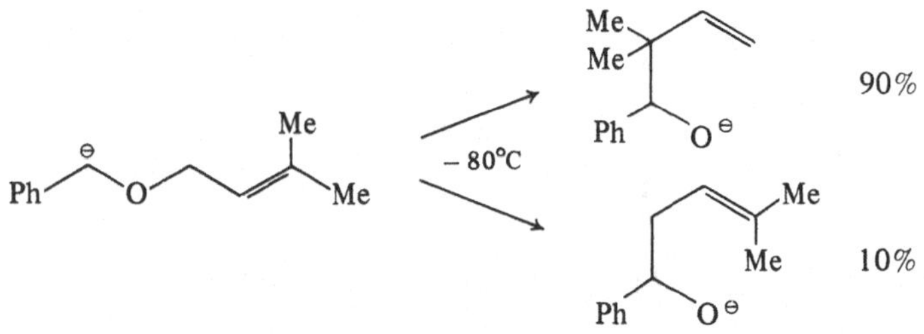

Me, Me, Ph, O⊖, 90%; Ph, O, Me, Me, −80°C; Me, Me, Ph, O⊖, 10%

Fig. 7.29. Competing rearrangements of benzyl dimethylallyl ether.

If the principle of orbital symmetry conservation is correct, then these minor reactions must be stepwise processes. One possible stepwise mechanism is a radical dissociation-recombination process, of the type proposed for 1,3-shifts and discussed in §7.2; it is shown in fig. 7.30 for the Stevens rearrangement

$$\mathrm{Ar\overset{\ominus}{C}H{-}\overset{\oplus}{N}R_3} \longrightarrow \mathrm{Ar\overset{\ominus}{C}H{-}\overset{\oplus}{\dot{N}}R_2}\ \cdot R \longleftrightarrow \mathrm{Ar\dot{C}H{-}NR_2}\ \cdot R \longrightarrow \mathrm{ArCHR{-}NR_2}$$

Fig. 7.30. Radical dissociation–recombination mechanism for the Stevens rearrangement.

As for the 1,3-shifts, the best diagnostic test for such a mechanism is the observation of CIDNP in the n.m.r. spectrum. This has been observed in Stevens rearrangements. The rearrangement of the ylide (**69**) has been followed by n.m.r., and a CIDNP signal appears during the reaction at the same place as the quartet of the methine proton in the product (fig. 7.31).[56]

Signal (*a*) is the quartet of the methine proton of an independently synthesised sample of *N*-methyl-*N*-(α-phenethyl)aniline. Signal (*b*) is the enhanced signal produced about 30 seconds after generation of the ylide (**69**), and (*c*) is the signal 10 minutes later, showing the normal quartet of the proton in the product, but at much lower intensity than that of (*b*). Signal (*b*) thus shows

† See footnote page 284.

two features which characterise it as a CIDNP signal: emission as well as absorption, and enhanced intensity.

Similarly, radical intermediates have been detected by CIDNP in a simple Meisenheimer rearrangement, that of *N,N*-dimethylbenzylamine oxide (fig. 7.32).[57]

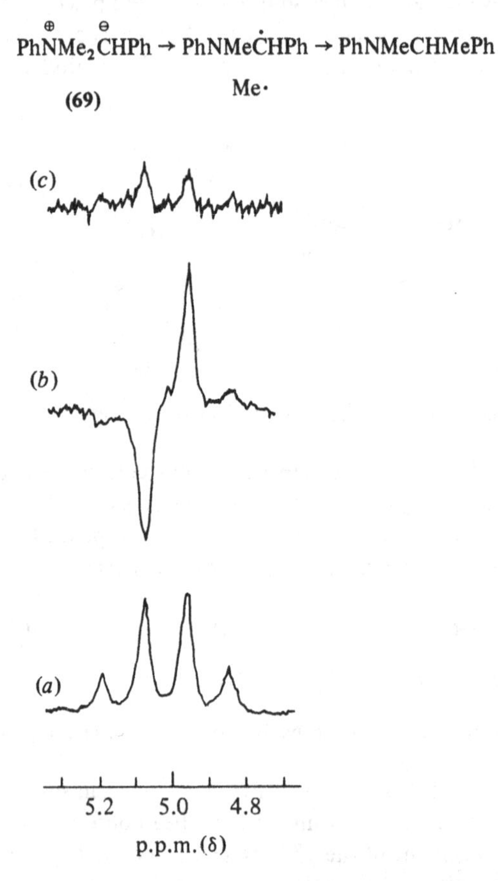

Fig. 7.31. Chemically induced dynamic polarisation signal in the Stevens rearrangement. Reprinted from *J. Am. chem. Soc.*, **91**, 1237 (1969).

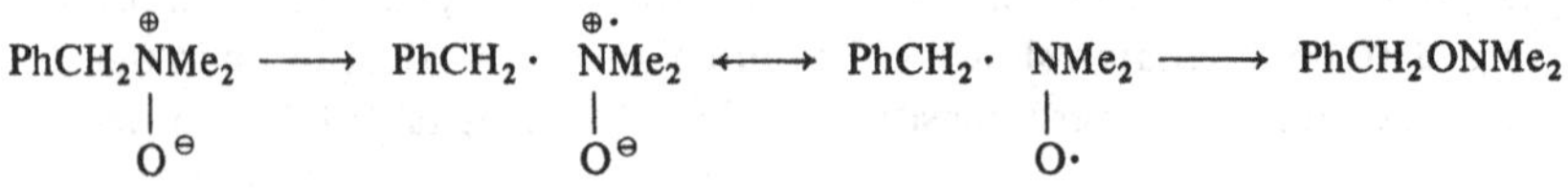

$$PhCH_2\overset{\oplus}{N}Me_2(O^{\ominus}) \longrightarrow PhCH_2\cdot \ \overset{\oplus\cdot}{N}Me_2(O^{\ominus}) \longleftrightarrow PhCH_2\cdot \ NMe_2(O\cdot) \longrightarrow PhCH_2ONMe_2$$

Fig. 7.32. Radical pairs in the Meisenheimer rearrangement.

In practical terms it can be very difficult to distinguish between radical-pair and concerted mechanisms for 1,2-anionic shifts. The migrations can go with a considerable degree of retention of stereochemistry and can have other characteristics normally associated with a concerted process. Thus, in the Wittig rearrangement of the anion derived from benzhydryl hex-5-enyl ether, a major proportion of the products contains the hexenyl group unchanged, and only a minor proportion contains a methylcyclopentyl group, the expected product of rearrangement of a hex-5-enyl radical (7.72).[49] The radical, if involved at all in the formation of the major products, cannot have an independent existence: the radical pair must react without ever having separated.[58]

$$Ph_2\overset{\ominus}{C}O(CH_2)_4CH{=}CH_2 \xrightarrow{\text{major}} Ph_2C(O^{\ominus})(CH_2)_4CH{=}CH_2$$

$$\xrightarrow{\text{minor}} Ph_2\dot{C}O + \dot{C}H_2\text{–cyclopentyl} \longrightarrow \text{Products} \qquad (7.72)$$

7.6. Other sigmatropic shifts

There are a few examples of other types of allowed sigmatropic shifts involving six- or ten-electron transition states, but these are not common reactions. A [3,4] shift is observed in competition with a [1,2] shift in cations such as **(70)** derived from cyclohexadienols.[59] This is a cationic equivalent of the Cope rearrangement.

(70) (Me, $CH_2C{\equiv}CH$ cyclohexadienyl cation) $\xrightarrow{[1,2]}$ 1-Me, 2-$CH_2C{\equiv}CH$ benzene

$\xrightarrow{[3,4]}$ $H_2C{=}C{=}CH$–C_6H_4–Me $\qquad (7.73)$

Examples of ten-electron processes are the [5,5] shift which takes place when the ether **(71)** is heated[60] and the [5,4] anionic shift which occurs in competition with a six-electron [1,4] shift when the quaternary oxide **(72)** is generated at 0 °C.[61]

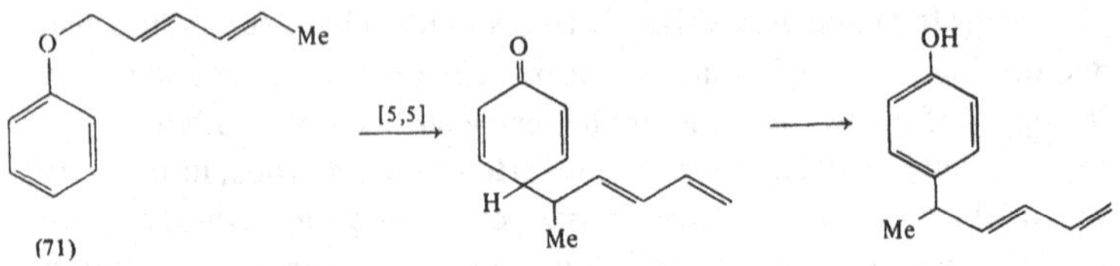

(71) [5,5] (7.74)

(72) [1,4] [5,4] (7.75)

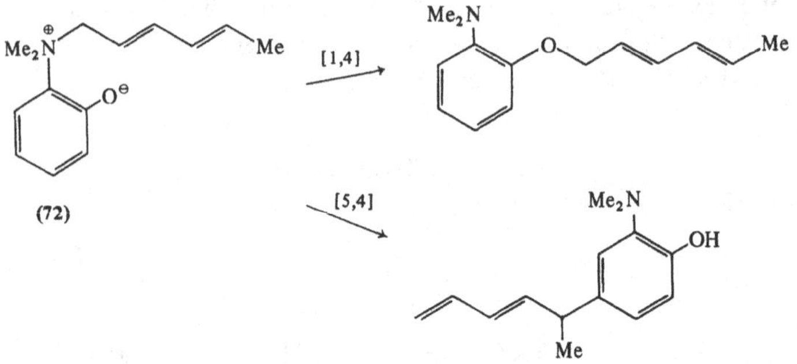

7.7. Photochemical rearrangements

Selection rules for photochemical sigmatropic shifts can be devised, starting from the assumption that for the first excited states, $4n$ electron systems of the Hückel type and $(4n + 2)$ electron systems of the Möbius type are the stable ones–the reverse of the ground state situation. Thus, sigmatropic shifts which are unfavourable in the ground state should be favourable in the first excited state, and vice versa.

A suprafacial [1,3] alkyl shift with retention of configuration provides an example. The transition state contains four electrons and is of the Hückel type (see fig. 7.8), so the reaction is unfavourable in the ground state. In the first excited state the four-electron Hückel transition state is favoured and the reaction should be allowed. Many photochemical reactions do give the products expected from a [1,3] sigmatropic shift; for example, cyclic unsaturated ketones of the general type **(73)** are found to rearrange photochemically to the four-membered ring isomers **(74)**.

This system illustrates a problem with the simple orbital symmetry rationalisation. The carbonyl group is necessary for the reaction to take place, and the initial photochemical excitation probably involves this group. The group could thus simply provide a means of absorbing energy for a concerted, excited state rearrangement not directly involving the carbonyl function, or the reaction could be a stepwise one. Like other photochemical reactions, the mechanism may be more complex than the simple theory suggests.

(73) $\xrightarrow{h\nu}$ (74) (7.76)

Metal catalysis might also be a method of bringing about [1,3] shifts; indeed, transition metal carbonyls have catalysed formal [1,3] hydrogen shifts in allylic systems, but the available evidence points to a stepwise mechanism, with metal hydride intermediates for these reactions.[62] It has not been conclusively shown that metal catalysts can reverse the symmetry rules for thermally forbidden rearrangements, and make them symmetry-allowed processes.

Photochemical analogues of several thermal processes are known; for example the 'photo-Claisen rearrangement.' This reaction has been shown to go by a radical-pair mechanism.[63]

7.8. Other six-electron pericyclic processes

There are several other pericyclic processes which combine the characteristics of cycloadditions with those of sigmatropic rearrangements, but do not strictly fit into either category.[64] The most important of these are intermolecular additions in which a σ bond (usually a bond to hydrogen) is broken, and the reverse of these processes. Some of the ways in which such reactions might occur as pericyclic six-electron reactions are shown in fig. 7.33.

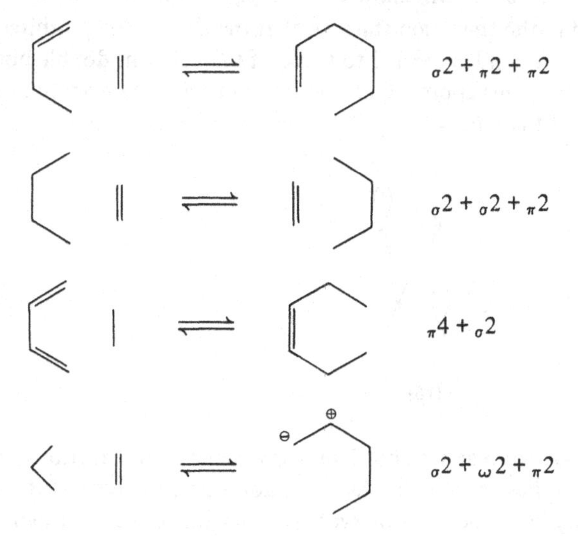

Fig. 7.33. Six-electron pericyclic processes.

The first and last of these are illustrated respectively by the ene reaction and by Cope eliminations and related reactions. The second and third types are exemplified by the addition and elimination of hydrogen in π bonded systems. These reactions are described in more detail below.

The ene reaction

The addition of an olefin containing an allylic hydrogen atom to a π bond, in the manner illustrated in fig. 7.34 for propene and ethylene, is the *ene reaction.*[65]

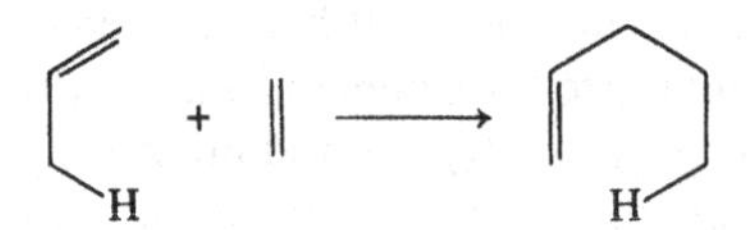

Fig. 7.34. The ene reaction.

As its name implies it is related to the diene cycloaddition reaction, the Diels–Alder reaction. The 'ene' component, the olefin with the allylic hydrogen, takes much the same part as the diene in the Diels–Alder reaction. The two reactions often compete in a particular system. Mechanistically, however, the reaction is much more closely related to [1,5] sigmatropic shifts.

The most likely geometry of approach for the components of an ene reaction is as shown, (75). This is very similar to the approach in the Diels–Alder reaction, and will lead to a 'boat-like' transition state. The extra flexibility of the ene system compared to the diene will allow the hydrogen atom to swing down towards the π bond in the transition state, so that the developing p orbital on the carbon of the C—H bond is parallel to those of the adjacent double bond, as in (76). The result of this approach geometry will be a *suprafacial* (*cis*) addition to the termini of the π bond.

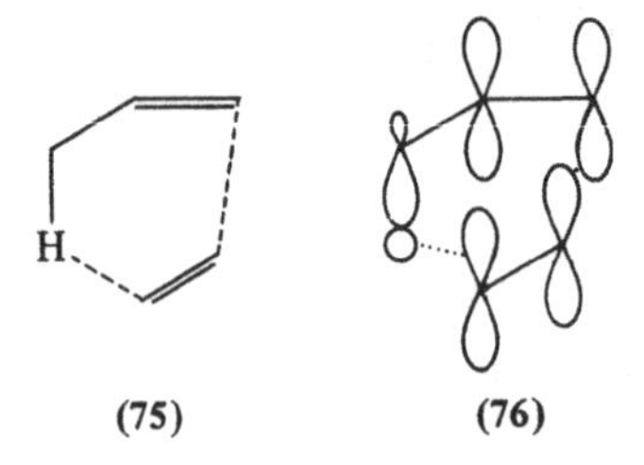

(75) (76)

Unlike the Diels–Alder reaction, but like many sigmatropic reactions, the ene reaction does not have a symmetrical transition state, and correlation diagrams are not readily constructed. However, we can deduce that it is thermally allowed as a concerted reaction by analogy with closely related systems: its transition state is 'aromatic' in the sense that it is a Hückel system and involves a suprafacial interaction of six electrons (four from the π bonds and two from the σ bond). In the general terminology of Woodward and Hoffmann it can be regarded as a ${}_{\sigma}2_{s} + {}_{\pi}2_{s} + {}_{\pi}2_{s}$ reaction.

The reactivity of the components of the ene reaction often parallels their reactivity in the Diels-Alder reaction. Like the Diels-Alder, the reaction goes best with electron-rich ene components and electron-deficient enophiles. Maleic anhydride is a good enophile, for example. Typical conditions involve heating it with the ene component in refluxing trichlorobenzene (at 210 °C) for 12-24 hours. Other very good enophiles include acetylenic esters, azodicarboxylic esters, nitrosobenzene, benzyne and singlet oxygen. Simple, unactivated olefins participate only under forcing conditions, unless the components are constrained within the same molecule; thus the diene **(77)** gives the products **(78)** and **(79)** in the ratio 14 : 1. This example illustrates two general features of such intramolecular ene reactions: C—C bond formation takes place predominantly between the closest olefinic termini, giving five-membered rings; and the reactions show considerable stereoselectivity.

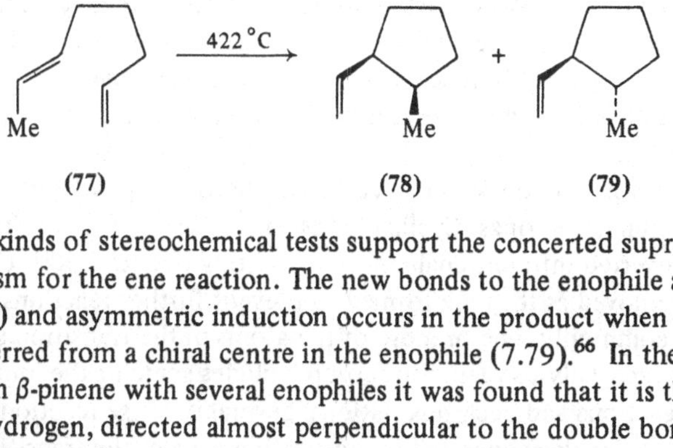

Me 422 °C Me + Me (7.77)

(77) **(78)** **(79)**

Two kinds of stereochemical tests support the concerted suprafacial mechanism for the ene reaction. The new bonds to the enophile are formed *cis* (7.78) and asymmetric induction occurs in the product when the hydrogen is transferred from a chiral centre in the enophile (7.79).[66] In the reaction of the olefin β-pinene with several enophiles it was found that it is the *trans*-allylic hydrogen, directed almost perpendicular to the double bond, which is exclusively transferred, and this is consistent with the transition state **(76)** for the reaction (7.80).[67]

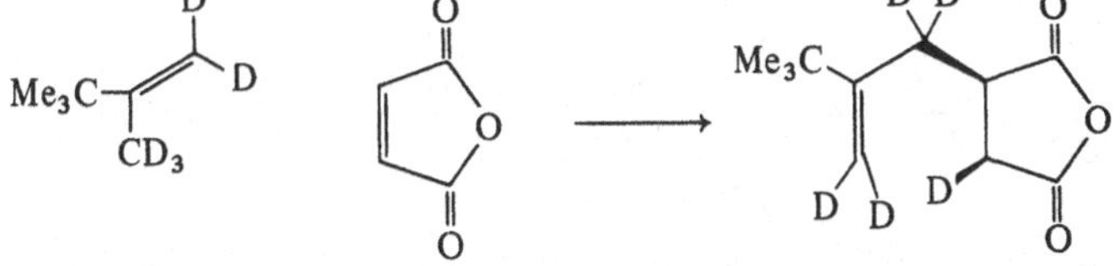

Me_3C, D, D, CD_3, O, O, O → Me_3C, D D, O, O, O, D, D, D (7.78)

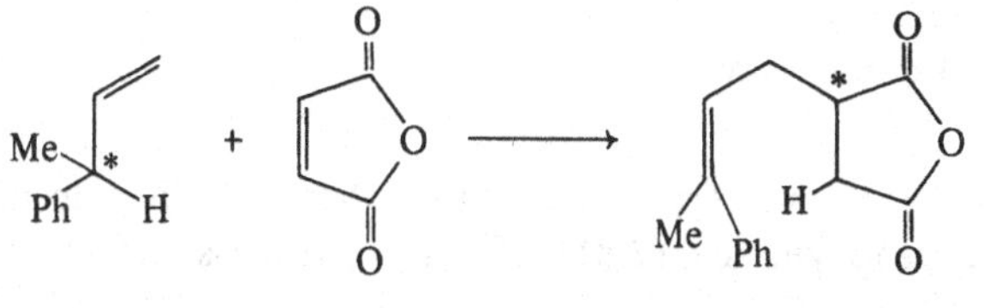

Me, *, Ph, H + O, O, O → *, O, O, O, H, Me, Ph (7.79)

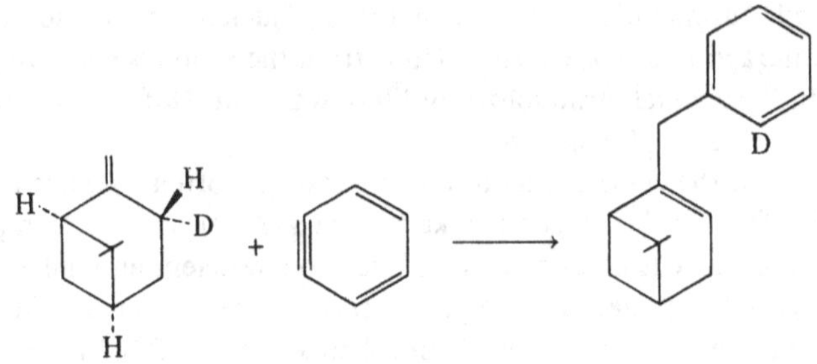

(7.80)

The ene reaction is relatively insensitive to changes in solvent polarity. A deuterium isotope effect k_H/k_D of 2.8 to 4.1 is observed in the ene reaction of diethyl azodicarboxylate with 1,4-dihydronaphthalene at 60–80 °C, indicating some C—H bond breaking in the transition state, and again consistent with a concerted mechanism.

The retro-ene reaction

This fragmentation essentially involves transfer of hydrogen through a six-membered cyclic transition state.[65,68] It bears a close resemblance to the [1,5] sigmatropic shift, the main difference being that in the retro-ene reaction, a σ bond breaks instead of a π bond. The transition state is 'aromatic' in that it involves six electrons.

The reaction is not well known in all-carbon systems except where they form part of medium-sized rings. Cyclic olefins with rings of eight to eleven atoms can be converted into terminal dienes by heating to about 500 °C, the products being removed as they are formed to prevent further reactions fig. 7.35). More commonly, one or more of the atoms in the transition state is a heteroatom, especially oxygen. This group includes some of the important olefin-forming reactions and decarboxylations. Several of these reactions which seem most likely to involve concerted fragmentation via a cyclic transition state will be described briefly.

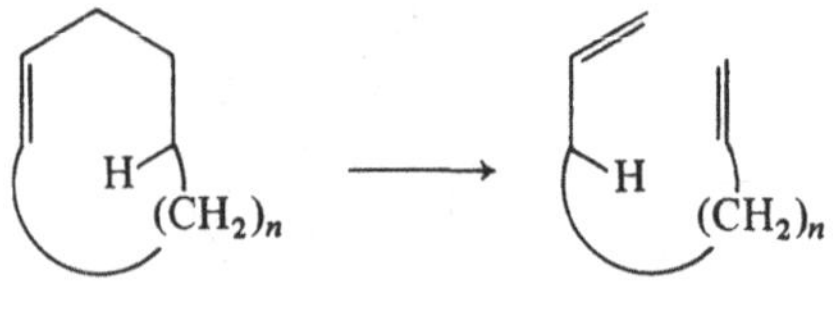

(n = 3 to 6)

Fig. 7.35. A retro-ene reaction.

1. *Pyrolysis of esters.* When esters are heated to 300–500 °C, they decompose to an olefin and a carboxylic acid (7.81). The ester pyrolysis is stereospecific

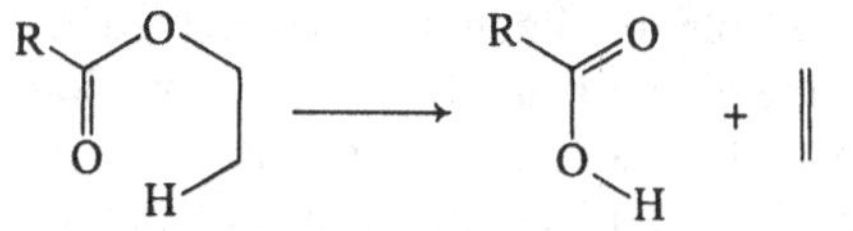

(7.81)

in the *cis* sense, unimolecular, shows a deuterium isotope effect and, generally, a negative entropy of activation. These characteristics are all to be expected for a concerted cyclic elimination. The *cis* stereospecificity has been demonstrated in several systems. An example (fig. 7.36) is the pyrolysis of *threo*- and *erythro*-3-deuterio-2-butyl acetates, (**80**) and (**81**). Both give *cis*- and *trans*-2-butenes. From the *threo*-isomer (**80**), the *cis*-butene contains deuterium and the *trans*-butene does not; from the *erythro*-isomer (**81**) the *cis*-butene is undeuterated and the *trans*-butene is deuterated, as expected for a *syn*-elimination.[69]

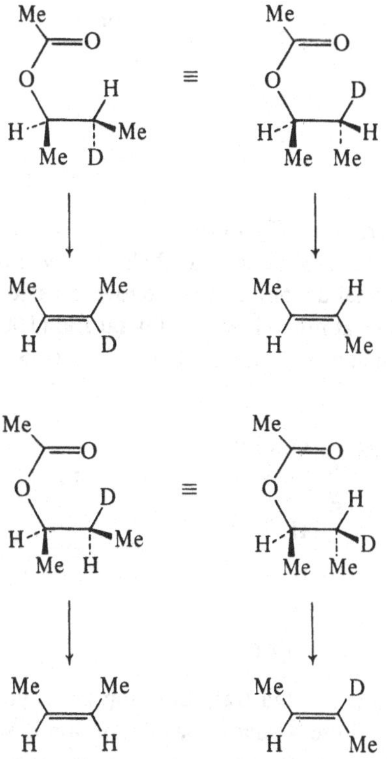

(±)-*threo* (**80**)

(±)-*erythro* (**81**)

Fig. 7.36

Some 1-butene is also formed in each case, by alternative elimination of a primary hydrogen from the α-methyl group. In cases such as this where there is more than one way in which the elimination can go, the ratio of olefins formed usually depends mainly on a statistical effect; that is, the number of hydrogens of a particular type available for elimination. Occasionally, other factors, such as the stabilities of the olefins being formed, become dominant. In cyclic systems, conformational effects are important: if the ester group occupies an axial position in a cyclic system, for example, then the hydrogen on the adjacent carbon must be equatorial for elimination to take place (fig. 7.37). If the leaving group is equatorial, then both *cis*- and *trans*-hydrogens on the adjacent carbon are sterically accessible for a cyclic transition state. However, a concerted elimination involving the *trans*-hydrogen will give a transition state in which the developing double bond is *transoid*. Again, therefore, *cis*-elimination is expected to be favoured energetically.

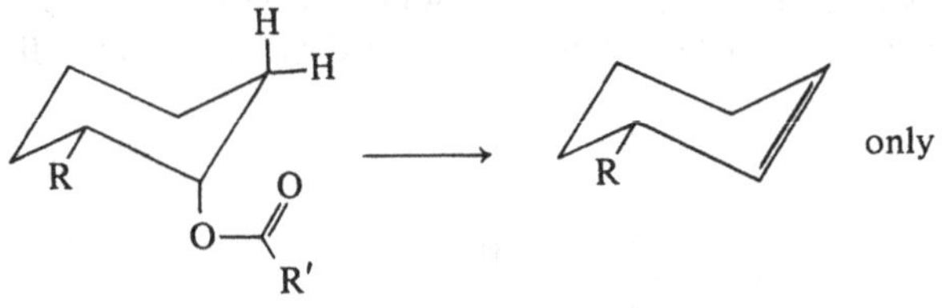

H H R O O R' → R only

Fig. 7.37

2. *Pyrolysis of xanthates* (*Chugaev reaction*).[68] Xanthate esters are prepared from alcohols by reaction with carbon disulphide and alkali, followed by methylation with iodomethane. Their thermal decomposition follows a pattern similar to that of carboxylate esters but goes at much lower temperatures (100–250 °C), probably because of the energy gained in the reorganisation $-O-\underset{|}{C}=S \rightarrow O=\underset{|}{C}-S-$ (7.82).

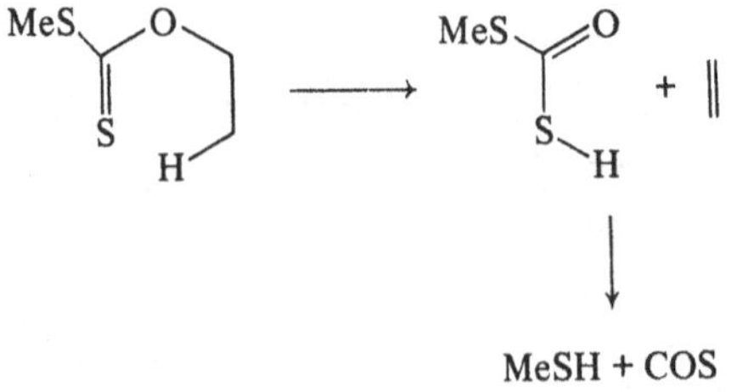

MeS O S H → MeS O S H + ‖ (7.82)

↓

MeSH + COS

This reaction probably involves a concerted fragmentation via a cyclic transition state, but the evidence is less definite. Usually the elimination is stereospecifically *syn*, but *anti*-eliminations have occasionally been observed, especially when there is an electron-withdrawing group at the β-carbon. The transition state may be highly polarised, so that suitable substituents can change the mechanism of fragmentation to a stepwise one.

3. *Decarboxylation of β-keto-acids.*[70] It is well known that β-keto-acids are thermally unstable, and lose carbon dioxide when heated (7.83). Since these

acids are intramolecularly hydrogen bonded even in the solid state a cyclic mechanism seems most likely for the fragmentation.

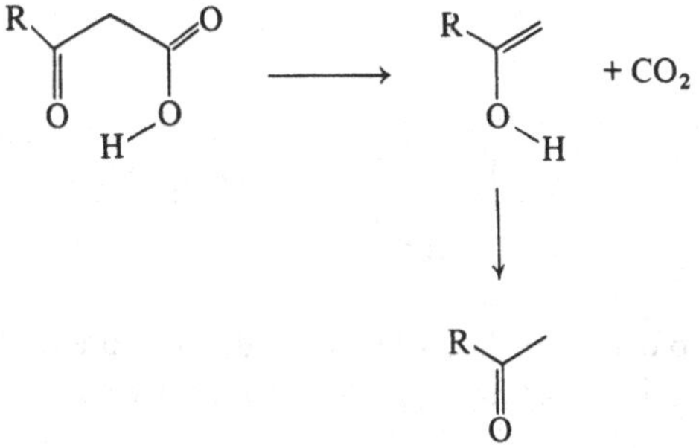

R O O H O → R O H + CO_2 → R O

(7.83)

The fact that different solvents have little effect on the rate argues against a dipolar intermediate in the reaction. However, the transition state probably does have some polar character; both deuterium isotope studies and measurements of volumes of activation suggest that there is some transfer of the acid proton to the ketone oxygen as the transition state is approached (fig. 7.38).

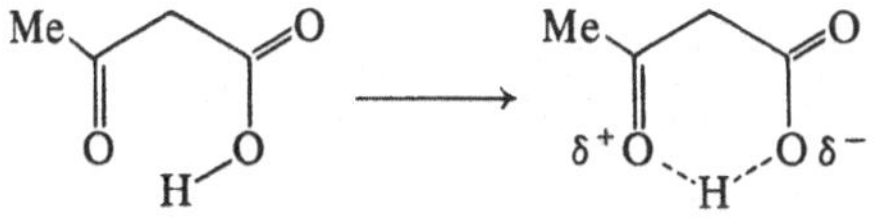

Me O O H O → Me O δ^+O H O δ^-

Fig. 7.38

4. *Pyrolysis of β-hydroxy-ketones.*[71] Like β-keto-acids, a six-membered cyclic hydrogen-bonded structure can be drawn for β-hydroxy-ketones. When they are heated to 200–250 °C, a thermal retrograde aldol reaction takes place. The vapour phase pyrolysis of diacetone alcohol gives acetone, for example (7.84); the reaction has a negative entropy of activation and an activation energy of 130 kJ mol^{-1}. A cyclic transition state, probably with some dipolar character, is likely. β-Hydroxy-esters can react in a similar way.[72]

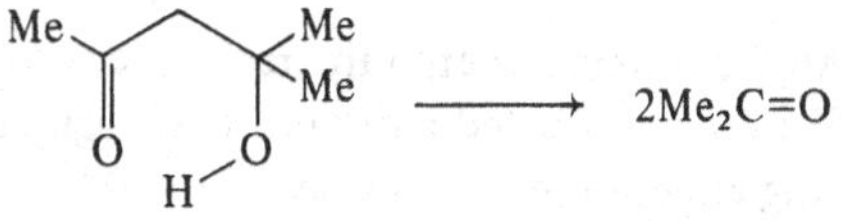

Me Me Me O H O → $2Me_2C{=}O$ (7.84)

5. *Decarboxylation of β,γ-unsaturated acids.*[73] β,γ-Unsaturated carboxylic acids lose carbon dioxide when heated above about 300 °C (7.85). The reaction is first order, and has a negative entropy of activation. A concerted fragmentation is the most probable mechanism. The product is an olefin in which the double bond has migrated through one carbon-typical of the ene and retro-ene reactions. Cyclohexenylacetic acid, for example, gives methylenecyclohexane (7.86).

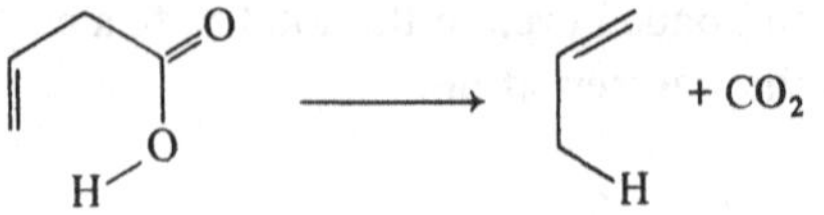

O, O, H $\longrightarrow$ H $+ CO_2$ (7.85)

CH_2CO_2H $\longrightarrow$ CH_2 $+ CO_2$ (7.86)

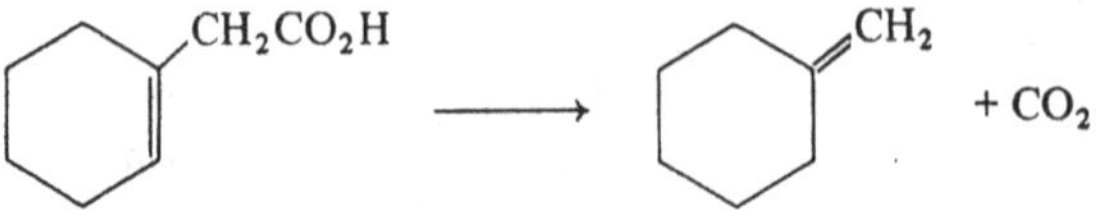

α,β-Unsaturated acids also decarboxylate when heated, but more slowly. They probably first rearrange (by a [1,5] sigmatropic shift) to the β,γ-unsaturated isomers, which then decarboxylate (7.87). The reaction is very similar to the decarboxylation of β-keto-esters but the reaction temperature is higher, presumably because of the absence of intramolecular hydrogen bonding in the unsaturated acids.

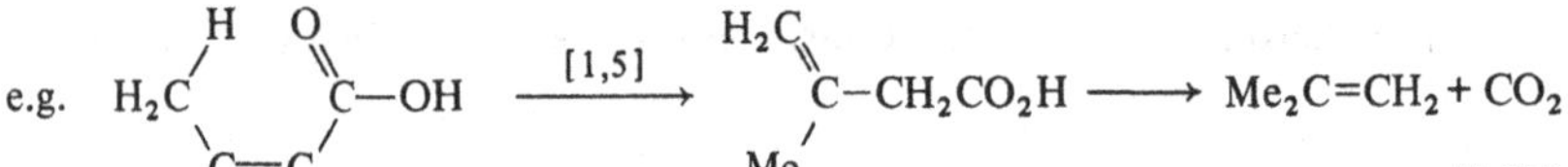

e.g. $H_2C(H)(Me)C{=}CH{-}C(=O){-}OH \xrightarrow{[1,5]} H_2C{=}C(Me){-}CH_2CO_2H \longrightarrow Me_2C{=}CH_2 + CO_2$ (7.87)

6. *Pyrolysis of β-hydroxy-olefins.*[74] β-Hydroxy-olefins fragment at about 500 °C to carbonyl compounds and olefins. A cyclic mechanism seems likely. The hydrogen transfer is intramolecular; this has been shown by deuterium labelling (7.88). A particular class of β-hydroxy-olefin pyrolysis involves a retro-ene reaction in competition with the oxy-Cope rearrangement (§7.4).

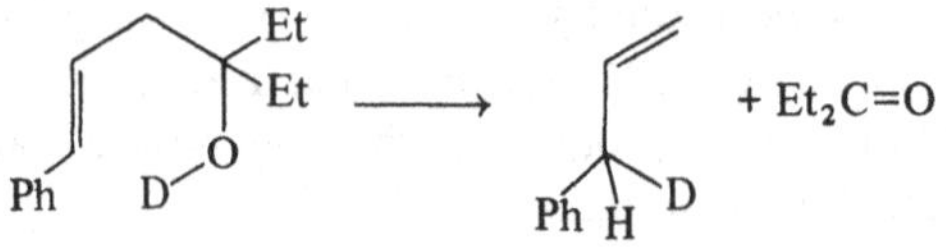

Et, Et, O, Ph, D $\longrightarrow$ Ph, H, D $+ Et_2C{=}O$ (7.88)

7. *Pyrolysis of allylic ethers.*[75] Allylic ethers undergo the retro-ene reaction when heated at 400–600 °C (7.89). The concerted nature of the fragmentation is supported by deuterium-labelling experiments, and since the yields in the fragmentation can be high, the reaction has been used as a synthetic route to α-deuterio-olefins.

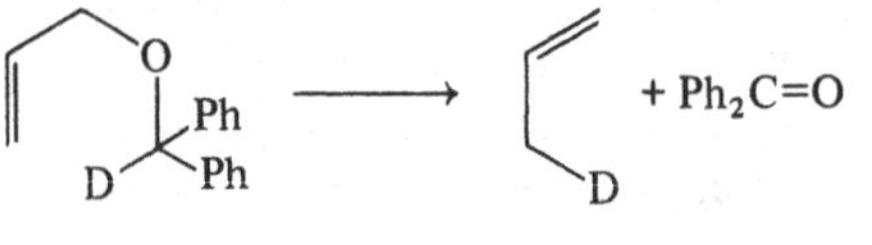

O, Ph, Ph, D $\longrightarrow$ D $+ Ph_2C{=}O$ (7.89)

The Cope elimination and related reactions[68]

There is a series of *syn*-elimination reactions which probably go concertedly via a five-membered cyclic transition state (**82**, 7.90); (X = SR or NR_2; Y = O, NR or CH_2). The most important of these is the Cope elimination (X = NR_2;

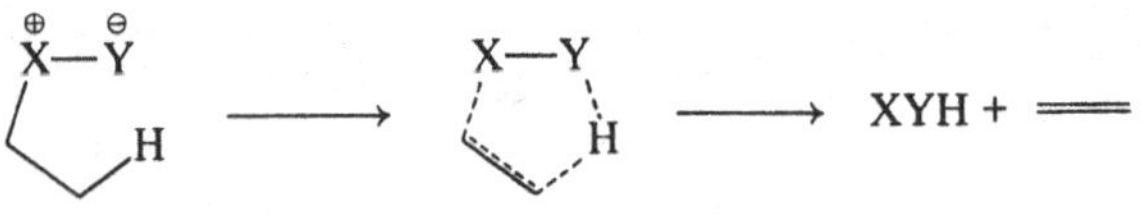

$$\overset{\oplus}{X}-\overset{\ominus}{Y} \longrightarrow X-Y \cdots H \longrightarrow XYH + = \qquad (7.90)$$

(**82**)

Y = O), which is a useful way of generating double bonds. Mechanistically, these eliminations bear the same relationship to the retro-ene reaction as do the ylide rearrangements and other [3,2] sigmatropic shifts (§7.5) to the well-established [3,3] shifts. A simpler analogy is to compare the transition states of these reactions with those of retro-ene reactions. Both involve the movement of six electrons in an 'aromatic' transition state. In this case the electrons involved are located in the C—H σ bond, the C—X σ bond and on Y (a lone pair). To draw a parallel, the retro-ene transition state is a six-membered 'benzene-like' one, whereas this transition state is a five-membered 'thiophene-like' one; both are aromatic. The elimination is therefore thermally allowed. On the Woodward-Hoffmann generalised picture it can be regarded as a ${}_{\sigma}2_s + {}_{\sigma}2_s + {}_{\omega}2_s$ reaction, as shown in fig. 7.39.

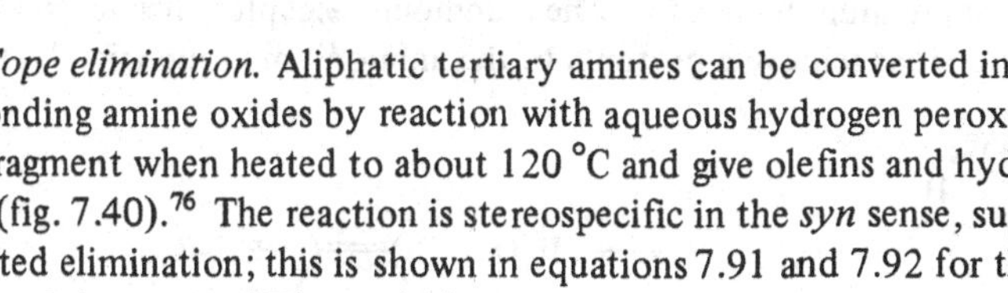

Fig. 7.39

1. *The Cope elimination.* Aliphatic tertiary amines can be converted into the corresponding amine oxides by reaction with aqueous hydrogen peroxide. The oxides fragment when heated to about 120 °C and give olefins and hydroxylamines (fig. 7.40).[76] The reaction is stereospecific in the *syn* sense, suggesting a concerted elimination; this is shown in equations 7.91 and 7.92 for the pyrolysis of the oxides (**83**) and (**84**).

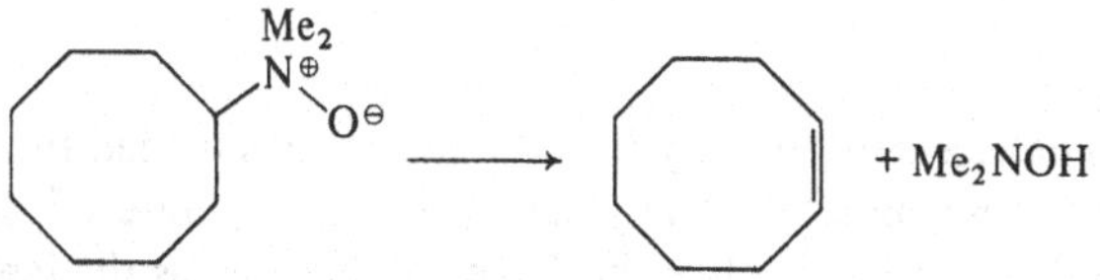

$$\longrightarrow + Me_2NOH$$

(*cis*)

Fig. 7.40. The Cope elimination.

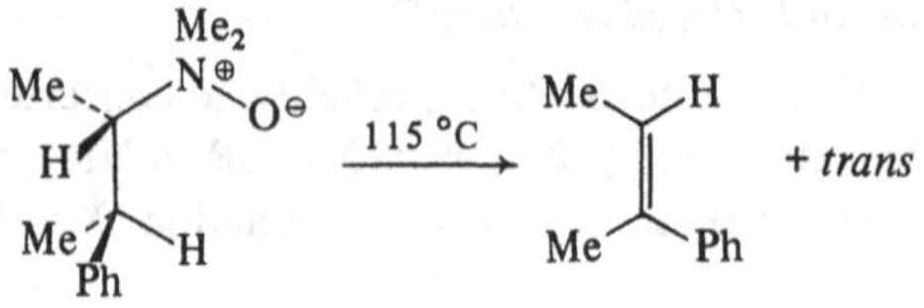

(83) $\xrightarrow{115\ °C}$ Me, H / Me, Ph alkene + *trans* (7.91)

93% 0.1%

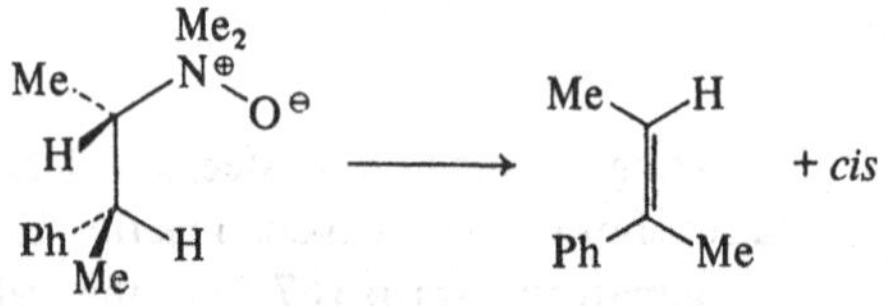

(84) $\longrightarrow$ Me, H / Ph, Me alkene + *cis* (7.92)

90% 2%

The reaction is thus very similar to the *syn*-eliminations of acetates and xanthates, but goes in even milder conditions. A similar reaction is observed when the imine (**85**) is heated (7.93); the product is predominantly *cis*-cyclooctene. A similar mechanism is probably involved.[77]

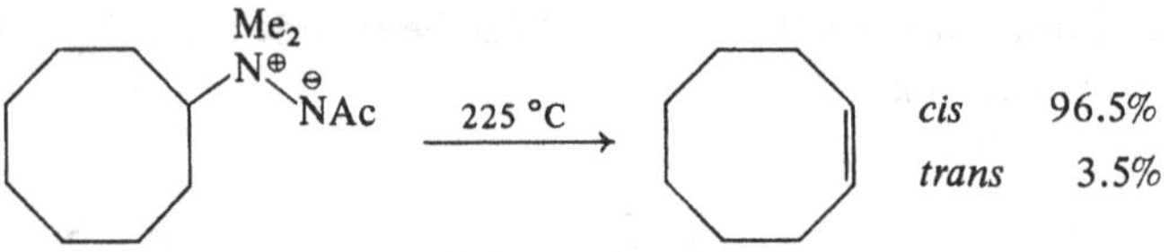

(85) $\xrightarrow{225\ °C}$ cyclooctene: *cis* 96.5%, *trans* 3.5% (7.93)

The series is completed by an ylide fragmentation which is a minor route of some Hofmann eliminations. The Hofmann elimination involves heating a quaternary ammonium hydroxide. The commonly accepted mechanism is that the hydroxide ions remove a proton β to the ammonium group (fig. 7.41).

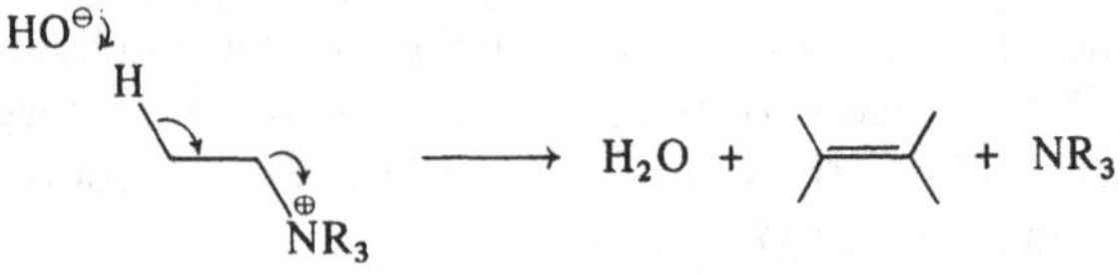

$HO^{\ominus}$ ... $\overset{\oplus}{N}R_3$ $\longrightarrow$ H_2O + alkene + NR_3

Fig. 7.41

However, in special cases where the β-hydrogen is sterically shielded from attack by an external base, an alternative ylide mechanism operates.[78] The elimination from the ylide is cyclic, like the Cope elimination, as the labelling experiment illustrated by fig. 7.42 shows: the trimethylamine produced is monodeuteriated.

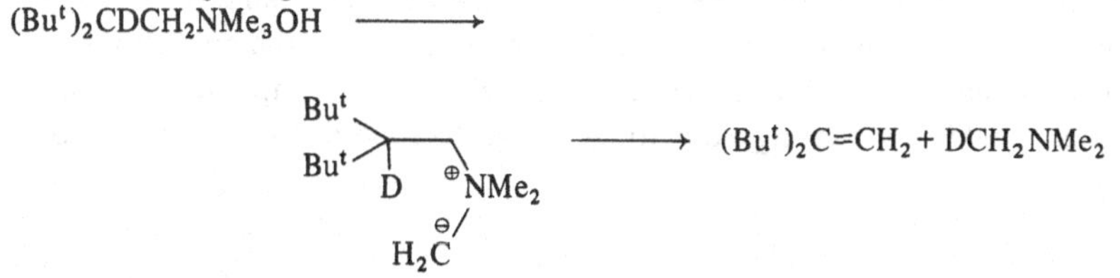

$(Bu^t)_2CDCH_2\overset{\oplus}{N}Me_3\overset{\ominus}{O}H \longrightarrow$

$\longrightarrow (Bu^t)_2C{=}CH_2 + DCH_2NMe_2$

Fig. 7.42

2. *Sulphoxide pyrolysis.*[68] When simple alkyl sulphoxides are heated to about 140 °C, fragmentation takes place and an olefin is formed (7.94). If the β-hydrogen is benzylic, the elimination goes at lower temperatures (80 °C and above). The reaction is stereoselective and insensitive to solvent changes at low

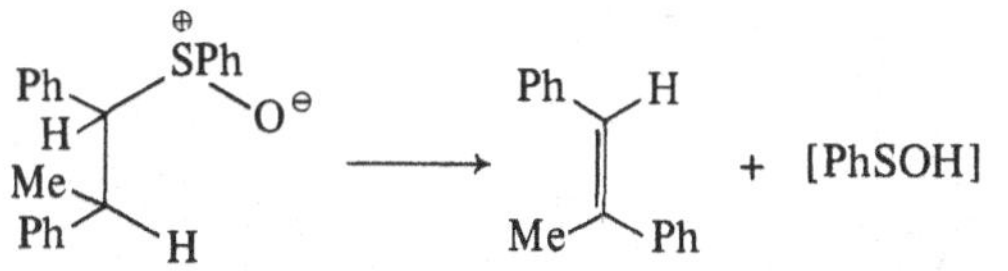

$\longrightarrow$ + [PhSOH] (7.94)

temperatures, and is probably concerted, like the Cope elimination. At higher temperatures the stereoselectivity is lost and other products are formed; a second, stepwise mechanism increasingly competes as the temperature is raised.

The reverse of this reaction, the addition of sulphenic acids to olefins, is also known. Both elimination and addition occur when the sulphoxide **(86)** is pyrolysed: elimination of 2-methylpropene gives the sulphenic acid **(87)** which then adds to the internal double bond to give the sulphoxide **(88)** in good yield.[79]

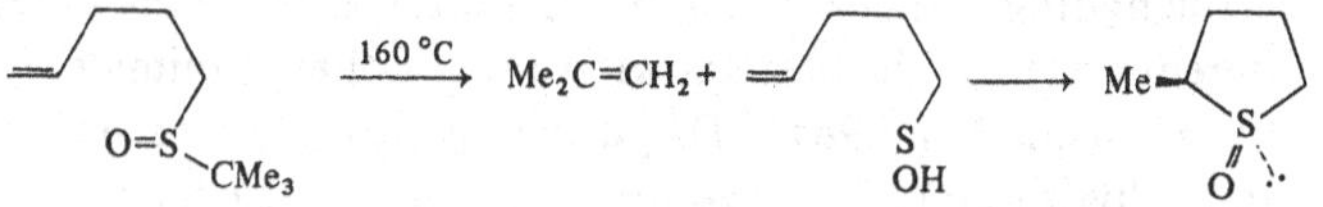

$\xrightarrow{160\,°C}$ $Me_2C{=}CH_2$ + $\longrightarrow$ (7.95)

(86) **(87)** **(88)**

The pyrolytic elimination is particularly important in the conversion of penicillin sulphoxides into cephalosporins (7.96). The sulphenic acid **(89)**, which is reversibly formed on heating the sulphoxide **(90)**, has been shown to be an intermediate in this conversion.[80]

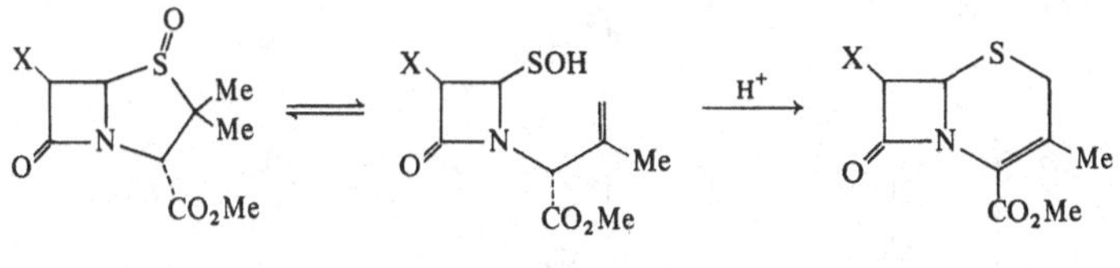

$\rightleftharpoons$ $\xrightarrow{H^+}$ (7.96)

(90) **(89)** (X = phthalimido)

3. *Selenoxide pyrolysis.*[81] The *syn*-elimination reactions of selenoxides (7.97) follow the same pattern as those of sulphoxides, except that the conditions are much milder: the elimination occurs at or below room temperature. Thus, oxidation of selenides bearing a β-hydrogen leads to spontaneous formation of olefins.

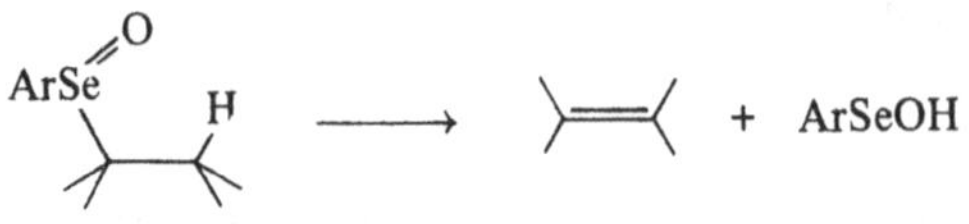

ArSe(=O)–C–C–H ⟶ C=C + ArSeOH (7.97)

This mild and stereospecific reaction has been exploited in a variety of synthetic transformations. An example is the conversion of ketones into α-β-unsaturated ketones by the sequence shown in fig. 7.43: the conditions are mild and the yields are generally very good.

$$R'COCHR^2CH_2R^3 \xrightarrow[\text{(ii) PhSeBr, (iii) Oxidation}]{\text{(i) LiNR}_2} R'COC(Se(=O)Ph)(R^2)-CH_2R^3 \longrightarrow R'COCR^2=CHR^3$$

Fig. 7.43. Introduction of unsaturation by selenoxide elimination.

Hydrogen-transfer processes

The simultaneous transfer of two hydrogen atoms in a pericyclic process can occur by either of the generalised mechanisms described earlier (fig. 7.33). The first type, stereospecific hydrogen transfer to a 2π acceptor, has been established for a number of different systems. The best known is the reaction of diimide with olefins or acetylenes (equation 7.98).[82] Dihydroaromatic molecules are similarly able to transfer hydrogen to 2π acceptors; an example of such a reaction is shown in equation 7.99.[83]

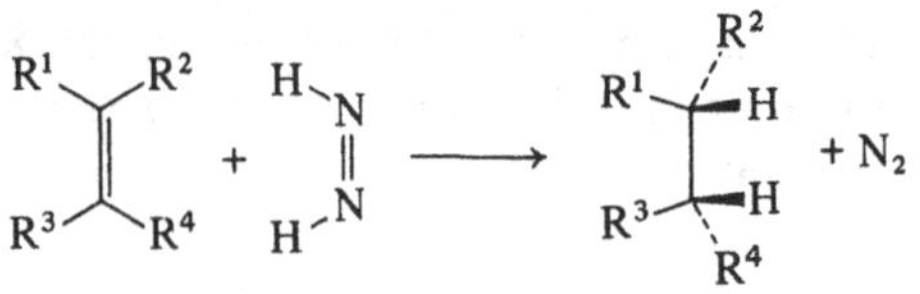

$$R^1R^2C=CR^3R^4 + HN=NH \longrightarrow R^1R^2CH-CHR^3R^4 + N_2 \quad (7.98)$$

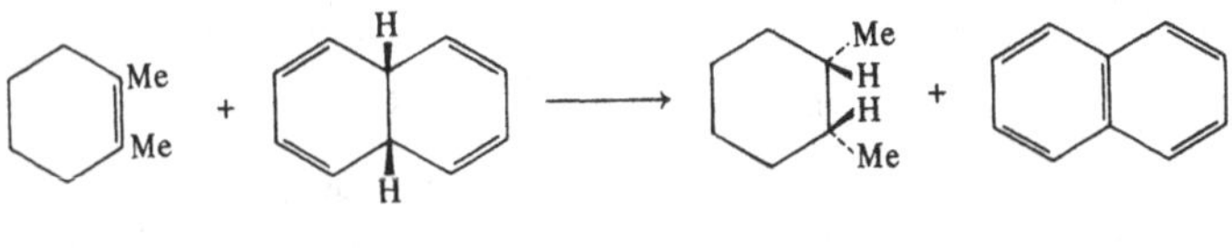

(7.99)

The second type of process is illustrated by the unimolecular stereospecific *syn*-1,4-elimination of hydrogen from dihydroaromatics.

A good illustration is provided by the pyrolysis of 1,4-cyclohexadiene to hydrogen and benzene. In contrast with the behaviour of 1,3-cyclohexadiene, the pyrolysis of the 1,4-diene is unimolecular, with an activation energy of 178 kJ mol^{-1}. Pyrolysis of the dideuteriated compound (**91**) gives HD and monodeuteriobenzene, showing that the elimination is *syn.*[84]

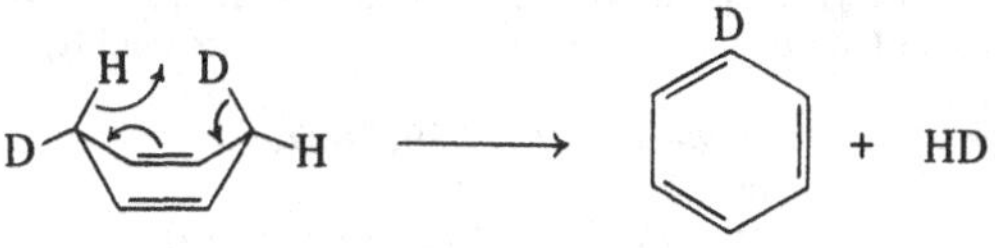

(91) → + HD (7.100)

7.9. Dyotropic reactions[85]

A pericyclic process in which two σ bonded groups migrate simultaneously has been called a *dyotropic reaction.* One type of such a reaction is illustrated in fig. 7.44. The process illustrated involves a four-electron transition state, and symmetry rules can be devised accordingly: thus, the reaction will be symmetry-forbidden if both groups migrate with retention, and symmetry-allowed if one group migrates with inversion. Alternative stepwise mechanisms may well compete in such a system.

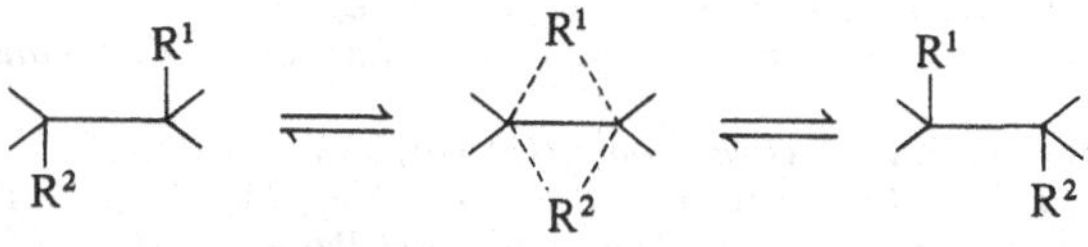

Fig. 7.44. A dyotropic reaction.

One way in which the pericyclic process can be facilitated is to increase the number of electrons in the transition state to six. This can be achieved, for example, by making one of the migrating groups an allyl group. The process is then symmetry-allowed when the remaining group migrates with retention. This has been shown to occur with the chiral silane (**92**), which rearranges with retention of configuration at silicon.

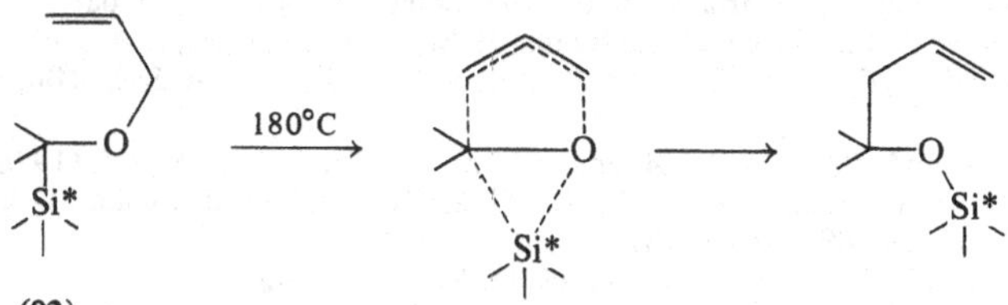

(92) →(180°C) → (7.101)

References

1. Review: Spangler, C. W., *Chem. Rev.*, **76**, 187 (1976).
2. Epiotis, N. D., *J. Am. chem. Soc.*, **95**, 1206 (1973).
3. Roth, W. R., König, J. and Stein, W., *Chem. Ber.*, **103**, 426 (1970).
4. Berson, J. A. and Aspelin, G. B., *Tetrahedron*, **20**, 2697 (1964).
5. McGreer, D. E. and Chiu, N. W. K., *Can. J. Chem.*, **46**, 2225 (1968).
6. Ando, W., *Tetrahedron Lett.*, 1969, 929. For an example of the reverse process, see Wilson, J. W. and Sherrod, S. A., *Chem. Communs*, 1968, 143.
7. Pleiss, M. G. and Moore, J. A., *J. Am. chem. Soc.*, **90**, 4738 (1968).
8. This effect has been described as 'subjacent orbital control': Berson, J. A., *Acc. chem. Res.*, **5**, 406 (1972); see also Baldwin, J. E., Andrist, A. H. and Pinschmidt, R. K., *Acc. chem. Res.*, **5**, 402.
9. Roth, W. R. and Friedrich, A., *Tetrahedron Lett.*, 1969, 2607.
10. Cookson, R. C. and Kemp, J. E., *Chem. Communs*, 1971, 385.
11. Baldwin, J. E. and Brown, J. E., *J. Am. chem. Soc.*, **91**, 3647 (1969).
12. W. D. Ollis and M. Rey, Personal communication.
13. Boersma, M. A. M., Haan, de J. W., Kloosterziel, H. and Ven, van de L. J. M., *Chem. Communs*, 1970, 1168.
14. Klärner, F.-G., *Angew. Chem., Int. Edn Engl.*, **13**, 268 (1974). For an explanation, see Baldwin, J. E. and Broline, B. M., *J. Am. chem. Soc.*, **100**, 4599 (1978).
15. Field, D. J., Jones, D. W. and Kneen, G., *J. chem. Soc. Perkin I*, 1978, 1050. An alternative view of relative rates of sigmatropic shifts, based on the donor–acceptor relationship of the migrating group and the migration framework, is given by Epiotis, N. D. and Shaik, S., *J. Am. chem. Soc.*, **99**, 4936 (1977).
16. Janssen, J. W. A. M., Habraken, C. L. and Louw, R., *J. org. Chem.*, **41**, 1758 (1976).
17. Patterson, J. M., Haan, de J. W., Boyd, M. R. and Ferry, J. D., *J. Am. chem. Soc.*, **94**, 2487 (1972); Gilchrist, T. L., Rees, C. W. and Thomas, C., *J. chem. Soc. Perkin I*, 1975, 12.
18. Franck-Neumann, M. and Dietrich-Buchecker, C., *Tetrahedron Lett.*, 1976, 2069, and references therein.
19. Klärner, F.-G. and Wette, M., *Chem. Ber.*, **111**, 282 (1978).
20. Gilchrist, T. L., Moody, C. J. and Rees, C. W., *J. chem. Soc. Chem. Communs*, 1976, 414.
21. Beggs, J. J. and Meyers, M. B., *J. chem. Soc. (B)*, 1970, 930.
22. Hart, H., Rodgers, T. R. and Griffiths, J., *J. Am. chem. Soc.*, **91**, 754 (1969).
23. Steinberg, H. and Sixma, F. L. J., *Recl. Trav. chim. Pays-Bas Belg.*, **81**, 185 (1962).
24. Review: Grovenstein, E., *Angew. Chem. Int. Edn Engl.*, **17**, 313 (1978).
25. Reviews: Rhoads, S. J. and Raulins, N. R., *Org. React.*, **22**, 1 (1975); Bennett, G. B., *Synthesis*, 1977, 589.
26. Doering, W. von E. and Roth, W. R., *Angew. Chem., Int. Edn Engl.*, **2**, 115 (1963).
27. Hill, R. K. and Gilman, N. W., *Chem. Communs*, 1967, 619.
28. Reviews: Schröder, G., Oth, J. F. M. and Merényi, R., *Angew. Chem., Int Edn Engl.*, **4**, 752 (1965); Paquette, L. A., *Angew. Chem., Int. Edn Engl.*, **10**, 11 (1971).
29. Cheng, A. K., Anet, F. A. L., Mioduski, J. and Meinwald, J., *J. Am. chem. Soc.*, **96**, 2887 (1974).
30. Hoffmann, R. and Stohrer, W.-D., *J. Am. chem. Soc.*, **93**, 6941 (1971).
31. Dewar, M. J. S. and Wade, L. E., *J. Am. chem. Soc.*, **99**, 4417 (1977). The detailed mechanism of the Cope rearrangement is the subject of continuing debate: see Dewar, M. J. S., Ford, G. P., McKee, M. L., Rzepa, H. S. and Wade, L. E., *J. Am. chem. Soc.*, 99, 5069 (1977); Gajewski, J. J. and Conrad, N. D., *J. Am. chem. Soc.*, **100**, 6268 (1978).
32. Gompper, R. and Ulrich, W.-R., *Angew. Chem., Int. Edn Engl.*, **15**, 299 (1976).
33. Viola, A., Ioro, E. J., Chen, K. K., Glover, G. M., Nayak, U. and Kocienski, P. J., *J. Am. chem. Soc.*, **89**, 3462 (1967).
34. Evans, D. A. and Golob, A. M., *J. Am. chem. Soc.*, **97**, 4765 (1975).
35. Rhoads, S. J. and Cockroft, R. D., *J. Am. chem. Soc.*, **91**, 2815 (1969).

36. Hansen, H.-J., in *Mechanisms of molecular migrations,* ed. B. S. Thyagarajan, vol. 3, p. 177, Interscience, New York, 1971.
37. Overman, L. E., Campbell, C. B. and Knoll, F. M., *J. Am. chem. Soc.,* **100,** 4822 (1978).
38. Vögtle, F. and Goldschmitt, E., *Chem. Ber.,* **109,** 1 (1976).
39. Hill, R. K. and Gilman, N. W., *Tetrahedron Lett.,* 1967, 1421.
40. Kwart, H. and Schwartz, J. L., *J. org. Chem.,* **39,** 1575 (1974).
41. Lewis, E. S., Hill, J. T. and Newman, E. R., *J. Am. chem. Soc.,* **90,** 662 (1968); see also Barton, D. H. R., Magnus, P. D. and Pearson, M. J., *J. chem. Soc. (C),* 1971, 2231.
42. Goldstein, M. J. and Judson, H. A., *J. Am. chem. Soc.,* **92,** 4119 (1970).
43. Garmaise, D. L., Uchiyama, A. and McKay, A. F., *J. org. Chem.,* **27,** 4509 (1962); see also Yamamoto, Y., Shimoda, H., Oda, J. and Inouye, Y., *Bull. chem. Soc., Jap.,* **49,** 3247 (1976).
44. Campbell, J. A., Mackay, D. and Sauer, T. D., *Can. J. Chem.,* **50,** 371 (1972); Kirby, G. W. and Mackinnon, J. W. M., *J. chem. Soc. Chem. Communs,* 1977, 23.
45. Reviews: Robinson, B., *Chem. Rev.,* **63,** 373 (1963); Robinson, B., *Chem. Rev.,* **69,** 227 (1969).
46. Review: Trost, B. M. and Melvin, L. S., *Sulfur ylides,* Academic Press, New York, 1975, p. 108.
47. Baldwin, J. E. and Urban, F. J., *Chem. Communs,* 1970, 165.
48. Reviews: Zimmerman, H. E. in *Molecular rearrangements,* ed. P. de Mayo, part 1, p. 345, Interscience, New York, 1963; Schöllkopf, U., *Angew. Chem., Int. Edn Engl.,* **9,** 763 (1970); Stevens, T. S. and Watts, W. E., *Selected molecular rearrangements,* Van Nostrand Reinhold, London, 1973.
49. Garst, J. F. and Smith, C. D., *J. Am. chem. Soc.,* **98,** 1526 (1976). See, however, evidence supporting the view that both the 1,2- and the 1,4-shifts involve radical pairs: Felkin, H. and Frajerman, C., *Tetrahedron Lett.,* 1977, 3485.
50. Reviews: Pine, S. H., *Org. React.,* **18,** 403 (1970); Lepley, A. R. and Giumanini, A. G., in *Mechanisms of molecular migrations,* ed. B. S. Thyagarajan, vol. 3, p. 297, Interscience, New York, 1971.
51. Kleinschmidt, R. F. and Cope, A. C., *J. Am. chem. Soc.,* **66,** 1929 (1944). Review: Johnstone, R. A. W., in *Mechanisms of molecular migrations,* vol. 2, p. 249, Interscience, New York, 1969.
52. Brindle, I. D. and Gibson, M. S., *J. chem. Soc. Perkin I,* 1979, 517.
53. Herriott, A. W. and Mislow, K., *Tetrahedron Lett.,* 1968, 3013; Farnham, W. B., Herriott, A. W. and Mislow, K., *J. Am. chem. Soc.,* **91,** 6878 (1969).
54. Rautenstrauch, V., *Chem. Communs,* 1970, 526; Bickart, P., Carson, F. W., Jacobus, J., Miller, E. G. and Mislow, K., *J. Am. chem. Soc.,* **90,** 4869 (1968).
55. Rautenstrauch, V., *Chem. Communs,* 1970, 4.
56. Lepley, A. R., *J. Am. chem. Soc.,* **91,** 1237 (1969).
57. Lepley, A. R., Cook, P. M. and Willard, G. F., *J. Am. chem. Soc.,* **92,** 1101 (1970).
58. Other work on the mechanisms of [1,2] shifts is described in Chantrapromma, K., Ollis, W. D. and Sutherland, I. O., *J. chem. Soc. Chem. Communs,* 1977, 97; Ollis, W. D., Rey, M., Sutherland, I. O. and Closs, G. L., *J. chem. Soc. Chem. Communs,* 1975, 543.
59. Heimgartner, H., Zsindely, J., Hansen, H.-J. and Schmid, H., *Helv. chim. Acta,* **55,** 1113 (1972).
60. Fráter, G. and Schmid, H., *Helv. chim. Acta,* **53,** 269 (1970).
61. Ollis, W. D., Somanathan, R. and Sutherland, I. O., *J. chem. Soc. Chem. Communs,* 1973, 661.
62. Review: Mango, F. D., *Coord. chem. Rev.,* **15,** 109 (1975).
63. Adam, W., Fischer, H., Hansen, H.-J., Heimgartner, H., Schmid, H. and Waespe, H.-R., *Angew. Chem., Int. Edn Engl.,* **12,** 662 (1973).
64. A review listing all the formally possible types of six-electron transition states is: Hendrickson, J. B., *Angew. Chem., Int. Edn Engl.,* **13,** 47 (1974).
65. Reviews: Hoffmann, H. M. R., *Angew. Chem., Int. Edn Engl.,* **8,** 556 (1969); Oppolzer, W. and Snieckus, V., *Angew. Chem., Int. Edn Engl.,* **17,** 476 (1978).

66. Friedrich, L. E., Kampmeier, J. A. and Good, M. *Tetrahedron Lett.*, 1971, 2783; see also Stephenson, L. M. and Mattern, D. L., *J. org. Chem.*, **41**, 3614 (1976).
67. Garsky, V., Koster, D. F. and Arnold, R. T., *J. Am. chem. Soc.*, **96**, 4207 (1974).
68. Review: Saunders, W. H. and Cockerill, A. F., in *Mechanisms of elimination reactions*, p. 377, Wiley, New York, 1973.
69. Skell, P. S. and Hall, W. L., *J. Am. chem. Soc.*, **86**, 1557 (1964). It has been suggested that many ester pyrolyses are not concerted, but involve surface-stabilised carbonium ions; Wertz, D. H. and Allinger, N. L., *J. org. Chem.*, **42**, 698 (1977).
70. Swain, C. G., Bader, R. F. W., Esteve, R. M. and Griffin, R. N., *J. Am. chem. Soc.*, **83**, 1951 (1961); Brower, K. R., Gay, B. and Konkol, T. L., *J. Am. chem. Soc.*, **88**, 1681 (1966).
71. Smith, G. G. and Yates, B. L., *J. org. Chem.*, **30**, 2067 (1965).
72. Yates, B. L. and Quijano, J., *J. org. Chem.*, **35**, 1239 (1970).
73. Smith, G. G. and Blau, S. E., *J. phys. Chem.*, **68**, 1231 (1964).
74. Spencer, H. K. and Hill, R. K., *J. org. Chem.*, **41**, 2485 (1976).
75. Cookson, R. C. and Wallis, S. R., *J. chem. Soc. (B)*, 1966, 1245.
76. Bach, R. D., Andrzejewski, D. and Dusold, L. R., *J. org. Chem.*, **38**, 1742 (1973).
77. Morris, D. G., Smith, B. W. and Wood, R. J., *Chem. Communs*, 1968, 1134.
78. Coke, J. L. and Cooke, M. P., *J. Am. chem. Soc.*, **89**, 6701 (1967).
79. Jones, D. N. and Lewton, D. A., *J. chem. Soc. Chem. Communs*, 1974, 457.
80. Chou, T. S., Burgtorf, J. R., Ellis, A. L., Lammert, S. R. and Kukolja, S. P., *J. Am. chem. Soc.*, **96**, 1609 (1974).
81. Reich, H. J., Renga, J. M. and Reich, I. L., *J. Am. chem. Soc.*, **97**, 5434 (1975). Review: Clive, D. L. J., *Tetrahedron*, **34**, 1049 (1978).
82. Review: Hünig, S., Müller, H. R. and Thier, W., *Angew. Chem., Int. Edn Engl.*, **4**, 271 (1965).
83. Doering, W. von E. and Rosenthal, J. W., *J. Am. chem. Soc.*, **89**, 4534 (1967).
84. Fleming, I. and Wildsmith, E., *Chem. Communs*, 1970, 223.
85. Reetz, M. T., *Tetrahedron*, **29**, 2189 (1973); Reetz, M. T., *Tetrahedron Lett.*, 1976, 817.

Index